This comprehensive text is a revised and greatly expanded second edition of a book first published in 1987, and provides a thorough introduction to the complex fields of signal coding and signal processing. As well as giving a detailed description of the theoretical aspects of these areas, emphasizing the particular algorithms currently employed, the book also discusses the general principles of system design, and frequent reference is made to relevant practical examples from the fields of broadcasting, mobile radio, magnetic recording, and satellite technology.

Among the key topics covered are: the fundamentals of pulse code modulation, including various aspects of sampling theory; modern data compression techniques; block and convolutional error control codes; transmission coding techniques, including partial response signalling and scrambling; digital filters; infinite and finite impulse response filters; adaptive filters, complete with computer simulation examples; and fast Fourier transform theory, implementation and applications.

Complete with problems and solutions, and containing over 230 diagrams, this textbook will be invaluable to third and fourth year undergraduates in electronic or communication engineering. It will also act as a useful reference for anyone working in this technologically important field.

Signal coding and processing

Signal coding and processing

SECOND EDITION

GRAHAM WADE

*Principal Lecturer and Head of Communication and
Information Engineering Group,
School of Electronic, Communication, and Electrical
Engineering, University of Plymouth*

CAMBRIDGE
UNIVERSITY PRESS

Published by the Press Syndicate of the University of Cambridge
The Pitt Building, Trumpington Street, Cambridge CB2 1RP
40 West 20th Street, New York, NY 10011–4211, USA
10 Stamford Road, Oakleigh, Melbourne 3166, Australia

First published by Ellis Horwood Limited 1987
Second edition published by Cambridge University Press, 1994

Printed in Great Britain at the University Press, Cambridge

A catalogue record for this book is available from the British Library

Library of Congress cataloguing in publication data
Wade, Graham.
Signal coding and processing / Graham Wade. – 2nd ed.
      p.   cm.
Includes bibliographical references (p.   ) and index.
ISBN 0-521-41230-7 (hardback). – ISBN 0-521-42336-8 (pbk.)
1. Digital communications. 2. Signal processing–Digital
techniques. 3. Coding theory. I. Title.
TK5103.7.W33   1994
621.382'2–dc20   93–37133   CIP

ISBN 0 521 41230 7 hardback
ISBN 0 521 42336 8 paperback

KT

To Jenny
Dominic, Alistair, and Oliver

The fear of the Lord is the beginning of knowledge
Proverbs 1 *v* 7

CONTENTS

PREFACE

Readership

This book has been written for students who are in the middle or final years of a degree in electronic or communication engineering and who are following a course in signal coding and signal processing techniques. It should also be helpful to engineers and managers in industry who need an introductory or refresher course in the field.

About the book

Many textbooks are devoted to either signal coding, e.g. error control, or to signal processing, e.g. digital filters, simply because there is great breadth and depth in each area. On the other hand, practical systems invariably employ a combination of these two fields and a knowledge of both is often required. For example, a knowledge of digital filtering, fast Fourier transforms (FFTs) and forward error control would often be required when designing a satellite data modem. Similarly, a knowledge of discrete transforms is fundamental to the understanding of some video compression schemes (transform coding), and basic digital filter theory is required for some speech codecs. Also, many undergraduate courses give an introduction to both fields in the middle and final years of a degree.

The philosophy behind this book is therefore to provide a single text which introduces the major topics in signal coding and processing, and to illustrate how these two areas interact in practical systems. The book is a blend of theory and modern practice and is the result of some 12 years lecturing and research experience in these two fields. Much of the material has been used on the communication engineering degree and industrial short courses run at the University of Plymouth. Sufficient mathematical detail is given to provide a sound understanding of each topic but every effort has been made to provide a practical context. To this end, the book examines coding and filtering aspects of practical systems taken from the broadcasting, satellite, mobile radio and magnetic recording fields, and is illustrated with over 230 diagrams and well over 100 worked examples. Generally speaking, only equations of broad significance are numbered in order to highlight important results.

Chapter 1 introduces the most fundamental coded system (PCM) and the basic signal processing concepts of sampling and quantization. It assumes a knowledge of the Fourier transform and so treats these topics in a generalized way. Important practical examples of sampling are highlighted by introducing decimation and interpolation (as found, for example, in high quality audio systems).

Chapter 2 (Source coding) takes a look at information theory and deals with the problem of redundancy reduction and the efficient coding of a signal source. This is a large field with applications from facsimile to disk file compression to satellite broadcasting to mobile radio, and an attempt has been made to survey the latest techniques. The first part of the chapter examines classical reversible techniques, such as variable length coding, and it concludes with a look at a modern technique for compressing computer data files. The second part of the chapter (non-reversible coding) embraces important philosophical and psychovisual/psychoacoustic concepts as well as signal processing since these are fundamental to many compression schemes. It commences with common compression techniques for speech and video (DPCM and transform coding) and concludes with a review of advanced signal processing techniques for videophone and digital mobile radio (e.g. GSM).

Chapter 3 (Channel coding) is probably the most theoretical chapter since error control coding has a strong mathematical basis (in linear algebra). On the other hand, the author has found that, once into the subject, many undergraduates find the subject stimulating and challenging. A theoretical basis of block and convolutional codes is provided together with practical illustrations from broadcasting, satellite systems and mobile radio. One of the most important areas is that of non-binary or Reed–Solomon (RS) codes, and some effort has been made to provide a readable introduction to this algebraically complex subject. To this end, Section 3.5 includes an introduction to Galois fields since this is sometimes overlooked in mathematics courses. The chapter includes the fundamental but important concepts of transmission coding, i.e. line codes, and concludes with the newer concept of partial response (PR) signalling (used particularly in magnetic recording), and scrambling theory.

Chapter 4 provides a fairly comprehensive introduction to digital filters and should be sufficient for most undergraduate courses. It also provides the theoretical basis for the filtering concepts discussed in other chapters. Only industrially relevant design techniques are covered and so some of the classical IIR filter design methods have been omitted. On the other hand, the chapter discusses some of the newer concepts in FIR filter design (for computation efficient filters) as well as the standard design techniques. Those new to the signal processing field will find the topic of adaptive filtering (Section 4.6) an interesting break from classical linear time-invariant filter theory. The mathematical basis and implementation issues are examined, and several adaptive systems are illustrated using computer simulation. Another break from classical filter theory is found in the subject of image filtering since these filters often use heuristic concepts and powerful non-linear operators.

Chapter 5 initially takes a broad look at discrete transforms since they are fundamental to a range of applications from spectral analysis to source coding. For instance, discrete transforms are the fundamental basis for the transform coding technique discussed in Chapter 2. The most important discrete transform is of course the DFT (FFT) and much of the chapter is concerned with FFT fundamentals and the use of the FFT for spectral analysis. More advanced applications of discrete transforms, such as fast digital filtering, fast correlation, and image transforms are discussed at the end of the chapter.

Acknowledgements

I am pleased to acknowledge the many hours spent reviewing the text by V. Bota and M. Borda, both from the Technical University of Cluj-Napoca, Romania. They provided detailed comments, corrections, and suggestions for Chapters 2 and 3. In addition, valuable comments and suggestions have been received from P. Farrell, University of Manchester, D. Pearson, University of Essex, D. Ingram, University of Cambridge, P. Lynn, T. Terrell, University of Central Lancashire, M. Beaumont and D. Cheeseman, BT Laboratories, Martlesham, and R. Stevens, GCHQ Cheltenham. I have also had very helpful discussions with colleagues at the University of Plymouth, namely, M. Tomlinson, A. Romilly, P. Sanders, M. Ali Abu-Rgheff, R. Gerry, P. Van Eetvelt, P. Smithson and P. Davey.

Graham Wade
University of Plymouth

ABBREVIATIONS

ADC   analogue-to-digital converter
ADPCM   adaptive differential pulse code modulation
AMI   alternate mark inversion
ARQ   automatic repeat request
ASK   amplitude shift keying
AWGN   additive white Gaussian noise
BCH   Bose–Chaudhuri–Hocquenghem
BER   bit-error rate
BnZS   bipolar n-zero substitution
BSC   binary symmetric channel
CCIR   International Radio Consultative Committee
CCITT   International Telegraph & Telephone Consultative Committee
CELP   codebook-excited linear predictive (coder)
codec   coder–decoder
cph(w)   cycles per picture height (width)
DAC   digital-to-analogue converter
DCT   discrete cosine transform
DED(C)   double error detecting (correcting)
DFT   discrete Fourier transform
DHT   discrete Hartley transform
DIF   decimation in frequency
DIT   decimation in time
DPCM   differential pulse code modulation
DWT   discrete Walsh transform
E   expectation
ECC   error control coding
ENBW   equivalent noise bandwidth

ETSI	European Telecommunications Standards Institute
FEC	forward error control
FFT	fast Fourier transform
FIR	finite impulse response
FSK	frequency shift keying
FWHT	fast Walsh–Hadamard transform
GF	Galois field
GSM	Groupe Speciale Mobile
HDBn	high-density bipolar n
IDFT	inverse discrete Fourier transform
IFFT	inverse fast Fourier transform
IIR	infinite impulse response
ISI	intersymbol interference
ITU	International Telecommunication Union
LMS	least mean square
LPC	linear predictive coding
LZW	Lempel–Ziv–Welch
mmse	minimum mean square error
NRZ	nonreturn to zero
NRZI	modified nonreturn to zero
NTSC	National Television Systems Committee
OFFT	offset FFT
1-D	one-dimensional
PAL	phase alternation line
PAM	pulse amplitude modulation
PCM	pulse code modulation
pdf	probability density function
PR	partial response (signalling)
PRBS	pseudo-random binary sequence
PSD	power spectral density
PSK	phase shift keying
QPSK	quadrature phase shift keying
RDS	radio data system
RLL	runlength-limited (code)
RLS	recursive least squares
RPE	regular-pulse excitation
RRS	recursive running sum
RS	Reed–Solomon
RZ	return to zero
SEC(D)	single error correcting (detecting)

SNR	signal-to-noise ratio
SOC	self-orthogonal code
SR	syndrome register
TC	transform coding
TED	triple error detecting

1

PCM signals and systems

This chapter examines theoretical and practical aspects of *pulse code modulation* (PCM) because of its direct relevance to the coding techniques presented in Chapters 2 and 3.

We start by comparing PCM with Shannon's ideal coding concept and by highlighting its noise immunity compared to uncoded systems. The chapter then examines the two fundamental aspects of PCM (sampling and quantizing) from theoretical and practical standpoints, and illustrates how sampling rates and quantizing noise can be 'adjusted' in order to convert efficiently analogue signals to PCM or vice-versa. Once in PCM form, a signal is usually processed by some *source coding* technique in order to satisfy a specific bit rate, and PCM network standards are outlined in Section 1.6.2. The chapter concludes by introducing the idea of *channel coding*, which is essentially a mechanism for matching the processed PCM signal to the transmission channel.

1.1 The ideal coding concept

Consider the receiver in Fig. 1.1 (essentially a demodulator or decoder) where S_i and N_i are the average signal and noise powers for the input, S_o and N_o are the average signal and noise powers for the output, B_T is the transmission bandwidth and B_m is the message bandwidth. In general (but not always) a practical receiver improves the signal-to-noise ratio (SNR) in exchange for signal bandwidth, so that by choosing a modulation or coding scheme such that $B_T \gg B_m$ it should be possible for the receiver to deliver

Fig. 1.1 Power–bandwidth exchange in a receiver.

1

a high-quality signal despite a low S_i/N_i ratio. Both coded and uncoded modulation schemes permit this type of exchange, and good examples of uncoded systems which do this are wideband frequency modulation (FM) and pulse position modulation (PPM).

The most efficient exchange can only be achieved using a coded system. In fact, it is reasonable to seek a coded system which will give arbitrarily small probability of error at the receiver output, given a finite value for S_i/N_i. To this end, it is helpful to have a broad view of coding which does not necessarily involve digital signals. Rather, we could view it as the transformation of a message signal or symbol into another signal or sequence of symbols for transmission. Using this broad concept of coding, it is possible to regard a coded message as a vector in multidimensional signal space and to deduce from a noise sphere packing argument that it is theoretically possible to reduce the probability of error at the receiver output to zero, despite finite channel noise. A rigorous treatment of this problem was given by Shannon in 1949 and, broadly speaking, his *ideal coding scheme* involves restricting the source output to a finite set of M different messages and assigning a particular message to a randomly selected signal or code vector of duration T s. For a binary digital system a particular message would be assigned to an n-digit codeword selected at random from the 2^n possible code vectors. The optimum receiver would involve correlation or matched filter detection and a minimum-distance (maximum-likelihood) decision criterion.

Unfortunately, it turns out that, to achieve arbitrarily small probability of error at the receiver output, it is necessary for T (or n) to extend to infinity! In turn, this means that to maintain a constant information rate of $(1/T) \log_2 M$ bits/s (see Section 2.1.2) then M must increase exponentially. These factors make such an ideal coding scheme impractical, although the concept is still extremely important since it defines the ultimate that can be achieved with respect to error-free transmission rate and bandwidth–SNR exchange. For his ideal coding scheme, Shannon showed that

$$C = B \log_2 \left(1 + \frac{S}{N} \right) \quad \text{bits/s} \tag{1.1}$$

where C is the *channel capacity*, B is the channel bandwidth (in Hz), S and N are the average signal and noise powers respectively and the noise is additive white Gaussian noise. Equation (1.1) is referred to as the *Shannon–Hartley* law and leads to Shannon's *noisy coding theorem*:

Theorem 1.1 *It is theoretically possible to transmit information through a noisy channel with an arbitrarily small probability of* *error provided that the information or source rate, R, is less than the channel capacity, i.e. R < C for reliable transmission.*

The Shannon–Hartley law underscores a general result for coded systems, namely, coding makes it possible to economize on power while consuming bandwidth. For example, letting $S/N = 15$ and $B = 3$ kHz gives $C = 12$ kbit/s. Alternatively, we could reduce S/N to 7 and increase B to 4 kHz to achieve the same capacity (noting that the noise power will increase by a factor 4/3). Similarly, in Chapter 3 (e.g. Fig. 3.20) we shall see that, in order to meet a specific decoded error rate, we can either increase the signal power, or increase the channel bandwidth by adding error control coding. We shall also see that, generally speaking, longer and longer coded sequences are required as we try and achieve lower and lower decoded error rate – as suggested by Shannon's ideal coding scheme.

1.2 Fundamentals of PCM

Shannon's work showed that every practical communication channel has a finite information capacity C owing to finite bandwidth and finite noise and signal powers. Also, according to Theorem 1.1, we must restrict the information rate R to C if we are to have a small probability of error at the decoder output. Following the concept behind Shannon's ideal coding scheme, we can restrict the information rate by restricting the source to a finite alphabet of M discrete messages or symbols transmitted every T s. In practice this *discrete source* is generated via *quantization* using an analogue-to-digital converter (ADC).

We will see in Section 2.1.2 that a discrete source of M symbols has a maximum information rate or *entropy rate* of $R = (1/T) \log_2 M$ bits/s and clearly this could be made less than the available channel capacity C. If this is the case, we can conjecture that virtually all the transmitted information will be received uncorrupted or, put another way, noise in the channel will have negligible effect upon the received signal quality. Clearly, the major impairment in such a system will then be the errors due to quantizing, and this is the fundamental concept behind PCM.

1.2.1 PCM bandwidth

A 'straight' PCM coder simply applies waveform sampling, quantizing and coding (usually binary number, offset binary, or Gray coding is used, although others are possible). We will see in Section 1.3 that if the analogue source is bandlimited to B_m then the signal can be sampled at the *Nyquist rate* of $2B_m$ (or greater), and all the information will be retained in the samples. To create a discrete source, we next quantize each sample to one of M levels and assume that the analogue signal forces each level to be equiprobable. As shown in Section 2.1.2, this assumption gives the maximum possible average information per sample ($\log_2 M$ bits) and so the discrete source will be generating information at the maximum rate of

$$R_{\max} = 2B_m \log_2 M \quad \text{bits/s} \tag{1.2}$$

Each quantized sample is now coded by assigning it to a code group as indicated in Fig. 1.2 and s-ary PCM is assumed where $s < M$ for a coded system. Since n s-ary pulses have s^n combinations, we require that $s^n = M$, corresponding to $\log_s M$ pulses per sample, and the serial transmission rate becomes $2B_m \log_s M$ pulses/s. At this point we can call upon Nyquist's bandwidth–signalling speed relationship for a transmission channel (Nyquist, 1924, 1928):

> **Theorem 1.2** *An ideal lowpass channel of bandwidth B can transmit independent symbols at a rate $R_s \leqslant 2B$ without intersymbol interference (ISI).*

From this theorem it follows that the minimum possible baseband transmission bandwidth for PCM is

$$B_T = B_m \log_s M \quad \text{Hz} \tag{1.3}$$

Example 1.1
For industrial closed-circuit television (CCTV) satisfactory monochrome images are obtained using a 6-bit ADC. This corresponds to $s = 2$, $M = 64$, and so at least six times the video bandwidth would be required for baseband serial transmission of the digitized signal.

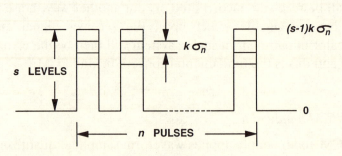

Fig. 1.2 PCM code group for s-ary pulses.

1.2.2 PCM channel capacity

If no information is to be lost during transmission (no decoding errors) then, according to Theorem 1.1 and Equation (1.2), the PCM channel capacity must be

$$
\begin{aligned}
C_{PCM} &= 2B_m \log_2 M \\
&= 2B_m \log_s M \log_2 s \\
&= B_T \log_2 s^2
\end{aligned}
\tag{1.4}
$$

For binary PCM($s = 2$) this means that the maximum transmission rate is simply $2B_T$ bits/s or 2 bits/s/Hz. In order to compare the power requirements of PCM with that of Shannon's ideal coding scheme, it is necessary to express s in terms of S_i/N_i and to assume that the channel is disturbed by additive white Gaussian noise. Let the s-ary levels be separated by $k\sigma_n$, where k is a constant, σ_n is the rms noise in the channel, and the normalized noise power is $N_i = \sigma_n^2$. Since the analogue signal is assumed to have a flat probability density function (pdf) it follows that the s levels are equiprobable, in which case the average normalized signal power at the demodulator input is

$$
\begin{aligned}
S_i &= \frac{1}{s} \sum_{r=0}^{s-1} (rk\sigma_n)^2 \\
&= \frac{(k\sigma_n)^2}{s} \sum_{r=0}^{s-1} r^2
\end{aligned}
$$

and this reduces to

$$
S_i = \frac{(k\sigma_n)^2 (s-1)(2s-1)}{6}
$$

Note that the s-ary pulses are unipolar and that the corresponding dc component has an average power

$$
S_{dc} = \frac{(k\sigma_n)^2 (s-1)^2}{4}
$$

Since this component does not convey useful information, we can resort to bipolar transmission of the s-ary pulses and so reduce S_i. In this case

$$
\begin{aligned}
S_i &= \frac{(k\sigma_n)^2 (s-1)}{24} [4(2s-1) - 6(s-1)] \\
&= \frac{(k\sigma_n)^2 (s^2-1)}{12}
\end{aligned}
$$

and so

$$\frac{S_i}{N_i} = \frac{k^2(s^2 - 1)}{12} \qquad (1.5)$$

Finally, substituting for s^2 in (1.4) gives

$$C_{PCM} = B_T \log_2\left(1 + \frac{12}{k^2}\frac{S_i}{N_i}\right) \qquad (1.6)$$

Note that by making this substitution we have introduced a practical coding scheme which, in turn, implies that the decoded error probability will not be arbitrarily small. In other words, whilst (1.1) assumes a complex coding technique yielding *arbitrarily small* error probability, (1.6) specifies the maximum rate at which information can be transmitted with *small* error. For example, considering a binary PCM system with $k = 10$, the bit-error probability $P(\varepsilon)$ for a Gaussian noise channel is approximately 3×10^{-7} (which turns out to be a reasonable objective for a PCM system handling broadcast quality video signals for example). Bearing this in mind, we can compare (1.1) and (1.6) for the same C/B ratios to give

$$\left(\frac{S_i}{N_i}\right)_{PCM} = \frac{k^2}{12}\left(\frac{S_i}{N_i}\right)_{ideal} \qquad (1.7)$$

This means that the PCM system needs to increase the signal power by a factor $k^2/12$ above that for Shannon's ideal coding scheme. For $k = 10$ this is an increase of 9.2 dB and even then the error probability for a binary system will be about 3×10^{-7} rather than zero. Increasing k will, however, reduce $P(\varepsilon)$ towards zero.

It is also important to note what happens as k, or, equivalently S_i/N_i, is reduced. Typically, a decrease of several dB in the SNR can increase $P(\varepsilon)$ by several orders of magnitude and $P(\varepsilon)$ can rapidly become significant! Clearly, in contrast to uncoded systems, if the SNR can be kept above this effective threshold then the PCM channel can be regarded as virtually transparent.

To summarize, we have highlighted the possibility of exchanging signal power for bandwidth in coded systems, and one way of achieving this is through forward error control (FEC). PCM in particular is extremely important because of its simple implementation, widespread use and versatility (it is not *signal specific*). Also, in Chapter 2 we shall see that PCM is used as a standard for comparison of data compression schemes.

1.3 Ideal sampling and reconstruction of signals

Conversion to PCM simply involves sampling and quantizing and, by implication, the coding of these quantizing levels. Invariably, the ADC carries out all three processes, and the reverse operations are carried out by the digital-to-analogue converter (DAC) and *reconstruction filter*. A simple PCM system is shown in Fig. 1.3. Note that the *bandlimitation* and reconstruction filters are crucial components of the system.

Sampling theory is neatly and powerfully handled using the Fourier transform and various convolutional relationships, and we shall adopt that approach here. In particular, *ideal sampling* can be neatly explained using the Fourier transform pairs in Fig. 1.4. Suppose that an arbitrary continuous time signal $x(t)$ is bandlimited to the frequency $\omega_m = 2\pi f_m$ and is sampled at the rate $\omega_s = 2\pi f_s$. If this signal is real, then the magnitude $|X(\omega)|$ of its Fourier transform is an even function as illustrated in Fig. 1.4(*a*). For ideal or *impulse* sampling, $x(t)$ is multiplied by the impulse sequence $\delta_T(t)$ to give the sampled signal $x_s(t)$ and the Fourier transform of $x_s(t)$ follows from the frequency convolution theorem:

$$X_s(\omega) = \frac{1}{2\pi}[X(\omega)*\omega_s\delta_{\omega_s}(\omega)]$$

$$= \frac{1}{T}\left[X(\omega)* \sum_{n=-\infty}^{\infty} \delta(\omega - n\omega_s)\right]$$

where n is an integer. Using the distributive convolution law

$$X_s(\omega) = \frac{1}{T}\sum_{n=-\infty}^{\infty} X(\omega)*\delta(\omega - n\omega_s)$$

$$= \frac{1}{T}\sum_{n=-\infty}^{\infty} X(\omega - n\omega_s) \tag{1.8}$$

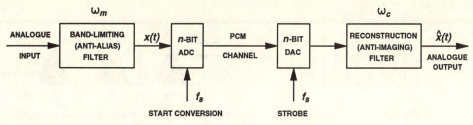

Fig. 1.3 A PCM system.

Apart from scaling, Equation (1.8) represents the original spectrum $X(\omega)$ plus *images* of $X(\omega)$ repeated in undistorted form at intervals of $\pm\, n\omega_s$, as shown in Fig. 1.4(c).

According to Fig. 1.4(c), $X(\omega)$ and therefore $x(t)$ can be recovered precisely by passing $x_s(t)$ through an ideal lowpass filter of transfer function $G_{2\omega_c}(\omega)$ with cutoff frequency ω_c, provided that

$$\omega_m \leqslant \omega_c \leqslant \omega_s - \omega_m \tag{1.9}$$

If this is true, then spectral overlap or *aliasing* is avoided and the sampling theorem is satisfied. In the above context, this theorem can be stated as follows:

> **Theorem 1.3** *A signal bandlimited to ω_m rad/s is completely determined by its instantaneous sample values uniformly spaced in time and corresponding to a sampling frequency $\omega_s \geqslant 2\omega_m$ rad/s. The minimum sampling frequency $\omega_s = 2\omega_m$ is called the* Nyquist rate.

Turning to the problem of reconstructing $x(t)$ from its samples, it is interesting to examine how an ideal lowpass filter interpolates between samples in order to generate the required continuous time signal. Let this filter have a transfer function

$$H(\omega) = G_{2\omega_c}(\omega)$$

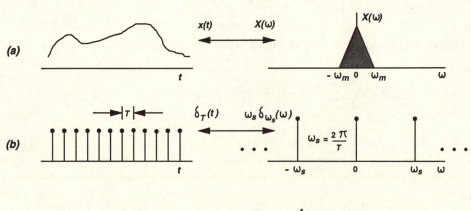

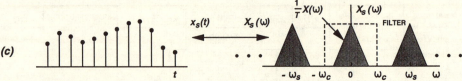

Fig. 1.4 Fourier transform pairs for ideal sampling (transform phase is omitted for clarity).

where $|G_{2\omega_c}(\omega)| = 1$ over the passband. Clearly, this filter will remove unwanted images to leave, from (1.8)

$$X(\omega) = X_s(\omega) T G_{2\omega_c}(\omega) \tag{1.10}$$

Fourier transform tables for a gate function give

$$\frac{\omega_c}{\pi} Sa(\omega_c t) \Leftrightarrow G_{2\omega_c}(\omega) \tag{1.11}$$

where

$$Sa(x) = \frac{\sin x}{x} \tag{1.12}$$

so that, using the time convolution theorem,

$$x(t) = Tx_s(t) * \frac{\omega_c}{\pi} Sa(\omega_c t)$$

$$= \frac{2\omega_c}{\omega_s} x_s(t) * Sa(\omega_c t)$$

$$= \frac{2\omega_c}{\omega_s} \sum_{n=-\infty}^{\infty} x(nT)\delta(t - nT) * Sa(\omega_c t)$$

$$= \frac{2\omega_c}{\omega_s} \sum_{n=-\infty}^{\infty} x(nT)Sa[\omega_c(t - nT)] \tag{1.13}$$

For Nyquist rate sampling this reduces to

$$x(t) = \sum_{n=-\infty}^{\infty} x(nT)Sa\left[\frac{\pi}{T}(t - nT)\right] \tag{1.14}$$

The sampling function $Sa[\pi(t - nT)/T]$ has zeros for all $t = kT \neq nT$ and is weighted by the value of $x(t)$ at $t = nT$. A graphical interpretation of (1.14) is shown in Fig. 1.5. Effectively, the ideal lowpass filter is replacing each sample $x(nT)$ with an interpolating function $x(nT)Sa[\pi(t - nT)/T]$ and the value of $x(t)$ at any point in time is the sum of contributions from all interpolating functions. Note that the cutoff frequency of this filter is $1/2T$ and effectively it is transmitting $(\sin x)/x$-shaped pulses at a rate of $1/T$. This illustrates a limiting form of data transmission and we shall return to this concept in Section 3.8.1. Also, it should now be clear why the bandlimiting filter in Fig. 1.3 is sometimes referred to as an *anti-alias* filter and why the reconstruction filter is sometimes referred to as an *anti-imaging* filter.

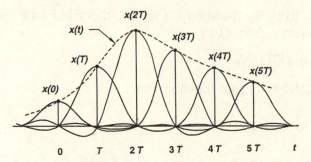

Fig. 1.5 Reconstruction of a signal $x(t)$ from its Nyquist rate samples.

1.3.1 Bandpass sampling

Sometimes it is necessary to apply digital signal processing (DSP) to a bandpass signal $v(t)$ with a spectrum centred on $f_c = (f_1 + f_2)/2$ (Fig. 1.6(*a*)). According to Theorem 1.3 we could treat this as a lowpass signal and sample at a rate $f_s \geq 2f_2$ although, clearly, f_s will often be ridiculously high! A better approach is to use a narrowband signal model and represent $v(t)$ as

$$v(t) = x(t)\cos(\omega_c t) - y(t)\sin(\omega_c t) \tag{1.15}$$

Here, $x(t)$ and $y(t)$ are independent, low-frequency (baseband) signals each bandlimited to $B_T/2 = (f_2 - f_1)/2$. Given $v(t)$ in this form it is apparent that $x(t)$ and $y(t)$ can be recovered using the quadrature carrier system in Fig. 1.6(*b*), which, in turn, means that each low-frequency signal can be sampled at a rate B_T. The net effective sample rate for $v(t)$ is therefore

$$f_s = 2B_T \tag{1.16}$$

It is also possible to sample the bandpass signal directly at a similar sample rate provided f_s is carefully selected to avoid spectral aliasing. By considering image spectra arising from sampling it can be shown that aliasing is avoided if

$$2B_T \frac{k}{N} \leq f_s \leq 2B_T \frac{k-1}{N-1} \tag{1.17}$$

where $k = f_2/B_T$, $k \geq N$ and N is a positive integer. Equation (1.17) defines a set of frequency bands suitable for direct bandpass sampling and the *minimum possible* sampling frequency can be written

$$f_s = 2B_T \frac{f_2/B_T}{\lfloor f_2/B_T \rfloor} \tag{1.18}$$

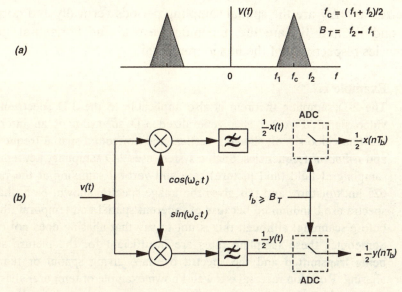

Fig. 1.6 Bandpass sampling: (*a*) spectrum of a bandpass signal $v(t)$; (*b*) quadrature sampling of $v(t)$.

where $\lfloor x \rfloor$ denotes the greatest integer in x. Clearly, if f_2 is an integer multiple of B_T, or if $f_2 \gg B_T$, then the minimum sample rate is $2B_T$, whilst the worst case occurs when $f_2 \approx 2B_T$ corresponding to a minimum rate of $4B_T$. We can therefore state a bandpass sampling theorem as follows:

Theorem 1.4 *The minimum sampling frequency for a bandpass signal of bandwidth B_T lies in the range $2B_T \leqslant f_s \leqslant 4B_T$.*

A practical example of bandpass sampling is found in sub-band coding (Section 2.2.4).

1.3.2 *Multidimensional sampling*

So far we have considered one-dimensional (1-D) forms of the sampling theorem for baseband and bandpass signals. However, it is important to realize that signals can exist in several dimensions and that we need to obey similar rules for such signals. Consider the two-dimensional (2-D) spatial sampling of a static image $f(x, y)$. In this case a form of the 2-D sampling theorem could be expressed as

$$\frac{1}{\Delta x} \geqslant 2W_u \qquad \frac{1}{\Delta y} \geqslant 2W_v \qquad\qquad (1.19)$$

where Δx and Δy are the spatial sampling periods vertically and horizontally, and W_u and W_v are the maximum vertical and horizontal *spatial* frequencies respectively of the image transform.

Example 1.2

The 2-D sampling theorem is also applicable to the 3-D spectrum of a video signal. Fig. 1.7 is a generalized 3-D spectrum of an interlaced 625-line video signal, showing vertical and horizontal spatial frequencies, and *temporal* frequencies. Such a system uses 2-D sampling, i.e. temporal sampling at field (and picture) rate and vertical sampling at the rate of 625 lines/picture, and this gives the image spectral shown. Note that the spectra are bandlimited because of inherent spatial and temporal filtering before scanning, although this is not to say that aliasing does not occur. Sometimes these sampling rates are insufficient for the picture source being transmitted and the spectra overlap, giving spatial or temporal aliasing. Rotation reversal is a well-known example of temporal aliasing.

As might be expected, the reconstruction process is similarly non-ideal. In practice, the images tend to be filtered out by the receiver display limitations and by the 3-D spatial–temporal lowpass filter provided by the human visual system.

1.4 Practical sampling and reconstruction

In this section we discuss the sampling and reconstruction aspects of practical systems, noting at the outset that both impulse sampling and ideal lowpass filtering are physically unrealizable.

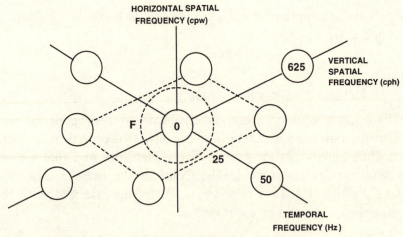

Fig. 1.7 3-D spectrum of an interlaced 625-line video signal. Reconstruction is aided by the 3-D lowpass filter F of the human visual system.

In practice, the anti-alias filter in Fig. 1.3 should not generate significant distortion of the input signal, and it should reduce alias-generating components to an acceptable level. For example, the filter cutoff frequency could be selected at, say, 20 times the fundamental frequency of a complex periodic signal, or at roughly the reciprocal of the pulse width for a single event. Filter stopband attenuation is determined by the basic need to keep alias generating components below the quantization threshold of the ADC, i.e. below $\frac{1}{2}$LSB. For an n-bit ADC this is $V_{FS}/2^{n+1}$ where V_{FS} is the ADC full-scale input, so that, for example, a 10-bit ADC would need a filter attenuation $> 2^{11}$ or at least 66 dB near $f_s/2$.

Sometimes practical systems can come surprisingly close to the theoretical ideal, as for example in the sampling process of an ADC. Ideally we wish to define a precise sampling instant. However, even if the sampling pulse applied to the ADC is absolutely jitter free, there is still uncertainty in the sample instant arising from the ADC itself and this time uncertainty is called the ADC *aperture uncertainty* or *aperture jitter* t_a. Over time t_a a sinusoidal signal of frequency f Hz occupying the full quantizing range of an n-bit ADC can change by as much as

$$\delta V = \pi f \Delta 2^n t_a \tag{1.20}$$

where Δ is the quantum interval, i.e. $V_{FS}/2^n$. It is obviously of interest to reduce this voltage uncertainty or *aperture error* to an acceptable level and usually we require $\delta V < \Delta/2$, in which case

$$t_a < (\pi f 2^{n+1})^{-1} \tag{1.21}$$

For a video system we might have $n = 8$ and $f = 5\,\text{MHz}$, giving $t_a < 124\,\text{ps}$. This is easily achieved by video ADCs or by preceding a slower ADC with a sample-and-hold circuit, and so it is reasonable to assume that sampling in a PCM system can closely approximate to uniform, instantaneous sampling.

In contrast, a significant deviation from ideal theory can occur in the practical reconstruction process. The DAC usually generates relatively broad pulses, far removed from impulses, and this can give a significant loss at high frequencies. Assuming the coding process to be transparent, then the sample levels from the DAC correspond to samples of the input signal $x(t)$, as shown in Fig. 1.8, and the pulse at $t = 0$ can be considered as arising from the convolution

$$x(t)\delta(t)*G_\tau(t)$$

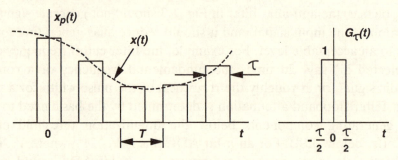

Fig. 1.8 PAM signal generated by a DAC.

where $G_\tau(t)$ is a gate function of duration τ. Similarly, the PAM signal generated by the DAC can be expressed as

$$x_p(t) = [x(t)\delta_T(t)]*G_\tau(t)$$

where $\delta_T(t)$ is the impulse train in Fig. 1.4(*b*). Since

$$G_\tau(t) \Leftrightarrow \tau\, Sa\left(\frac{\omega\tau}{2}\right)$$

and

$$x(t)\delta_T(t) \Leftrightarrow \frac{1}{T}\sum_{n=-\infty}^{\infty} X(\omega - n\omega_s)$$

then

$$x_p(t) \Leftrightarrow \frac{\tau}{T} Sa\left(\frac{\omega\tau}{2}\right)\sum_{n=-\infty}^{\infty} X(\omega - n\omega_s) \qquad (1.22)$$

Usually, $\tau \approx T$ in which case

$$x_p(t) \Leftrightarrow Sa\left(\frac{\omega T}{2}\right)\sum_{n=-\infty}^{\infty} X(\omega - n\omega_s) \qquad (1.23)$$

Finally, passing $x_p(t)$ through the reconstruction filter to remove all image spectra gives the continuous signal

$$\hat{x}(t) \Leftrightarrow Sa\left(\frac{\omega T}{2}\right)X(\omega) \qquad (1.24)$$

Equation (1.24) shows that the ideal spectrum $X(\omega)$ is weighted by a $(\sin x)/x$ function and so high baseband frequencies may suffer significant attenuation. For example, Table 1.1 shows the $(\sin x)/x$ loss for several

Table 1.1. *Maximum (sin x)/x loss for a sampling frequency of 13.5 MHz.*

Frequency (MHz)	$Sa(\omega T/2)$	Loss (dB)
1	0.991	0.08
3	0.921	0.72
5	0.789	2.06

video frequencies when the sampling frequency is 13.5 MHz. For broadcast quality video systems this sort of loss is sufficient to warrant the use of an anti-imaging filter with a $(\sin x)/x$ corrected passband response.

1.4.1 Decimation

Many signal processing applications require a change of sample rate, and this invariably requires the use of a *multirate* filter. This class of *digital* filters convert a set of input samples to another set of samples that represent the same analogue signal sampled at a different rate. Typical applications include professional audio systems, sub-band coding for speech processing, narrowband lowpass and bandpass filters, transmultiplexers for time-division multiplexing to frequency-division multiplexing translation, and multiband spectral analysis (Section 5.5.7).

Suppose we wish to reduce the (angular) sample rate by an integer factor M, from $\omega_s = 2\pi$ to $\omega'_s = 2\pi/M$. Ideal sampling theory tells us that this can be achieved without danger of aliasing provided the sampled signal is first bandlimited to a maximum frequency

$$\omega_m \leq \pi/M \tag{1.25}$$

In other words, to satisfy the sampling theorem the input signal must be bandlimited to one-half the *final* sample rate. Typical magnitude Fourier transforms illustrating the limiting case of a sample rate reduction process are shown in Fig. 1.9(a). Here the symbol denotes sample rate compression by an integer factor M, which means that sequence $y(m)$ is formed by extracting every Mth sample from sequence $w(n)$, i.e.

$$y(m) = w(Mm) \tag{1.26}$$

Since bandlimitation is essential for sample rate reduction it is usual to associate an anti-alias lowpass filter with the compressor and to refer to the combination as a *decimation filter*, Fig. 1.9(b). As indicated, the role

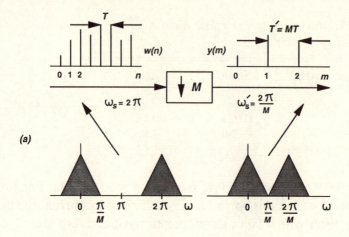

(a)

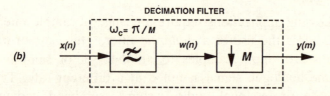

(b)

Fig. 1.9 Decimation filtering: (*a*) sampling rate compressor; (*b*) model for a decimation filter.

of the lowpass filter is to ensure components of $w(n)$ above $\omega = \pi/M$ are negligible. Note that the decimator is a time-varying system and so, unlike most of the filters we shall be dealing with, in general it cannot simply be described in terms of an impulse response or transfer function.

The structural design of a decimation filter can be understood with reference to the digital filter theory presented in Section 4.2. The anti-alias lowpass filter is usually implemented using a *transversal finite impulse response* (FIR) structure, as in Fig. 4.14, and for convenience this is redrawn as a signal–flowgraph in Fig. 1.10(*a*). Here, z^{-1} corresponds to a one (input) sample delay, and $h(0)$ to $h(N-1)$ denote N digital multipliers, the outputs of which are summed before being applied to the compressor. The significant point about Fig. 1.10(*a*) is that, for every M inputs to the compressor, $M-1$ are discarded – a clear waste of computational effort!

Fortunately, it can be shown using signal–flowgraph theory that the two branch operations of compression and multiplication (gain) *commute* (Crochiere and Rabiner, 1983), and this leads to the more computation

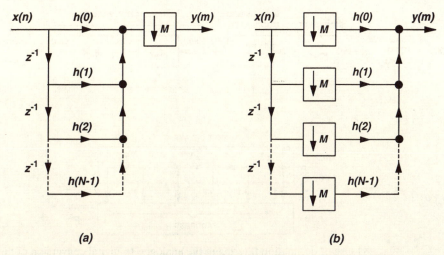

Fig. 1.10 Transversal realization of a decimation filter: (*a*) direct realization of Fig. 1.9 (*b*); (*b*) more efficient realization.

efficient structure in Fig. 1.10(*b*). This structure reduces the computation rate by a factor M since the multipliers and adders are now working at the lower sample rate of $2\pi/M$. Although similar computation gains are claimed for some *recursive* (IIR) decimation structures (Ansari and Liu, 1983), the transversal arrangement in Fig. 1.10(*b*) is usually preferred since it is easily modified to generate an even more computation efficient filter when linear phase is required (see also Section 4.2.2).

Example 1.3

Decimation filters are widely used in professional audio systems in order to eliminate high-precision analogue components. Typically we require a 16-bit (preferably 20-bit) PCM audio signal at the compact disc sampling rate of $f_s = 44.1$ kHz, and for an audio bandwidth of 20 kHz the conventional approach would require a precision, high-order, analogue, anti-alias filter followed by a precision 16-bit ADC.

A better approach is shown in Fig. 1.11(*a*). This utilizes a relatively low-order, low-precision, anti-alias filter F_1, a delta-sigma modulator (1-bit coder), and a high-order, very sharp cutoff decimation filter. The purpose of the 1-bit coder (Goodman, 1969; Agrawal and Shenoi, 1983) is to provide simple quantization coupled with *noise shaping* such that virtually all the *quantization noise* falls between 20 kHz and $Mf_s/2$. Since, using a simple argument, it can be shown that the signal-to-quantizing noise ratio (SNR) at the modulator output is approximately 0 dB, this means that the stopband attenuation of filter F_2 essentially defines the

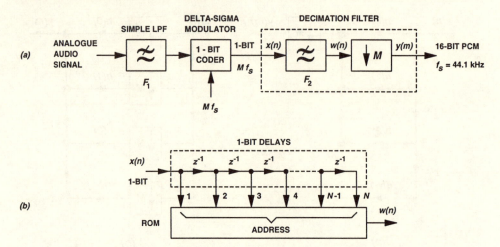

Fig. 1.11 Use of decimation filtering in the analogue-to-digital conversion of hi-fi audio signals (M is typically 72): (a) ADC system; (b) table look-up concept for filter F_2.

required SNR of the overall system. For instance, Example 1.6 shows that 16-bit PCM would require 98 dB stopband attenuation.

The need for high stopband attenuation and a narrow transition band demands a very high-order filter, typically with several thousand coefficients! Fortunately, this is still realistic since the filter input is only 1-bit wide and therefore each coefficient in the transversal structure of Fig. 1.10(a) is effectively multiplied by $+1$ or -1. In other words, the total sum entering the compressor is simply the addition and subtraction of N coefficients. There are 2^N possible values arising from such addition and subtraction and, in principle, these could be stored in a ROM. The (large) N-bit ROM address is then simply the last N bits held in the filter delay line, as indicated in Fig. 1.11(b). This basic structure can then be modified to incorporate compression by a factor M, which, in turn, reduces the ROM address size to a realistic value.

1.4.2 Interpolation

Another type of multirate filter uses interpolation or oversampling in order to increase the sample rate. Suppose we wish to increase the sample rate by an integer factor L. A model of the required process is shown in Fig. 1.12 and the corresponding signals for $L = 4$ are illustrated in Fig. 1.13. For a bandlimited signal $x(n)$ sampled at rate $f_s = 1/T$, ideal sampling theory gives the spectrum $X(e^{j\omega T})$ shown in Fig. 1.13(a). Here,

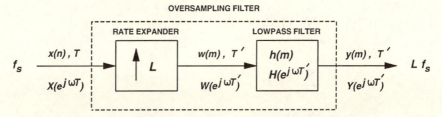

Fig. 1.12 Model for increasing the sample rate by an integer factor L.

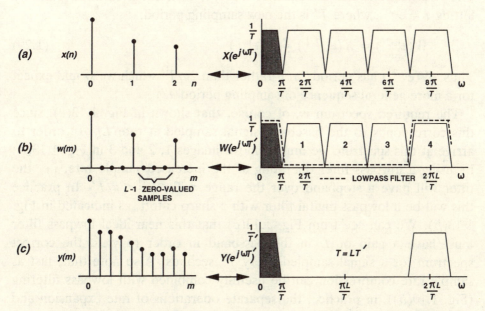

Fig. 1.13 Sequences and their transforms for interpolation by $L = 4$ (diagram shows transform magnitudes).

$X(e^{j\omega T})$ is the Fourier transform of sequence $x(n)$, as defined in (4.22). The *rate expander* now inserts $L - 1$ zero-valued samples after each input sample (Fig. 1.13(b)) but, intuitively, we might expect the spectrum to remain the same since essentially nothing has been added to $x(n)$.

For those familiar with z-transform notation (Section 4.1) we can formally determine the Fourier transform of $w(m)$ as follows. Sequence $w(m)$ can be expressed

$$w(m) = \begin{cases} x\left(\dfrac{m}{L}\right) & m = 0, \pm L, \pm 2L, \ldots \\ 0 & \text{otherwise} \end{cases} \quad (1.27)$$

Taking the z-transform of sequence $w(m)$ gives

$$W(z) = \sum_{m=-\infty}^{\infty} x\left(\frac{m}{L}\right) z^{-m}$$

$$= \sum_{m=-\infty}^{\infty} x(m) z^{-Lm}$$

$$= X(z^L) \tag{1.28}$$

The Fourier transform of sequence $w(m)$ is then obtained from (1.28) by letting $z = e^{j\omega T'}$, where T' is the new sampling period:

$$W(e^{j\omega T'}) = X(e^{j\omega T'L}) = X(e^{j\omega T}) \tag{1.29}$$

Thus, $W(e^{j\omega T'})$ has period $2\pi/T$ rather than $2\pi/T'$ which we would expect for a more general sequence of sampling period T'.

The required spectrum is, of course, that shown in Fig. 1.13(c), since this corresponds to the baseband signal sampled at rate Lf_s. In order to arrive at this spectrum we must remove images 1, 2 and 3 in Fig. 1.13(b) by using a lowpass filter operating at the increased sample rate, i.e. the filter will have a stopband over the range $\pi/T < |\omega| < \pi/T'$. In practice this will be a lowpass digital filter with a sharp cutoff, as indicated in Fig. 1.13(b). We can see from Fig. 1.13(c) that this near-ideal lowpass filter must have a gain of L in the passband in order to yield the correct spectrum for a signal sampled every T' seconds. Also note that, just as sample rate compression can be usefully combined with lowpass filtering (Fig. 1.10(b)), in practice, the separate operations of rate expansion and lowpass filtering are usually combined into a single oversampling filter, as indicated in Fig. 1.12. This avoids the storage and processing of zero-valued samples.

Example 1.4

A good example of interpolation is found in the DAC processing of a compact disc system, and a simple four-times oversampling configuration is shown in Fig. 1.14. Just as for decimation, the digital filter typically has a transversal structure, as in Fig. 1.10(a). However, in order to perform interpolation, each sample shifted into the filter is now multiplied by four different coefficients in turn, rather than by a single coefficient. The products are also summed four times during each sample period, T, and passed to the output, thereby giving an output rate of $4f_s$. As indicated in Fig. 1.14, the filter is of high order (typically 100 or more coefficients), cutting off sharply at the audio limit of 20 kHz so that images 1, 2 and 3 are rejected. The hold circuit following the DAC deliberately generates

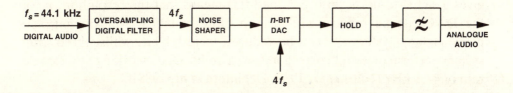

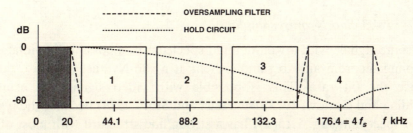

Fig. 1.14 A simple four-times oversampling system for digital-to-analogue conversion in compact disc systems.

$(\sin x)/x$ distortion and, according to (1.24), gives a first zero at the output sampling rate, i.e. at $4f_s = 176.4$ kHz. The $(\sin x)/x$ response, together with that of a simple lowpass analogue filter is sufficient to remove image 4.

The significant point about Fig. 1.14 is that it avoids the need for a sharp cutoff analogue lowpass filter. This filter would also need a linear phase response in the passband to avoid impairment of pulsed sounds. In contrast, linear phase, sharp cutoff and high stability are relatively easy to achieve using the oversampling filter.

Noise shaping

Configurations similar to Fig. 1.14 are also used to simplify the DAC considerably. First note that the wordlength of the oversampling filter output is typically 28 bits (corresponding to a 16-bit input and 12-bit coefficients) and that this must be rounded off to n-bits at the DAC input. In practice this is achieved by a *noise shaper* which, in its first-order form, is essentially a quantizer with the rounding error fedback and subtracted from its input.

Clearly, since the quantizing error will tend to be similar for successive samples of a low-frequency signal, subtraction of the error from the next sample will tend to reduce the average quantizing noise at low frequencies. More specifically, since the signal bandwidth is considerably less than half the sampling frequency due to oversampling, this process enables the noise density to be shaped such that it is low at low frequencies and high outside the signal bandwidth. The low density over the signal bandwidth gives a signal-to-noise ratio *gain* and this means that the resolution, n, of the DAC can be reduced. In fact, in more complex oversampling and noise shaping systems, n can be reduced to just 1 bit, whilst the overall system performance still corresponds to a 16-bit DAC working at a sample rate of 44.1 kHz (Naus *et al*, 1987; Uchimura *et al*, 1988).

1.4.3 Sub-Nyquist sampling

For some signals with a well-defined spectrum it is possible to violate the sampling theorem and to sample them at a sub-Nyquist rate in order to reduce the bit rate. This is possible with video signals for example (Golding and Garlow, 1971; Devereux and Stott, 1978) because the typical video spectrum (Fig. 1.15(a)) has a strong line structure with most of the signal energy concentrated near line frequency harmonics, i.e. at mf_H,

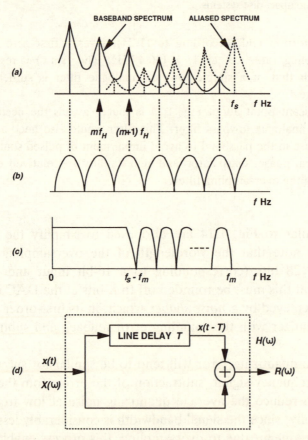

Fig. 1.15 Sub-Nyquist sampling of a video signal: (a) interleaved baseband and aliased spectra; (b) comb response for removal of aliased components; (c) required overall response; (d) comb filter for removal of components at $(m + \frac{1}{2})f_H$.

where m is an integer and f_H is the horizontal or line frequency. If components near $(m + \frac{1}{2})f_H$ can be assumed to be less important than components near mf_H, then it is possible to select a sub-Nyquist sampling frequency such that the major aliased components fall at these frequencies, and they can subsequently be removed by the *comb filter* shown in Fig. 1.15(b). On this assumption, the sub-Nyquist sampling frequency should be

$$f_s = (M + \tfrac{1}{2})f_H \tag{1.30}$$

where M is an integer, and a comb filter having the response shown in Fig. 1.15(c) should be inserted after the DAC. Note that comb filtering is only required over a frequency band disturbed by aliased components, i.e. from $(f_s - f_m)$ to f_m where f_m is the maximum video frequency. Also, other aliased components can be reduced by additional filtering. For example, a baseband component at a video frequency

$$f = (m + \tfrac{1}{2})f_H$$

appears as an aliased component at

$$f = (M + \tfrac{1}{2})f_H - (m + \tfrac{1}{2})f_H = Nf_H$$

where N is an integer. This will not be removed by the above comb filter but it can be removed by preceding the ADC by an identical comb filter.

Example 1.5

Suppose that we require a simple analogue comb filter which will remove *all* the frequency components at $f = (m + \frac{1}{2})f_H$ in a monochrome signal. Since these components are 180° out of phase on adjacent scan lines of the same field, they can be cancelled by adding delayed and undelayed lines as shown in Fig. 1.15(d). Since

$$x(t - T) \Leftrightarrow X(\omega)e^{-j\omega T}$$

then

$$R(\omega) = H(\omega)X(\omega) = (1 + e^{-j\omega T})X(\omega)$$

Therefore

$$H(\omega) = 2\cos^2\left(\frac{\omega T}{2}\right) - j2\sin\left(\frac{\omega T}{2}\right)\cos\left(\frac{\omega T}{2}\right)$$

$$= 2\cos\left(\frac{\omega T}{2}\right)e^{-j\omega T/2}$$

This is a comb response for all ω, with minima at $(m + \frac{1}{2})f_H$ as required.

1.5 Quantization

This section first illustrates typical quantizing and coding characteristics. It then takes a general theoretical look at *quantizing noise* in terms of its power and spectral content, and concludes by computing the SNR for some simple PCM systems.

1.5.1 Quantizing characteristics and coding

We saw in Section 1.2 that amplitude quantization is fundamental to the success of PCM, and that in a well-designed PCM system quantizing error or quantizing noise should be the dominant impairment.

There are essentially two ways of approximating an input sample x to a quantized value or *representative level* y_k $k = 1, 2, \ldots, M$. The usual method is to select the quantizing characteristic such that, if x lies between

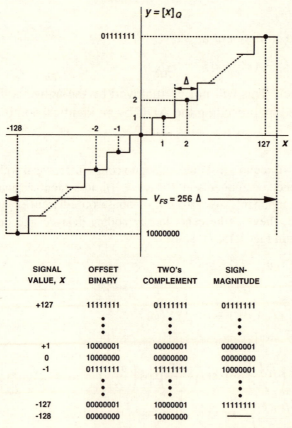

SIGNAL VALUE, X	OFFSET BINARY	TWO's COMPLEMENT	SIGN-MAGNITUDE
+127	11111111	01111111	01111111
⋮	⋮	⋮	⋮
+1	10000001	00000001	00000001
0	10000000	00000000	00000000
-1	01111111	11111111	10000001
⋮	⋮	⋮	⋮
-127	00000001	10000001	11111111
-128	00000000	10000000	——

Fig. 1.16 Typical uniform, bipolar quantizing characteristic and coding of an 8-bit ADC.

$(m - \frac{1}{2})\Delta$ and $(m + \frac{1}{2})\Delta$, where Δ is the *quantum interval* (1 LSB), then x is *rounded* to $m\Delta$. This approach minimizes the quantizing noise power and is illustrated in Fig. 1.16, e.g. $x = 1.8$ is rounded to 2. If the characteristic were to be shifted $\Delta/2$ to the right then an input lying between $m\Delta$ and $(m + 1)\Delta$ would be *truncated* to $m\Delta$, and the quantization noise would increase. In Fig. 1.16 the full-scale input is defined as $V_{FS} = 256\Delta$ or $2^n\Delta$ for an n-bit ADC, so if $V_{FS} = 10$ V we have $\Delta \approx 39$ mV for $n = 8$ and $\Delta \approx 153 \ \mu V$ for $n = 16$. Therefore, despite this rather large input range a 16-bit ADC still requires an exceptionally stable circuit and very-high-precision voltage dividers, and in this case there is good reason to resort to alternative techniques to achieve analogue-to-digital conversion (see Example 1.3).

The quantizer output y is a quantized version of x, i.e. $y = [x]_Q$, and it can use one of a number of codes. For example, if we wish to use straight binary coding for a bipolar input it is necessary to 'offset' it, as shown in Fig. 1.16. Whilst this is a popular coding technique for ADCs, difficulties can arise when performing arithmetic operations. In particular, if the offset binary values are multiplied by a factor, the offset (10000000) is also multiplied, which causes an unwanted shift of the signal zero position. As discussed in Appendix B, most arithmetic is performed using two's complement representation, where a negative number is formed by bit-by-bit inversion of the magnitude, plus 1 LSB (in practice this is often obtained by simply inverting the MSB of the offset binary representation). For the characteristic in Fig. 1.16 a two's complement number can be written

$$y = [x]_Q = 128\left(-b_0 + \sum_{i=1}^{7} b_i 2^{-i}\right)$$

where bit b_i, $i = 0, 1, \ldots, 7$, is 1 or 0, which also means that zero signal is conveniently represented by the all-zero code. Finally, note that sign-magnitude coding can sometimes be advantageous in applications involving symmetrical positive and negative signals since only the sign bit (the MSB) changes between positive and negative values of a number.

The foregoing discussion and Fig. 1.16 illustrate the common case of mapping from analogue to digital domains through the use of *uniform quantization*. This is appropriate for digitizing video signals for example, where the signal pdf $p(x)$ tends to be 'flat'. In this case we usually assume that the quantizing noise power is independent of the input signal, which, in turn, means that the SNR will decrease with decreasing signal level. On the other hand, when $p(x)$ is non-uniform it is possible to quantize the input with a smaller error variance by using a *non-uniform* quantizing

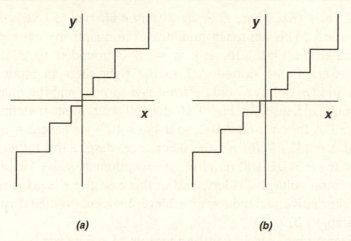

Fig. 1.17 Non-uniform quantizing characteristics: (*a*) midrise; (*b*) midtread.

characteristic. Typically the signal pdf $p(x)$ peaks at $x = 0$ (as in the case of speech signals) and the corresponding optimized quantizing character-istic will be of the form shown in Fig. 1.17. The general idea is to use more quantizing levels where $p(x)$ is relatively high and vice-versa. Clearly, this type of *pdf optimized quantizer* is susceptible to mismatch with respect to both pdf shape and input variance and this implies that the SNR is only optimal for one specific variance. A good example of pdf optimized quantizers and non-uniform quantizing is discussed in Section 2.2.2 under DPCM. Here it will be seen that the quantizer input x is a *digital difference* signal with a pdf peak at $x = 0$. A similar idea is found in *digital companding* systems for sound signals (Gilchrist, 1984).

1.5.2 *Quantizing noise*

The quantizing process for an arbitrary waveform $x(t)$ is illustrated in Fig. 1.18. In this diagram an input in the range

$$d_k \leqslant x < d_{k+1}$$

(where d_k is a *decision level*) is quantized to the representative level y_k and the quantum interval corresponding to y_k is Δ_k. Over this quantum interval the instantaneous quantizing error or quantizing noise is

$$q_n = x - y_k \qquad (1.31)$$

and we are interested in computing its mean square value for a particular quantizing problem. Note that, whilst q_n is obviously a function of time, it

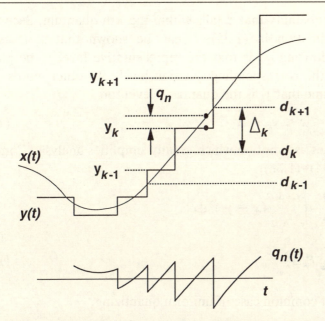

Fig. 1.18 Quantizing error for a non-uniform characteristic.

is also a function of the instantaneous signal amplitude x. Put more formally, the random variable Q_n corresponding to quantizing noise is a function of the random variable X since

$$Q_n = X - Y \tag{1.32}$$

where Y is the discrete-valued random variable representing the quantizer output levels. Having recognized this, we can now apply simple statistics to evaluate the mean square value of q_n. Recalling that, for a continuous random variable X with a pdf $p(x)$,

$$E[f(X)] = \int_{-\infty}^{\infty} f(x)p(x)\,\mathrm{d}x \tag{1.33}$$

it follows that the mean square quantizing noise for L quantizing levels is

$$\overline{q_n^2} = E[Q_n^2] = \sum_{k=1}^{L} \int_{d_k}^{d_{k+1}} (x - y_k)^2 p(x)\,\mathrm{d}x \tag{1.34}$$

This can be simplified by making several assumptions. Firstly, provided that quantizing is not too coarse, then $p(x)$ can be approximated by a constant $p(y_k)$ over the kth quantum interval, i.e. over this interval,

$$p(x) = p(y_k) = P_k/\Delta_k \tag{1.35}$$

where P_k is the probability that x falls within the kth quantum. Secondly, using the approximation in (1.35), it can be shown that a necessary condition for minimizing $\overline{q_n^2}$ is that the representative level y_k be placed midway between the corresponding decision levels. In other words, it is reasonable to assume that y_k is the quantized level for

$$y_k - \Delta_k/2 \leqslant x < y_k + \Delta_k/2 \tag{1.36}$$

This is an important assumption which greatly simplifies analysis. Combining Equations (1.34)–(1.36),

$$\overline{q_n^2} = \sum_{k=1}^{L} \frac{P_k}{\Delta_k} \int_{y_k-\Delta_k/2}^{y_k+\Delta_k/2} (x - y_k)^2 \, dx$$

$$= \frac{1}{12} \sum_{k=1}^{L} P_k \Delta_k^2 \tag{1.37}$$

For the special but common case of uniform quantizing,

$$\overline{q_n^2} = \Delta^2/12 \tag{1.38}$$

where Δ is the quantum interval, and we conclude that, if quantizing is uniform and reasonably fine, then $\overline{q_n^2}$ is approximately independent of the pdf of the input signal. Note that in the case of fine uniform quantizing we could assume that

$$p(q_n) = \begin{cases} 1/\Delta & -\Delta/2 \leqslant q_n \leqslant \Delta/2 \\ 0 & |q_n| > \Delta/2 \end{cases} \tag{1.39}$$

and use a more direct calculation for $\overline{q_n^2}$ from (1.33):

$$\overline{q_n^2} = \int_{-\Delta/2}^{\Delta/2} q_n^2 p(q_n) \, dq_n = \Delta^2/12 \tag{1.40}$$

Also, since the pdf $p(q_n)$ has zero mean, then

$$\overline{q_n^2} = E[Q_n^2] = \sigma_q^2 = \Delta^2/12 \tag{1.41}$$

where σ_q^2 is the quantizing noise variance or normalized noise power.

System considerations

At this point it is useful to model the principal operations of a PCM system in order to illustrate how the quantizing noise propagates to the system output, and a suitable model (ignoring coding) is shown in Fig. 1.19(*a*).

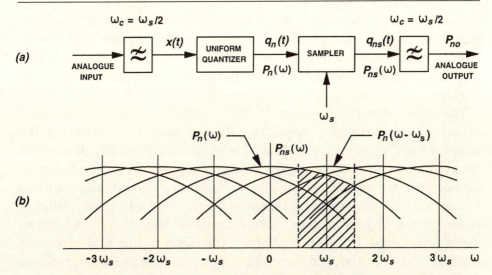

Fig. 1.19 Quantizing noise: (*a*) model of a PCM system; (*b*) quantizing noise spectra.

We shall assume fine uniform quantizing and for mathematical convenience this is placed before (ideal) sampling. In practice, sampling precedes quantizing but these two operations commute and so they can be represented as shown. Also, the lowpass filters are assumed to be ideal with cutoff frequency $\omega_c = \omega_s/2$. Intuitively, it appears from Fig. 1.18 that quantizing noise is a relatively wideband signal with a *power spectral density (PSD)* $P_n(\omega)$ which extends well above the sampling frequency ω_s. Assuming this to be the case, then the PSD $P_{ns}(\omega)$ of the sampled quantizing noise will be a highly aliased spectrum (Fig. 1.19(b)) and the noise power P_{no} at the output of the reconstruction filter will be

$$P_{no} = \frac{1}{2\pi} \int_{-\omega_s/2}^{\omega_s/2} P_{ns}(\omega)\, d\omega \tag{1.42}$$

$$= \frac{1}{2\pi} \sum_{k=-\infty}^{\infty} \int_{-\omega_s/2}^{\omega_s/2} P_n(\omega - k\omega_s)\, d\omega$$

$$= \frac{1}{2\pi} \sum_{k=-\infty}^{\infty} \int_{k\omega_s - \omega_s/2}^{k\omega_s + \omega_s/2} P_n(\omega)\, d\omega$$

For $k = 1$ the integral corresponds to the shaded region in Fig. 1.19(b) and so summation over all k must give the area under $P_n(\omega)$. Therefore,

$$P_{no} = \frac{1}{2\pi} \int_{-\infty}^{\infty} P_n(\omega)\, d\omega = \frac{\Delta^2}{12} \tag{1.43}$$

by definition of the PSD. This interesting result states that all the quantizing noise power generated by the quantizer appears in the band $0-\omega_s/2$ at the analogue output. In practice, $\omega_s > 2\omega_m$ and $\omega_c < \omega_s/2$ (see Fig. 1.4(*c*)) and so the output noise power will be somewhat less than $\Delta^2/12$.

The actual output noise power P_{no} for $\omega_c < \omega_s/2$, and the characteristics of the output quantizing noise, can be estimated by considering the sampled noise signal $q_{ns}(t)$ (a train of impulses in our model). Except for special cases (periodic or slowly changing inputs), it is reasonable to assume that the samples of quantizing noise are uncorrelated and this means that the autocorrelation function of $q_{ns}(t)$ over a large time window tends to an impulse function. According to the *Wiener–Khinchine theorem*, the PSD $P_{ns}(\omega)$ is the Fourier transform of this autocorrelation function and so $P_{ns}(\omega)$, and therefore the output noise spectrum, tends to be flat. This is readily confirmed for fine quantizing, say, 10 bits/sample or more for uniformly quantized sound signals and 6 bits/sample or more for uniformly quantized video signals. In these cases, quantizing noise is similar to white Gaussian noise, although the noise spectra become progressively more peaked as the digital resolution is reduced. Subjectively, these deviations from white noise characteristics are referred to as *granular noise* for small sound signals and *contouring* (as in a map) for slowly varying video signals.

Assuming a flat PSD and a reconstruction filter with a cutoff frequency $\omega_c < \omega_s/2$, it follows from the definition of PSD in (1.42) that

$$P_{no} = 2\sigma_q^2 \frac{\omega_c}{\omega_s}; \quad \sigma_q^2 = \frac{\Delta^2}{12} \tag{1.44}$$

1.5.3 SNR in PCM systems

Knowing the quantizing noise power we can now compute a signal-to-quantizing noise ratio (SNR) at the output of a PCM system (after the reconstruction filter). In doing so, we must be careful to state how the SNR is defined, together with the exact terms of measurement. For instance, the overall SNR depends upon ω_c and ω_s (Equation (1.44)), and upon any increase in noise level due, say, to multiple coding–decoding operations in tandem or to the addition of a *dither* signal at the ADC input. Also, in practice a weighting factor (in dB) is often added to the basic computed SNR to account for the subjective effect of different noise spectra.

Example 1.6

The SNR in digital audio systems is often defined as the ratio of full-scale signal power to the noise power. Taking the full-scale signal at the output of the DAC as $2^n\Delta$, then the power of a digitized sinusoidal signal is simply $(2^{n-1}\Delta/\sqrt{2})^2$. Assuming fine uniform quantizing and that all the quantizing noise power $(\Delta^2/12)$ is available at the reconstruction filter output, we have

$$\begin{aligned} \text{SNR} &= 10\log_{10}\left[\frac{\text{signal power}}{\text{noise power}}\right] \quad \text{dB} \\ &= 20\log_{10}[2^{n-1}\sqrt{6}] \\ &= 6n + 1.8 \quad \text{dB} \end{aligned} \qquad (1.45)$$

With the above assumptions, this means, for example, that the SNR in a uniformly quantized 16-bit PCM audio system is limited to about 98 dB.

This example highlights several important points. First, the SNR changes by 6 dB/bit, and this is generally true for other definitions of SNR. Secondly, in general, SNR calculations using (1.45) or a similar expression cannot be directly compared with signal-to-random-noise measurements in an analogue system. This is because the *subjective effect* of quantizing noise in audio and video systems differs from that of truly random noise, particularly for coarse quantization.

Now consider a more general calculation of SNR for a PCM system using uniform quantization. The system may consist of a number of identical ADC–DAC pairs (codecs) interspaced with analogue signal processing, as in Fig. 1.20(*a*), and the overall SNR has to be computed. We shall ignore the analogue processing and also assume that each codec

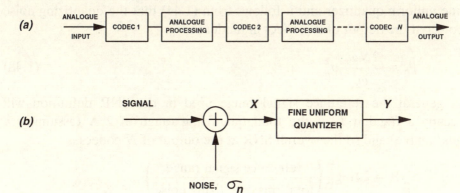

Fig. 1.20 Combining noise sources in a PCM system: (*a*) multiple codecs in an analogue/digital environment; (*b*) a model for computing total output noise variance.

incorporates an input and output lowpass filter with cutoff frequency $\omega_c < \omega_s/2$. The system is then a tandem connection of codecs and this raises the question of how the noise variance at the input to a codec affects the noise variance at its output. To answer this, consider the random variables X and Y in Fig. 1.20(b); as before, we can write

$$Y = X - Q_n$$

where Q_n is the random variable representing the quantizing noise. The variance of the quantizer output is then

$$\sigma_y^2 = E[(Y - \mu_y)^2]$$

where $\mu_y = E[Y]$. Noting that $E[Y] = E[X] = \mu_x$, since $E[Q_n] = 0$, this reduces to

$$\sigma_y^2 = \sigma_x^2 + \sigma_q^2 - 2E[(X - \mu_x)Q_n]$$

If we now make the reasonable assumption that Q_n is independent of X (as implied by (1.38)) we have

$$\sigma_y^2 = \sigma_x^2 + \sigma_q^2 \tag{1.46}$$

(Note that this assumption is not true and Equation (1.46) does not hold when the quantizer is optimized to the input pdf.) Finally, if the noise component of X is uncorrelated with the signal component of X, then the total output noise power from the quantizer is

$$P_n = \sigma_n^2 + \sigma_q^2 \tag{1.47}$$

where σ_n^2 is the input noise variance and $\sigma_q^2 = \Delta^2/12$. In other words, to a good approximation we can add noise powers at the input and output of a fine uniform quantizer and it follows from (1.44) that the quantizing noise power at the output of N codecs in tandem is

$$P_N = \frac{2N\omega_c}{\omega_s}\sigma_q^2 \tag{1.48}$$

In general the reference signal range used in the SNR definition will occupy a fraction k of the total quantizing range of $2^n\Delta$ (assuming n bits/sample) and so the overall SNR at the output of N codecs is

$$\text{SNR} = 20\log_{10}\left(\frac{\text{reference signal range}}{\text{total rms quantizing noise}}\right)$$

$$= 20\log_{10}\left(\frac{k2^n\Delta}{\sqrt{P_N}}\right)$$

$$= 6n + 10.8 + 10\log_{10}\left(\frac{k^2\omega_s}{2N\omega_c}\right) \text{dB} \qquad (1.49)$$

Example 1.7

The CCIR (International Radio Consultative Committee) recommendation (ITU, 1982) for the digital coding of 625-line 50-fields/s luminance video signals (the 4:2:2 standard) gives the following parameters:

digital resolution n	8 bits/sample, uniform quantization
sampling frequency	13.5 MHz (864 × line frequency)
reference signal range	220 Δ

Note that for video signals the reference range is between black and white levels and in the 4:2:2 coding standard this is 220/256 of the total quantizing range. Typical luminance SNR estimates found using (1.49) and the above parameters are shown in Table 1.2.

Table 1.2. *Overall SNR for luminance video signals and N codecs in tandem. Coding is to the 4:2:2 standard.*

f_c(MHz)	SNR (dB)		
	$N = 1$	$N = 2$	$N = 3$
5.0	58.8	55.8	54.0
5.5	58.4	55.4	53.6

1.6 Source and channel coding

So far we have concentrated on 'straight PCM', i.e. waveform sampling, quantizing, and binary coding. In communications systems this is usually only the first step in what is often a complex signal processing sequence prior to transmission or data storage. It is usual to split this processing into *source coding* and *channel coding*, as shown in Fig. 1.21, and we could regard the straight PCM coder as the 'front end' of the source coder.

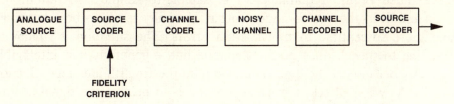

Fig. 1.21 A coding model for digital communications systems.

1.6.1 Source coding

Source coding techniques are discussed at length in Chapter 2 and so here we will outline the basic philosophy. We could identify the following important concepts associated with source coding.

Data compression

The broad objective of source coding is to exploit or remove 'inefficient' redundancy in the PCM source and thereby achieve a reduction in the overall source rate R. According to Theorem 1.1, information can be reliably transmitted provided $R < C$, which, in turn, means that source coding provides a way of minimizing the required channel capacity C, and hence the cost of the link. For example, it would be uneconomic to transmit directly a conventional broadcast quality video signal for video-conferencing, and for this application source coding techniques have resulted in more cost effective bit rates of the order of 1 Mbit/s.

Signal fidelity

A *fidelity* or *quality criterion* (Fig. 1.21) is an essential concept behind source coding. Sometimes we require data compression without information loss (as in the compression of computer data files) and this leads to a branch of source coding called *reversible* or *lossless* coding. Alternatively, we may be satisfied with a non-reversible (lossy) source coding scheme which, by definition, results in an acceptable amount of signal *distortion*. Non-reversible source coding is the basis of most speech and video compression schemes since slight information loss can go undetected by the human observer.

Robustness

A consequence of efficient coding of the PCM source will be that the source coder output tends to be a sequence of *independent* (and therefore *uncorrelated*) symbols. Put another way, we might regard the source coder as attempting to generate a random signal at its output. Unfortunately, since much of the redundancy, e.g. sample-to-sample correlation, has been removed the source coder output generally tends to be less robust and more susceptible to transmission errors. The situation is similar to listening to speech on a noisy telephone line – generally it is intelligible but only because of the high redundancy in the English language (Section 2.1.3). As we shall see, the usual solution is to protect the signal against transmission errors using channel coding.

1.6.2 Coding standards

In practice the target bit rate for a source coder is often determined by the need to meet a specific CCITT bit rate. Consider the standardized, gross (total package) rates for digital networks in Table 1.3. For Europe, level 1 corresponds to CCITT 30-channel PCM, i.e. 30 voice channels and 2 framing/signalling channels. The 32 8-bit time slots span 125 μs giving a gross rate of 2.048 Mbit/s and a voice channel rate of 64 kbit/s, i.e. 8 kHz sampling and 8 bits/sample. A speech rate of 64 kbit/s achieves *toll-quality* (commercial telephony quality – which is higher than communication quality) by using non-uniform quantizing or digital companding, and in Europe this is an approximation to *A*-law companding. For the USA (and generally for Japan), level 1 corresponds to CCITT 24-channel PCM i.e. 24 8-bit time slots plus 1 framing bit span 125 μs and gives a gross rate of 1.544 Mbit/s (signalling being provided by occasional 'bit stealing' from 8-bit time slots). In this system digital companding is an approximation to μ-law companding and the gross channel rate of 64 kbit/s includes signalling. Also note that 64 kbit/s is the capacity of an *Integrated Services Digital Networks* (*ISDN*) B channel.

We might regard conversion to 64 kbit/s *A*-law or μ-law PCM as the 'front end' processing of the source coder in Fig. 1.21. For example, the 64 kbit/s PCM data stream could then be applied (after first conversion to uniform PCM) to an *adaptive differential PCM (ADPCM)* system in order to reduce the bit rate from 8 bits/sample to 5, 4, 3 and 2 bits/sample under CCITT Rec. G721, G723 and G726. These compressed rates are useful for speech storage and low-bit-rate telephony, e.g. at 2 bits/sample (16 kbit/s) an 8-minute conversation could be stored on a 1 Mbyte DRAM. The ADPCM system is referred to as a *transcoder* between the uniform PCM

Table 1.3. *Standardized digital network hierarchies (rates in* Mbit/s).

	Level			
	1	2	3	4
Europe	2.048	8.448	34.368	139.264
USA	1.544	6.312	44.736	274.176
Japan	1.544	6.312	32.064	97.728

system and the channel coder. Similarly, a *linear predictive coder (LPC)* for speech compression is also referred to as a transcoder between uniform PCM and the channel.

Coding targets can also be identified for video systems. Consider first broadcast (or studio) quality signals and in particular the CCIR 4:2:2 coding standard (ITU, 1982). In this case the recommended sampling frequencies for the luminance and two-colour difference signals are 13.5 MHz and 6.5 MHz respectively, and both are uniformly quantized to 8 bits/sample. This gives a raw bit rate of 216 Mbit/s which can be reduced to an effective 166 Mbit/s if the video blanking intervals are removed before coding. We therefore need a compression ratio of 1.2:1 to achieve transmission at the European 140 Mbit/s level and a compression ratio of 4.9:1 to achieve transmission at the 34 Mbit/s level, see Table 1.4. Transmission at 34 Mbit/s is in fact quite feasible using some initial sample rate reduction followed by differential pulse code modulation (DPCM), and possibly *variable-length coding*. Note that 34 Mbit/s is the 'total package' rate embracing video error correction, sound and ancillary services, and so the actual compressed video rate will be around 30 Mbit/s.

For videoconferencing we can accept lower spatial resolution in the picture, and transmission at the European 2 Mbit/s level can be achieved using a combination of *intraframe* and *interframe* DPCM coding followed by variable-length coding. More significantly, CCITT Rec. H.261 defines a source coding standard for videoconference/videophone codecs operating from 64 kbit/s to 2 Mbit/s, in integer multiples of 64 kbit/s ('$p \times$ 64 kbit/s'). To enable interchange between world standards, Rec. H.261 specifies a 'common intermediate format' (CIF), and Table 1.4 shows the compression ratios required to achieve videoconference and videophony rates for this format.

Table 1.4. *Video compression ratios required for European transmission rates.*

	Transmission rate				
	64 kbit/s	384 kbit/s	2 Mbit/s	34 Mbit/s	140 Mbit/s
Broadcast television			83	4.9	1.2
Videoconferencing (CIF, 30 pictures/s)		95	18.2		
Videophony (0.25 CIF, 20 pictures/s)	95				

1.6.3 Channel coding

Channel coding can be considered as the matching of the source coder output to the transmission channel, and usually it involves two distinct processes:

(1) error control coding,

(2) transmission coding.

Consider first the concept of error control coding. In general, the 'noisy channel' in Fig. 1.21 represents a digital modulator, a *physical analogue channel*, and a digital demodulator, and one of the main aims of a channel coder is to combat the effects of noise in the analogue channel. Paradoxically, whilst the source coder is removing 'inefficient' redundancy, the channel coder reinserts 'efficient' redundancy in the form of *parity* or *check* symbols in order to protect the compressed signal from channel noise. Fortunately, the added redundancy is usually only a small fraction of that which is removed by the source coder. For example, in broadcast quality digital video systems, a source coder will typically achieve 3:1 compression whilst error control coding in the corresponding channel coder will increase the bit rate by typically 5–10%.

The second major role of channel coding (i.e. transmission coding) is to modify the transmitted spectrum so that it is more suitable for the transmission channel. This often involves additional coding techniques and these are discussed in detail in Section 3.8.

2

Source coding

In Section 1.6 we saw that source coding is the process whereby redundancy is removed in order to reduce the bit rate (or to compress a data file prior to storage). It was also pointed out that a data compression technique can usually be classified as either reversible or non-reversible, as shown in Fig. 2.1. This diagram highlights practical techniques which have found application in the communications and computing fields.

2.1 Reversible coding

Practical reversible coding techniques can be classified as either *statistical coding*, or what may loosely be termed *string coding*. If transmission is

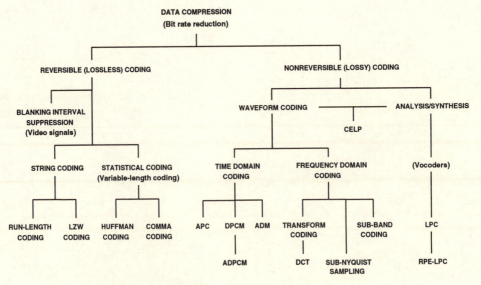

Fig. 2.1 Useful data compression techniques.

error-free, either approach implies that the decoder output is *exactly* the same as the coder input – an obvious requirement for computer data storage, for example.

Statistical coding achieves compression by exploiting the *statistical* redundancy in the signal, in contrast to, say, DPCM which exploits *correlation* redundancy in the signal. It is also referred to as *variable-length coding* or *entropy coding* and so it relies heavily upon the concept of *source entropy*. This means that statistical coding generally requires *a priori* knowledge of the source (in the form of source symbol probabilities) and, clearly, this may not always be available. In contrast, string coding techniques such as *run-length coding* and the *LZW* (Lempel–Ziv–Welch) algorithm do not require *a priori* knowledge of the source. On the other hand, the LZW algorithm still exploits statistical redundancy, and an underlying reason why LZW coding is successful can be explained by variable-length coding theory. Also, in some applications, coding of run-length makes use of variable-length coding.

2.1.1 Run-length coding

This simple reversible technique converts a string of repeated characters into a compressed string of fewer characters. A general format for run-length coding is shown in Fig. 2.2(*a*), together with several examples. A special character S is sent first to indicate that compression follows and X and C denote the repeated character and character count respectively. Clearly, S should not naturally occur in the data string, and the format in Fig. 2.2(*a*) is only useful for strings of four or more repeated characters.

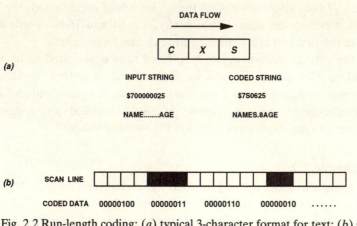

Fig. 2.2 Run-length coding: (*a*) typical 3-character format for text; (*b*) simple fixed wordlength coding for two-tone graphics.

Alternatively, if S denotes a specific character, such as the 'space' character, the format reduces to S and a space-count character, C. In other words, space compression in text requires just two characters.

Run-length coding is also used extensively for compressing two-tone (black and white) images prior to transmission. Typical images are pages of text and drawing (as found in facsimile) and cartoon-type images for low bit rate visual communication. In this application run-length coding operates in one dimension (horizontally) and exploits the correlation between pixels along a line. The simplest scheme uses fixed wordlength coding and Fig. 2.2(b) illustrates this for a single scan line and 8-bit words. The transmitted sequence is simply the run-lengths 4, 3, 6, 2, . . . and each new symbol denotes a tone change. A facsimile system scanning at the CCITT rate of 1728 samples/line is quite capable of generating (usually white) runs of the order of 100 or larger, and so the use of 8-bit words can achieve significant compression, i.e. a run of up to 255 will be represented by only 8-bits. On the other hand, for a facsimile document short runs are usually more probable than long ones and this statistical redundancy is often exploited by applying variable-length coding to the run-lengths. Run-length/variable-length coding for two-tone graphics is discussed in Example 2.6.

Example 2.1

Multispectral (visible and infrared) image data from weather satellites can be clustered into three homogeneous regions representing a surface class (sea plus land) and two cloud classes. This format is suitable for image dissemination on the PSTN (Wade and Li, 1991) and the homogeneous regions lend themselves to run-length coding prior to transmission (Fig. 2.3). If the run-length $r < 64$ then 8-bit words are selected, the first two bits denoting the class. Occasionally, $r \geq 64$ and 16-bit words are used, with the first two bits indicating the extended wordlength.

For this application typically 94% of runs were found to have $r < 64$, and so 8-bit coding is used for most of the time (corresponding to an average wordlength of 8.5 bits/run). This simple byte-oriented run-length/variable-length coding scheme is easily handled by computer and achieved typically 4:1 compression for weather images.

Example 2.2

A transform coder applied to video signals converts an $N \times N$ pixel block of a digitized image into an $N \times N$ block of transform coefficients in the 'frequency' domain. After coefficient quantization the non-zero coefficients tend to be clustered around the origin, or DC, and much of the

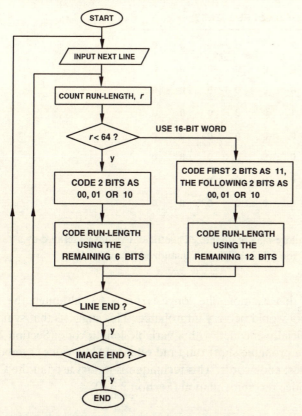

Fig. 2.3 Use of an 8/16 run-length code for weather image compression (00, 01, and 10 code three different classes in the image).

block is taken up with zero-valued coefficients. Some types of transform coder (see Section 2.3.3) exploit this redundancy in the coefficient block in order to increase the efficiency of coefficient transmission.

Consider the simple 8×8 example in Fig. 2.4. If the coefficients are transmitted by addressing them in a zig zag way, as shown, then the non-zero coefficients are concentrated at the start of the transmitted sequence and the efficiency of run-length coding is improved. Each non-zero coefficient can be regarded as a run terminator, and so the run-length followed by the value of the non-zero coefficient can be treated as a 'source symbol', a_j, $j = 1, 2, 3, \ldots$. For example, the zig zag scanning in Fig. 2.4 yields the symbols

$$a_1 = (0, 9) \qquad a_2 = (0, 6) \qquad a_3 = (0, 4)$$
$$a_4 = (2, 2) \qquad a_5 = (4, 1) \qquad a_6 = (7, 1)$$

and the remaining zero-valued coefficients can then be denoted by simply transmitting an end-of-block character, i.e. coefficients after the last non-zero one are not transmitted.

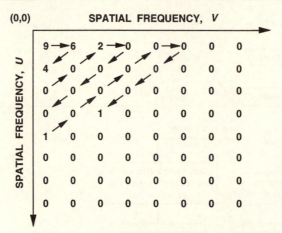

Fig. 2.4 Zig zag addressing of quantized coefficient values generated by a transform coder (the DC coefficient is quantized to a value 9).

Clearly, for this example the longest run is 63, although the corresponding symbol would be very improbable. This suggests that symbols a_j could be beneficially encoded with a variable-length code (Section 2.1.4), where the more probable short run (and relatively low value) symbols are given the shortest codewords. This technique is in fact used in the CCITT H.261 video codec recommendation (Section 2.3.4).

2.1.2 *Information measure and source entropy*

Before discussing variable-length coding we need a quantitative description of the discrete source in terms of its *information rate* and this is usually given by the source entropy. Consider a discrete source having a finite alphabet A of M different symbols a_j, $j = 1, 2, \ldots, M$ each with probability $P(a_j)$ of being generated. Clearly,

$$P(a_j) \geq 0, \quad j = 1, 2, \ldots, M$$

and

$$\sum_{j=1}^{M} P(a_j) = 1$$

The less likely that a symbol a_j will be generated, the more surprised we are when it is generated and we could interpret this as giving us more information than that conveyed by a commonly occurring symbol. This and similar arguments lead to the familiar definition for *self information* of a discrete source, namely for symbol a_j

$$I(a_j) = \log\left[\frac{1}{P(a_j)}\right] \tag{2.1}$$

This is the information conveyed when the source output assumes symbols a_j at some point in time. The units of information depend upon the base and it is standard practice in information theory to use base 2 and so quantify information in *bits*.

As far as communication systems are concerned, it is more useful to describe the source in terms of the *average* information that it generates per symbol rather than the information per symbol and this is simple to calculate if we assume all M symbols to be statistically independent. In this case the jth symbol occurs about $NP(a_j)$ times in a long message of N symbols and the total information is approximately

$$-\sum_{j=1}^{M} NP(a_j)\log P(a_j) \quad \text{bits}$$

The source entropy $H(A)$ is the statistical average information per symbol:

$$H(A) = -\sum_{j=1}^{M} P(a_j)\log P(a_j) \quad \text{bits/symbol} \tag{2.2}$$

As in statistical mechanics, $H(A)$ is a measure of the uncertainty or disorder of the source or system and it is low for an ordered source. In the limit, if $P(a_1) = 1$ and $P(a_j) = 0$, $j = 2, 3, \ldots, M$, then $H(A) = H(A)_{\min} = 0$. However, for a very disordered source, all symbols would be equiprobable $(P(a_j) = 1/M)$ and it can be shown that this leads to the maximum possible average information per symbol:

$$H(A)_{\max} = \sum_{j=1}^{M} \frac{1}{M}\log M$$

$$= \log M \tag{2.3}$$

Therefore,

$$0 \leqslant H(A) \leqslant \log M \tag{2.4}$$

Finally, for a symbol rate k we can define an average information rate or *entropy rate* of $R = kH(A)$ bits/s.

It is important to note the limitations of (2.2). First, $H(A)$ will have little meaning if the $P(a_j)$ vary with time, i.e. if the source is *non-stationary* and so we might expect difficulty with video signals for example.

Secondly, (2.2) accurately represents the average information content of the source only when the symbols are statistically independent, i.e. when $P(a_j|a_i) = P(a_j)$. We say the source must have *zero-memory* and that (2.2) is the *zero-order entropy*. For some practical sources this is a reasonable approximation. For example, successive run-lengths in two-tone facsimile are often assumed statistically independent and so, regarding the runs as symbols, (2.2) can be used to give a good measure of the source entropy (in bits/run). Very often, it is also reasonable to apply (2.2) to a (quantized) difference source, as found in DPCM systems. On the other hand, for many information sources the zero-memory assumption in (2.2) is inadequate and it is necessary to account for the fact that the occurrence of a given symbol depends on preceding symbols (Section 2.1.3). In other words, we may wish to use a better source model.

Example 2.3

A simple PCM coder is modelled in Fig. 2.5 and it is assumed that the analogue source statistics result in the indicated probabilities of the four quantizing levels. Taking the quantizer output as a discrete source with independent symbols, the entropy of the discrete source is

$$H(A) = -\sum_{j=1}^{4} P(a_j) \log P(a_j)$$

$$= \tfrac{1}{8}\log 8 + \tfrac{1}{2}\log 2 + \tfrac{1}{4}\log 4 + \tfrac{1}{8}\log 8$$

$$= 1.75 \text{ bits/symbol}$$

Clearly, the *information rate* of 1.75 bits/symbol is less than the *bit rate* (or *data rate*) of 2 bits/symbol. The significant point here is that, in general, raw bit rates will not necessarily represent information rates! In fact, the role of source coding is to reduce the raw PCM bit rate so that the final average bit rate is close to the average information rate $H(A)$ found from some source model (and not necessarily from (2.2)). This is stated formally in Shannon's *noiseless coding theorem*, Theorem 2.1.

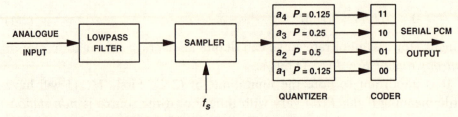

Fig. 2.5 A simple PCM coder.

Theorem 2.1 *A discrete source of entropy H(A) bits/sample or symbol can be binary coded with an average bit rate R_b that is arbitrarily close to H(A) i.e.*

$$R_b = H(A) + \varepsilon \quad \text{bits/sample} \tag{2.5}$$

where ε is an arbitrarily small positive quantity.

Shannon's theorem implies that source entropy, as found using some source model, can be used as a target for data compression schemes, and in practice this can be tackled through variable length coding (Section 2.1.4). It also implies that the maximum achievable compression ratio is

$$C_R = \frac{\text{average bit rate of source coder input}}{\text{average bit rate of source coder output}} \approx \frac{n}{H(A)} \tag{2.6}$$

(where the PCM source quantizes to n bits/sample) and that the entropy rate gives a lower bound on the required channel capacity (see Theorem 1.1).

Example 2.4

A pessimistic estimate can be made for the channel capacity C required by a broadcast quality television signal. Suppose the signal is sampled at 13.5 MHz and is quantized to 8 bits/sample ($M = 256$). For UK signals (System I) this corresponds to some 4×10^5 samples per picture (ignoring blanking) and so

$H(A)_{\text{max}} \approx 3.2 \text{ Mbits/picture}$

Therefore

$C \geqslant 25 \times 3.2 \quad (25 \text{ pictures/s})$

$= 80 \text{ Mbits/s}$

This is pessimistic because the maximum possible entropy has been assumed and the *actual* entropy will be less than this.

2.1.3 Conditional entropy

For maximum compression $H(A)$ in (2.5) and (2.6) needs to be an *accurate* measure of source entropy and, as previously pointed out, the use of (2.2) at an arbitrary point in the digital system, e.g. the ADC output, may be unrealistic. For many information sources the current sample or symbol has statistical dependence upon preceding samples and so we need a better source model, i.e. one with memory. A common approach is to model the source as an *mth-order Markov source*, where the occurrence of a sample depends on m immediately preceding samples (an mth-order Markov process is also referred to as an mth-order autoregressive (AR) process). This inherently involves *conditional probabilities* and the resulting entropy can be termed a *conditional entropy* $H_c(A)$.

Let the m preceding samples define a state s_i and let the next source sample be a_j, where a_j is drawn from the alphabet $A = \{a_1, a_2, \ldots, a_M\}$. The occurrence of a_j is then defined by a conditional probability $P(a_j|s_i)$ and for state s_i we can define a conditional entropy

$$H(s_i) = -\sum_{j=1}^{M} P(a_j|s_i) \log P(a_j|s_i) \tag{2.7}$$

By definition the source entropy is the statistical average of $H(s_i)$ over the $q = M^m$ possible states, and so

$$H_c(A) = \sum_{i=1}^{q} P(s_i) H(s_i) \tag{2.8}$$

$$= -\sum_{i=1}^{q} \sum_{j=1}^{M} P(s_i) P(a_j|s_i) \log P(a_j|s_i) \tag{2.9}$$

The conditional entropy $H_c(A)$ will be a measure of the *actual* source entropy. For a zero-memory source a_j is independent of preceding symbols ($m = 0$) and there is only one possible state, i.e.

$$q = 1, \quad P(s_i) = 1, \quad P(a_j|s_i) = P(a_j)$$

In this case (2.9) reduces to (2.2), as expected. For a first-order Markov source ($m = 1$) each state s_i is determined by only one previous symbol, say a_i, and so $P(a_j|s_i) = P(a_j|a_i)$. In this case (2.9) reduces to

$$H_c(A) = -\sum_{i=1}^{M} \sum_{j=1}^{M} P(a_i, a_j) \log P(a_j|a_i) \tag{2.10}$$

Example 2.5

The importance of conditional entropy can be illustrated using English text. This has $M = 27$ symbols if the space symbol is included and so, from (2.3)

$$H(A)_{\text{max}} = \log 27 = 4.8 \text{ bits/symbol}$$

Of course not all letters are equiprobable and so $H(A)$ will be less than this. A measure of letter frequencies may, for example, yield

$P(\text{space}) \approx 0.2$

$\quad P(z) \approx 0.001$

from which we conclude that thespacegiveslittleinformation! Taking estimated probabilities of all 27 symbols (Shannon, 1951) and applying (2.2) gives $H(A) \approx 4.1$ bits/symbol, although the actual entropy must be lower than this because we know that the symbols are not statistically independent. For example, if letter Q occurs it is virtually certain to be followed by U, i.e. $P(U|Q) \approx 1$. Taking into account the full statistical dependence between symbols (which can extend over many symbols) the actual entropy of English text has been estimated at around 2 bits/symbol.

Example 2.5 highlights the concept of *redundancy*, e.g. a letter U following letter Q is really redundant since from a knowledge of Q we can fairly safely predict the next symbol. We can express this redundancy formally as a ratio

$$E = 1 - \frac{H(A)}{\log M} \tag{2.11}$$

or as

$$E = \log M - H(A) \quad \text{bits/sample or symbol} \tag{2.12}$$

where $H(A)$ is a measure of the source entropy using some assumed source model. If, for example, we take $H(A) = 4.1$ bits/symbol in Example 2.5 this exposes a *statistical redundancy* of 0.7 bits/sample. On the other hand, a better model for English text is the Markov model in (2.9) and typically this might yield $H_c(A) \approx 2$ bits/symbol, implying that the actual or *correlation redundancy* is about 2.8 bits/symbol. Put another way, the redundancy in English text is roughly 50%.

Differential transmission

The conditional entropy in (2.9) is a measure of the remaining uncertainty about the next sample given m preceding samples. Intuitively, if there is high correlation between samples then this uncertainty or entropy will be relatively low, as demonstrated in Example 2.5. A practical way of estimating this entropy is to process a correlated source such that the source coder output symbols are approximately statistically independent. In fact, statistical independence of the source coder output is a necessary condition for optimal coding. Using (2.2) on the pdf of the practical source coder output would then give a good estimate of the actual source entropy and a good target for data compression. A simple but effective way of achieving this is to use differential transmission.

Consider the system in Fig. 2.6. The analogue input is assumed to be a monochrome video signal and this is linearly quantized to create a discrete source at the PCM coder output. The discrete pdf $P(a_j)$ of this source will generally be 'flat' but in detail non-uniform, as indicated. If we assume the 64 grey level samples a_j to be statistically independent then (2.2) can be used to calculate the source entropy. A typical monochrome image might give $H(A) \approx 5$ bits/sample (Kretzmer, 1952), corresponding to a statistical redundancy of

$$E = \log M - H(A) = 1 \quad \text{bit/sample}$$

Some form of optimum coding scheme could be applied to yield a bit rate near 5 bits/sample. However, we know that the image (PCM output) possesses significant sample to sample correlation and so the actual source entropy will usually be less than 5 bits/sample. The source coder must decorrelate the PCM signal and in this case the simple previous sample predictor in Fig. 2.6 is quite effective. For monochrome signals we could assume that the difference samples d_j are approximately independent and that inserting the probabilities $P(d_j)$ into (2.2) will give a good estimate of the actual entropy of the difference output. The interesting point is that this will also be very close to the conditional entropy of the PCM signal.

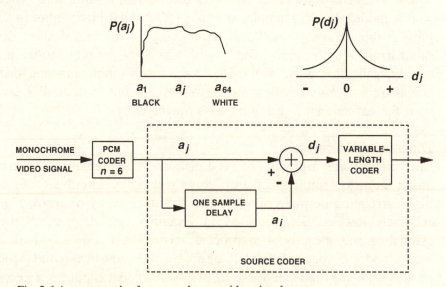

Fig. 2.6 A source coder for monochrome video signals.

Using probabilities $P(d_j)$ rather than $P(a_j)$ in (2.2) typically yields $H(A) \approx 3$ bits/sample (Kretzmer, 1952), indicating that the actual average information in the image is nearer to 3 bits/sample rather than to the 6 bits/sample assumed by the PCM coder. Efficient coding of the difference signal using variable-length coding would then give a bit rate and bandwidth compression of approximately 2:1.

The concept of differential transmission can be extended to colour television signals. As before, the basic assumption is that the zero-order entropy of the difference signal is lower than the zero-order entropy of the PCM signal and so compression can be achieved by coding each difference sample independently. The zero-order entropy of the difference signal closely represents the minimum average bit rate needed to convey the information source without loss. Suppose that a PAL colour signal is sampled at N times the colour sub-carrier frequency and that the 8-bit PCM output is applied to the predictor in Fig. 2.6 (the delay would be increased to N samples to ensure that samples of the same sub-carrier phase are differenced). Typical pdfs for $N = 3$ and 9-bit difference samples (but ignoring blanking) are shown in Fig. 2.7, and using (2.2) it was found that the entropy for PAL difference signals is typically in the range 3.5–6.5 bits per difference sample (Reid, 1974). Also, by removing blanking (see Fig. 2.1), the effective sampling rate reduces to about 10 MHz and so a better estimate for the required channel capacity for broadcast quality video signals (Example 2.4) is 35–65 Mbit/s. In fact, PAL transmission at the CCITT 34 Mbit/s level is perfectly feasible, although the compression techniques to achieve this are usually non-reversible (Ratliff and Stott, 1983).

Fig. 2.8 illustrates a possible coding scheme for colour video signals. The variable-length coder generates a variable bit rate and so for a fixed bit rate system a buffer store is required to absorb this varying rate and to regenerate it at a constant bit rate for transmission. An additional problem is that the variable-length coder must assume a representative set of difference probabilities but, as Fig. 2.7 shows, these can vary significantly for video sources. In fact, a video source is often regarded as non-stationary and, if this technique were to be used for a fixed bit rate system, the transmission rate would be fixed just above a typical entropy rate. Fig. 2.8 assumes that the typical zero-order entropy of the difference signal is 4.5 bits/sample. Note the need for feedback from the buffer store in order to minimize buffer overflow or underflow. In practice, variable-length coding is usually used to code the output of a DPCM quantizer, rather than in the context of pure reversible coding.

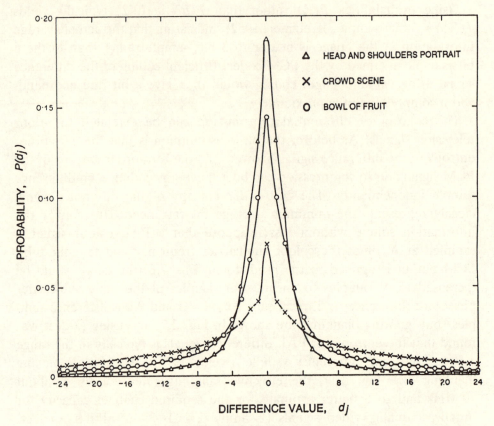

Fig. 2.7 Pdf of difference samples for PAL colour signals (reprinted by permission, BBC Research Department, 1974/37).

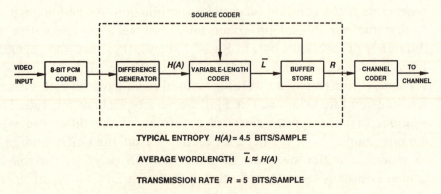

TYPICAL ENTROPY $H(A) = 4.5$ BITS/SAMPLE

AVERAGE WORDLENGTH $\bar{L} \approx H(A)$

TRANSMISSION RATE $R = 5$ BITS/SAMPLE

Fig. 2.8 Transmission of colour video signals using reversible coding.

2.1.4 Variable-length coding

The previous section introduced the concept of applying variable-length coding to the difference signal. By this we mean that variable length codewords would be applied to non-uniform discrete pdfs such as those in Fig. 2.7, and intuitively we should assign the shorter codewords to the most probable difference samples.

Consider the following variable-length binary codes:

Source symbol	Code A	Code B
a_1	10	1
a_2	100	01
a_3	1000	001
a_4	1100	0001

For the source sequence $a_3a_4a_2a_1a_3$ we obtain the following coded sequences:

> code A   10001100100101000
> code B   0010001011001

Both codes are useful or *uniquely decodable* because each codeword can be unambiguously identified in the code sequence. However, for code A it is necessary to receive several codewords to decode one codeword, whilst code B can be decoded on a word-for-word basis (instantaneously). Code B is an example of an *instantaneous* code and a necessary and sufficient condition for a uniquely decodable code to be instantaneously decodable is that no codeword must be a prefix for another. Clearly, this is not true for code A.

The variable-length coding process can be formally stated as the mapping of a source-symbol sequence to a code-symbol sequence where A and V below are the source and code alphabets respectively and l_j is the length of codeword v_j:

$$A = \left\{ \begin{matrix} a_1 & a_2 & a_3 & \ldots & a_M \\ P(a_1) & P(a_2) & P(a_3) & \ldots & P(a_M) \end{matrix} \right\} \Rightarrow$$

$$V = \left\{ \begin{matrix} v_1 & v_2 & v_3 & \ldots & v_M \\ l_1 & l_2 & l_3 & \ldots & l_M \end{matrix} \right\}$$

Since the $P(a_j)$ are the *a priori* probabilities, the average length of the code is

$$\bar{L} = \sum_{j=1}^{M} P(a_j)l_j \quad \text{code symbols/coder input symbol} \qquad (2.13)$$

Having defined the average length of a code we can now restate Shannon's noiseless coding theorem in more precise terms:

> **Theorem 2.2** *For a code with c possible code symbols (c = 2 for a binary code) the average length of the codewords per source symbol can be made arbitrarily close to H(A)/log c by coding longer and longer sequences of source symbols rather than coding individual symbols. Expressed mathematically*
>
> $$\frac{H(A)}{\log c} \leqslant \frac{\bar{L}}{n} \leqslant \frac{H(A)}{\log c} + \frac{1}{n} \qquad (2.14)$$
>
> *where $\bar{L}$ is now the average length of the codeword associated with a sequence of n source symbols.*

This is referred to as *block* or *string* coding, since strings of *n* source symbols are coded, and the advantages of this type of coding are illustrated in Example 2.10. If the symbols are individually coded then Theorem 2.2 reduces to

$$\frac{H(A)}{\log c} \leqslant \bar{L} \leqslant \frac{H(A)}{\log c} + 1$$

where the usual base 2 logarithm applies. For binary codes this further reduces to

$$H(A) \leqslant \bar{L} \leqslant H(A) + 1 \qquad (2.15)$$

and the corresponding *coding efficiency* is

$$\eta = \frac{H(A)}{\bar{L}} \qquad (2.16)$$

Also, it is easily seen from (2.2) and (2.13) that, if the source symbols are statistically independent and if their probabilities obey

$$P(a_j) = 2^{-l_j} \quad j = 1, 2, 3, \ldots, M \qquad (2.17)$$

then the resulting binary code will be 100% efficient or *absolutely optimum*. An approximation to (2.17) is, in fact, more likely to occur for block coding rather than for coding individual symbols.

Example 2.6

The above concepts can be applied to facsimile (two-tone) image transmission. Lumping white and black run-lengths together for simplicity, we could obtain the run-length probabilities for a typical image:

Source symbol	Run-length r (pixels)	Run probability
a_1	1	P_1
a_2	2	P_2
a_3	3	P_3
.	.	.
.	.	.
.	.	.
a_M	M	P_M

Note that M now corresponds to the number of pixels in a scan line and so for standard facsimile this will be a *large* alphabet. Also, for this type of source it is usual to assume that successive symbols (run-lengths) are statistically independent (Meyr *et al*, 1974) and so we can find the *run-length entropy* as

$$H(A) = -\sum_{i=1}^{M} P_i \log P_i \quad \text{bits/run}$$

Using (2.15), an optimum variable-length binary code for the source will have an average wordlength $\bar{L}$ (in bits/run) bounded by

$$H(A) \leq \bar{L} \leq H(A) + 1$$

and the corresponding bit rate R_b (in bits/pixel) will be

$$\frac{H(A)}{\bar{r}} \leq R_b \leq \frac{H(A) + 1}{\bar{r}}$$

Here $\bar{r}$ is the average run-length

$$\bar{r} = \sum_{i=1}^{M} iP_i \quad \text{pixels/run}$$

For typical facsimile weather maps, $H(A)/\bar{r} \approx 0.2$ bits/pixel (Huang, 1977). This indicates that run-length/variable-length coding should yield an average bit rate close to 0.2 bits/pixel, in contrast to an uncoded rate of 1 bit/pixel, and that the typical compression ratio is about 5. It is interesting to note that $H(A)/\bar{r}$ for weather maps can also be well estimated using a binary first-order Markov model (see (2.10)), e.g. given a white pixel, it is highly probable that the next pixel is also white. This approach circumvents the need to measure run-length probabilities.

2.1.5 Huffman coding (Huffman, 1952)

Given the symbol probabilities $P(a_j)$, Huffman's basic method guarantees an optimum variable-length code in the sense that it will be instantaneously decodable and will have an average length less than or equal to that of all other instantaneous codes for the same coding problem. The coding procedure consists of a step-by-step reduction in the number of source symbols until there are c source symbols, corresponding to c code symbols. However, to do this, we must start with a particular number M of source symbols, since each reduction step reduces the number of symbols by $(c-1)$ and there must be an integral number α of steps. Therefore,

$$M = c + \alpha(c - 1) \tag{2.18}$$

If necessary, (2.18) can be satisfied by adding dummy symbols having $P(a_j) = 0$ and then discarding them at the end. The Huffman coding algorithm for a zero-memory source and c different code symbols is as follows:

(1) List the source symbols in decreasing probability.
(2) Combine the c smallest probabilities and reorder. Repeat α times to achieve an ordered list of c probabilities.
(3) Coding starts with the last reduction. Assign the first code symbol as the first digit in the codeword for all source symbols associated with the first probability of the reduced source. Assign the second code symbol as the first digit in the codeword for all the second probabilities, etc.
(4) Proceed to the next-to-last reduction. For the c new probabilities the first code digit has already been assigned and so assign the first code symbol as the second digit for all source symbols associated with the first of these c new probabilities. Assign the second code symbol as the second digit for the second of these c new probabilities, etc.
(5) Proceed through to the start of the reductions in a similar manner.

Example 2.7 illustrates Huffman coding for a binary code. Note that no short codeword is a prefix of a longer word, implying that the code is instantaneously decodable and no codeword synchronization or *comma* between codewords is required. The codewords can be concatenated in a comma-free way and can still be instantaneously decoded. For example,

Example 2.7 Example of Huffman coding

a_j	$P(a_j)$	Final code	Step 1		Step 2		Step 3		Step 4		Step 5		Step 6	
a_1	0.35	0	0.35	0	0.35	0	0.35	0	0.35	0	0.35	0	0.65	1
a_2	0.3	10	0.3	10	0.3	10	0.3	10	0.3	10	0.35	11	0.35	0
a_3	0.15	110	0.15	110	0.15	110	0.15	110	0.2	111	0.3	10		
a_4	0.08	1110	0.08	1110	0.08	1110	0.12	1111	0.15	110				
a_5	0.05	11110	0.05	11110	0.07	11111	0.08	1110						
a_6	0.03	111110	0.04	111110	0.05	11110								
a_7	0.03	1111111	0.03	111111										
a_8	0.01	1111110	0.03	1111110										

$\bar{L} = 2.43$ $H(A) = 2.34$ $\eta = 96.25\%$

using the code in Example 2.7, it is easily seen that the code sequence {1111010111111100111111111110} corresponds to the message sequence $(a_5a_2a_8a_1a_7a_4)$. Huffman decoders can be implemented using table-lookup or tree-follower methods.

Comma coding

At this point it is interesting to compare Huffman coding with a very simple variable-length coding procedure. A *comma code* simply has an obvious terminating symbol or comma at the end of each codeword.

Example 2.8

It was pointed out in Example 2.3 that source coding should be able to reduce the final average bit rate from 2 bits/symbol to nearer the source entropy of 1.75 bits/symbol. A step in this direction could be made by coding the PCM output in Fig. 2.5 according to the following comma code (the zero corresponds to the comma).

a_j	$P(a_j)$	v_j
a_2	0.5	0
a_3	0.25	10
a_1	0.125	110
a_4	0.125	1110

$$\therefore \bar{L} = \sum_j P(a_j)l_j$$

$$= 0.5 + 0.25 \times 2 + 0.125 \times 3 + 0.125 \times 4$$

$$= 1.875 \quad \text{bits/symbol}$$

For some applications it may be preferable to use a comma code rather than a Huffman code in order to simplify implementation and improve error resilience (in contrast to Huffman coding, a comma code is immediately self-resynchronizing after an error). These advantages are at the expense of a reduction in coding efficiency compared to Huffman coding, although the reduction can be marginal.

Consider the coding of video telephone signals (slow-scan television over low-capacity telephone lines). These signals can be satisfactorily coded using DPCM with 4 bits describing each difference sample (4-bit DPCM) and for most picture material the 4-bit code words will have a pdf of the form shown in Fig. 2.9. Since the pdf is non-uniform, it is

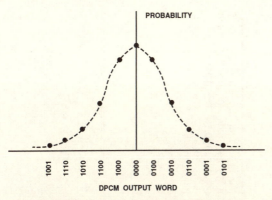

Fig. 2.9 Typical pdf for 4-bit DPCM codewords.

reasonable to apply variable-length coding to the DPCM output in order to achieve further compression (Chow, 1971; Nicol *et al*, 1980). To illustrate the advantage of doing this, Example 2.9 shows variable-length coding of the 11 binary number coded difference samples in Fig. 2.9 (typical probabilities for this sort of application having been assumed). Simply transmitting the 4-bit DPCM output gives a coding efficiency of only 60.9%, but Example 2.9 shows that comma coding can give an efficiency close to 95%, and that this is only marginally less than that for Huffman coding. This increased efficiency means that the update time for a typical slow-scan picture would be reduced by a factor of $2.576/4 = 0.64$ for comma coding compared with the time for straight transmission of the DPCM signal.

Modified Huffman code for facsimile
The Huffman coding procedure assumes independent source symbols and so it is applicable to facsimile transmission (see Example 2.6). On the other hand, the CCITT Group 3 facsimile standard (Hunter and Robinson, 1980) has a normal resolution of 1728 samples/line for A4 documents, which means that the maximum run-length and number of codewords is 1728. Such a large number of codewords rules out the use of a pure Huffman code and so a modified and *fixed* Huffman code was designed. Every run-length greater than 63 is broken into two run-lengths; one is a make-up run-length having a value $N \times 64$ (where N is an integer) and the other is a terminating run-length having a value between 0 and 63. In fact, since runs are approximately independent it is advantageous to code white and black runs independently since this leads to a lower bit rate. The final code is shown in Table 2.1.

Example 2.9 Comparison of Huffman and comma coding

a_j	$P(a_j)$	Comma code	Huffman code									
1	0.4	0	0	0.4 0	0.4 0	0.4 0	0.4 0	0.4 0	0.4 0	0.4 0	0.4 0	0.6 1
2	0.175	10	111	0.175 111	0.175 111	0.175 111	0.175 111	0.175 111	0.175 111	0.25 10	0.35 11	0.4 0
3	0.175	110	110	0.175 110	0.175 110	0.175 110	0.175 110	0.175 110	0.175 110	0.175 111	0.25 10	
4	0.09	1110	100	0.09 100	0.09 100	0.09 100	0.09 100	0.09 100	0.16 101	0.175 110		
5	0.09	11110	1011	0.09 1011	0.09 1011	0.09 1011	0.09 1011	0.09 1011	0.09 100			
6	0.028	111110	10100	0.028 10100	0.028 10100	0.028 10100	0.042 10101	0.07 1010				
7	0.028	1111110	101011	0.028 101011	0.028 101011	0.028 101011	0.028 10100					
8	0.005	11111110	1010100	0.005 10101011	0.009 1010101	0.014 101010						
9	0.005	111111110	10101011	0.005 1010100	0.005 1010100							
10	0.002	1111111110	101010101	0.004 10101010								
11	0.002	11111111110	101010100									

$H(A) = 2.435$ $\bar{L}_{comma} = 2.576$ $\eta_{comma} = 94.54\%$ $\bar{L}_{Huffman} = 2.499$ $\eta_{Huffman} = 97.45\%$

Redundancy removal via run-length/variable-length coding makes the transmitted signal more susceptible to transmission errors (a problem common to many compression schemes). In general, but not always, a single transmission error will generate error propagation as the decoder loses synchronization and generates a string of decoded runs different from those at the transmitter. Fig. 2.10(*a*) shows an arbitrary transmitted sequence derived from Table 2.1, and Fig. 2.10(*b*) shows the same sequence received with one error. In this particular case there is no loss of synchronization although there will be a slight displacement of the de-coded scan line. Fig. 2.10(*c*) illustrates typical error propagation and the generation of a string of false run-lengths. Note, however, that eventually the code *resynchronizes* (this being a property of most comma-free codes).

The modified Huffman code tries to ensure that errors do not propagate beyond the current line by transmitting a unique synchronizing word at the end of each coded line. This is the end-of-line (EOL) codeword in Table 2.1.

Adaptive coding

Huffman coding has often been applied to English text since the symbol probabilities are known and reasonably stable, and compression ratios of 1.5:1 are typical. On the other hand, for many applications the symbol probability distribution is unknown and so fixed or static Huffman coding is impractical. For example, the distributions for data files are very application specific and files such as object code and source code will have dissimilar distributions. One solution is to analyse each data block prior to coding in order to ascertain the distribution (requiring in all two passes over the data). The derived coding table could then be transmitted along with the compressed data.

An alternative approach is to make the coding and decoding tables adapt to changing message statistics by adding a *symbol count field*. The role of this field is to position the most frequently occurring symbols near the top of the coder (decoder) table – which corresponds to the shortest codewords – thereby generating an optimum table for the current mes-sage. Suppose that an initial Huffman code is established (Fig. 2.11) and that a message sequence $\{a_3a_4a_3a_2\}$ is applied to the coder. The count field is initialised to zero, as shown. Codeword v_3 is transmitted for symbol a_3 and then the count field and symbol ordering is updated. Similarly, symbol a_4 generates codeword v_4 and the table is again updated. The significant point is that at the end of the sequence the most frequently occurring symbols are at the top of the table. Clearly, the decoding table

Table 2.1. *Modified Huffman code table (Hunter and Robinson © 1980, IEEE).*

Terminating codewords

Run-length	White	Black	Run-length	White	Black
0	00110101	0000110111	32	00011011	000001101010
1	000111	010	33	00010010	000001101011
2	0111	11	34	00010011	000011010010
3	1000	10	35	00010100	000011010011
4	1011	011	36	00010101	000011010100
5	1100	0011	37	00010110	000011010101
6	1110	0010	38	00010111	000011010110
7	1111	00011	39	00101000	000011010111
8	10011	000101	40	00101001	000001101100
9	10100	000100	41	00101010	000001101101
10	00111	0000100	42	00101011	000011011010
11	01000	0000101	43	00101100	000011011011
12	001000	0000111	44	00101101	000001010100
13	000011	00000100	45	00000100	000001010101
14	110100	00000111	46	00000101	000001010110
15	110101	000011000	47	00001010	000001010111
16	101010	0000010111	48	00001011	000001100100
17	101011	0000011000	49	01010010	000001100101
18	0100111	0000001000	50	01010011	000001010010
19	0001100	00001100111	51	01010100	000001010011
20	0001000	00001101000	52	01010101	000000100100
21	0010111	00001101100	53	00100100	000000110111
22	0000011	00000110111	54	00100101	000000111000
23	0000100	00000101000	55	01011000	000000100111

Run-length	White	Black		Run-length	White	Black
24	0101000	00000010111		56	01011001	000000101000
25	0101011	00000011000		57	01011010	000001011000
26	0010011	000011001010		58	01011011	000001011001
27	0100100	000011001011		59	01001010	000000101011
28	0011000	000011001100		60	01001011	000000101100
29	00000010	000011001101		61	00110010	000001011010
30	00000011	000001101000		62	00110011	000001100110
31	00011010	000001101001		63	00110100	000001100111

Make-up codewords

Run-length	White	Black		Run-length	White	Black
64	11011	0000001111		960	011010100	0000001110011
128	10010	000011001000		1024	011010101	0000001110100
192	010111	000011001001		1088	011010110	0000001110101
256	0110111	000001011011		1152	011010111	0000001110110
320	00110110	000000110011		1216	011011000	0000001110111
384	00110111	000000110100		1280	011011001	0000001010010
448	01100100	000000110101		1344	011011010	0000001010011
512	01100101	0000001101100		1408	011011011	0000001010100
576	01101000	0000001101101		1472	010011000	0000001010101
640	01100111	0000001001010		1536	010011001	0000001011010
704	011001100	0000001001011		1600	010011010	0000001011011
768	011001101	0000001001100		1664	011000	0000001100100
832	011010010	0000001001101		1728	010011011	0000001100101
896	011010011	0000001110010		EOL	000000000001	000000000001

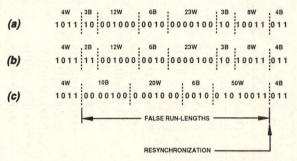

Fig. 2.10 Effect of a single transmission error upon the modified Huffman code: (*a*) transmitted sequence; (*b*) and (*c*) received sequences.

MESSAGE	a_3	a_4	a_3	a_2
TRANSMITTED CODEWORD	v_3	v_4	v_1	v_4

TIME →

INITIAL CODE AND FIELD

v_1	a_1	0	a_3	1	a_3	1	a_3	2	a_3	2
v_2	a_2	0	a_1	0	a_4	1	a_4	1	a_4	1
v_3	a_3	0	a_2	0	a_1	0	a_1	0	a_2	1
v_4	a_4	0	a_4	0	a_2	0	a_2	0	a_1	0

Fig. 2.11 Adaptive coding.

at the receiver must keep track of coder changes and so it is updated in the same way as the coding table each time a codeword is received. Occasionally both count fields may need scaling down to prevent overflow.

Block coding

According to Theorem 2.2 coding efficiency will be improved if we code blocks or strings of *n* source symbols rather than individual symbols. This is true even for independent symbols and so is easily demonstrated.

Example 2.10

Assume a statistically independent source with

$$A = \{a_1, a_2\} \quad P(a_1) = 0.9 \quad P(a_2) = 0.1$$

Using (2.2) the source entropy is

$$H(A) = -(0.9 \log 0.9 + 0.1 \log 0.1) = 0.469 \quad \text{bits/source symbol}$$

In this case, Huffman coding of individual symbols is trivial and corresponds to $\bar{L} = 1$ (see (2.13)) and a coding efficiency of $\eta = 46.9\%$. Now suppose the coder input comprises strings of $n = 2$ symbols rather than individual symbols and that these strings are independent. Huffman coding gives

Coder input string	Probability	Final code	Step 1		Step 2	
a_1a_1	0.81	1	0.81	1	0.81	1
a_1a_2	0.09	00	0.10	01	0.19	0
a_2a_1	0.09	011	0.09	00		
a_2a_2	0.01	010				

The average length is now $\bar{L} = 1.29$ and so $\bar{L}/n = 0.645$ bits/source symbol, corresponding to a coding efficiency of $\eta = 72.7\%$. The reader should show that extending the concept to strings of three symbols gives

Coder input string	Probability	Final code
$a_1a_1a_1$	0.729	1
$a_1a_1a_2$	0.081	011
$a_1a_2a_1$	0.081	010
$a_2a_1a_1$	0.081	001
$a_1a_2a_2$	0.009	00011
$a_2a_1a_2$	0.009	00010
$a_2a_2a_1$	0.009	00001
$a_2a_2a_2$	0.001	00000

In this case $\bar{L}/n = 0.533$ and $\eta = 88\%$. It is interesting to note that the increased efficiency available through block or string coding is exploited by commercial string compression algorithms such as the LZW algorithm.

2.1.6 The LZW string compression algorithm

The LZW compression algorithm (Welch, 1984) is a variation on the basic LZ algorithm (Ziv and Lempel, 1977), and its relatively simple form makes it attractive for commercial computer applications. A significant practical point is that, like the LZ algorithm, the LZW algorithm adapts to the source statistics and so no *a priori* knowledge of the source is required. In other words, it is a *universal compression algorithm* whereby the coding (decoding) process is interlaced with a learning process as the coder (decoder) builds and dynamically changes its string table. This means that the LZW decoder does not require transmission of the decoding table since it builds an identical table as it receives compressed data. Also, like Huffman coding, the LZ and LZW algorithms exploit the statistical redundancy of the source (frequently used symbols and symbol patterns) rather than any positional redundancy. On the other hand, unlike Huffman coding the basic form of these algorithms employs fixed-length (typically 12-bit) codewords.

Basic coding algorithm

The algorithm parses a string $S = s_1 s_2 s_3 \ldots$ of source symbols into successive sub-strings S_i such that $S = S_1 S_2 S_3 \ldots$. Once a sub-string S_i is found (the longest recognized input string is parsed off each time) it is coded into a fixed-length uniquely decodable codeword v_i. No real attempt is made to parse string S optimally and so the LZW algorithm is less than optimum, but on the other hand it can be very fast.

initialise string table to contain single symbol strings

read first symbol s_1

$P = s_1$

for next input symbol s_k {

    $E = s_k$

    if (PE is in the string table) $P = PE$

    else {

        transmit code(P)

        add PE to string table

        $P = E$

    }

}

Fig. 2.12 Pseudo-code for LZW coding.

The LZW string table has a prefix property in that for every string in the table its *prefix string* is also in the table. For example, if the table has string 'and' it also has a prefix string 'an'. Let us generalize this by defining a string in the table as *PE* where *P* is a prefix string and *E* is an *extension symbol*. The basic LZW coding algorithm is then as shown in Fig. 2.12. Note that strings are assigned a unique codeword as they are added to the table.

Example 2.11

Let the source alphabet be $A = \{a, b, c\}$ so that the string table is initialized to

s_i	v_i
a	1
b	2
c	3

LZW coding for the input string

$S = ababcbababaaaaaaa$

is then as shown in Fig. 2.13. Coding starts by reading symbol a and equating P to a. Symbol b is then read, and, since string ab is not in the table, the code for symbol a is transmitted, string ab is added to the table, and P is equated to b (note the convenient storage format). The third symbol is now read and, since string ba is not in the table, the code for b is transmitted, ba is added to the table, and P is equated to a. For the fourth input symbol, ab is already in the table and so no codeword is transmitted and P is equated to ab. If the input is considered to be a string of 8-bit ASCII characters and 12-bit codewords are used, then the compression ratio for this particular input string is about 1.3:1. Clearly,

INPUT STRING	S	a	b	a	b	c	b	a	b	a	b	a	a	a	a	a	a	a
PREFIX STRING	P	a	b	a	ab	c	b	ba	b	ba	bab	a	a	aa	a	aa	aaa	a
OUTPUT CODE			1	2		4	3		5			8	1		10			11
TABLE ADDITION			ab	ba		abc	cb		bab			baba	aa		aaa			aaaa
ADDED CODEWORD v_i			4	5		6	7		8			9	10		11			12
STORED FORMAT			1b	2a		4c	3b		5b			8a	1a		10a			11a

Fig. 2.13 Example of LZW coding.

good compression occurs when the coder recognizes a long input string since it will then be replaced by a relatively short codeword, and for English text compression is typically 2:1. Similar compression can be achieved for colour images.

As might be expected, the decoding process is essentially the reverse of the coding process and uses the same string table as used for compression. For Example 2.11, each received codeword is translated by the decoder's string table into a prefix string and an extension symbol (which is pulled off and stored), and this is repeated in a recursive way until the prefix string is a single symbol. For example, given input code '9' from Fig. 2.13, the decoder would perform the following steps:

$$9 \Rightarrow 8a \text{ (store } a) \Rightarrow 5b \text{ (store } b) \Rightarrow 2a \text{ (store } a) \Rightarrow b$$

giving the decoded string *baba*.

The principal concern when implementing the LZW algorithm is that of storing the string table. Clearly, the table size is limited by the preselected codeword length. If this is too small then the table rapidly fills, whilst if this is too large the compression overhead for single character strings is excessive. An enhanced LZW algorithm might therefore use variable-length codewords, starting with a short-length codeword and extending the length as the table fills. For example, when the input is a string of 8-bit ASCII characters, the wordlength could start at 9 bits and extend to 12 or 13 bits. Clearly, even then the string table will eventually overflow and when this happens one possible solution is to clear the table and rebuild it in order to keep the compression adaptive. Obviously, compression will be poor whilst the table is sparsely filled. Finally, note that, like other compression schemes, the LZW algorithm is susceptible to error propagation in the presence of channel errors. In the context of data storage on disk or tape, the usual solution is to employ *cyclic redundancy checks* (Section 3.3.4).

2.2 Non-reversible coding – differential PCM (DPCM)

Important non-reversible coding techniques are highlighted in Fig. 2.1. As the name implies, non-reversible coding gives compression at the expense of information loss. Often this loss is slight (virtually undetectable), as in some ADPCM and transform coding (TC) systems for video signals, but in other cases it can be significant. For example, in LPC, speech is compressed by assuming a parametric model of the vocal-tract transfer function, and this imposes a constraint on reproduced speech quality. On the other hand, compared with some direct waveform coding techniques, such as ADPCM, the bit rate can be relatively low (of the order of 1000 bits/s).

2.2.1 Principles of DPCM

In Section 2.1.3 we saw that if the signal has significant correlation redundancy then a more optimum source to transmit would be a differential one since the difference samples are less correlated. In Fig. 2.8 difference transmission and compression was achieved using variable-length coding, but an alternative approach is to requantize and recode the difference signal. The resulting system is then called differential PCM (DPCM). Both variable-length coding and DPCM exploit the non-uniform pdf of the difference signal and DPCM does this by coarsely quantizing rare samples and finely quantizing the most probable samples. In other words, basic DPCM schemes achieve compression by *non-uniform quant-izing* of the difference signal, and in doing this the quantizer loses information and the process becomes non-reversible. Note however, that some advanced DPCM schemes use linear quantization and achieve compression through a combination of quantizing and statistical coding.

The basic DPCM system is shown in Fig. 2.14 and the diagram highlights a number of fundamental points.

(1) The difference sample $e_n = x_n - \hat{x}_n$ depends upon a prediction $\hat{x}_n$ derived from k previous samples, where k is the order of the predictor, P.

(2) In the absence of transmission errors the coder develops an output signal which is identical to that at the receiver.

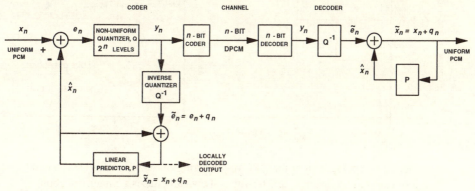

Fig. 2.14 Block diagram of a DPCM system (non-adaptive).

(3) The non-uniform quantizer, Q, is placed within the closed loop rather than outside it in order to avoid integration of quantizing errors at the receiver decoder. In this way, only the quantizing noise or error q_n associated with the current input sample x_n appears at the locally decoded output and previous quantizing errors are subtracted out.

(4) The non-uniform quantizing law could be a 'midriser' character- istic, as shown in Fig. 2.15, or it could be a 'midtread' character- istic with an output level at zero. Either way, it is helpful to discuss it in the context of a video input signal. For slowly changing signals, as might occur in low detail images, the quantizer input will be small and the quantizing error is determined by the fine quantizing levels, giving so-called *granular noise* (see Fig. 2.15). In fact, for quasi-flat regions of an image the vertical step at the origin of a 'midriser' characteristic causes the DPCM loop to oscillate between small positive and negative steps. For larger signals, such as edges in an image, the increased quantizing noise manifests itself as an *edge busyness*, whilst *slope overload* occurs when the input error is larger than the largest quantizing step.

(5) Error sample e_n in Fig. 2.14 is usually normalized and compressed prior to quantization, and so typically the actual quantizer input is of the form

$$\log_2 |e_n| - g_n$$

where g_n is a scaling factor. A practical normalized 15-level

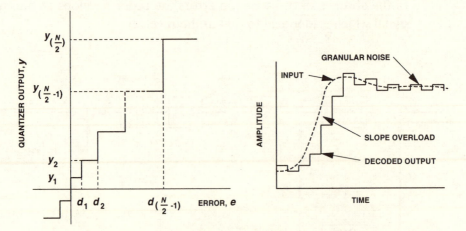

Fig. 2.15 General form of the DPCM quantizing characteristic and associated distortions (a 'midriser' characteristic is shown and this is symmetrical about $e = 0$).

Table 2.2. *Quantizer normalized input/output characteristic for 32 kbit/s 4-bit DPCM voice frequency signals (s is a sign bit) (CCITT Rec. G.721, 1988).*

Quantizer input $\log_2 \|e_n\| - g_n$	Quantizer output $y_n = \log_2 \|\tilde{e}_n\| - g_n$	Coder output v_n
3.12, ∞	3.32	s111
2.72, 3.12	2.91	s110
2.34, 2.72	2.52	s101
1.91, 2.34	2.13	s100
1.38, 1.91	1.66	s011
0.62, 1.38	1.05	s010
−0.98, 0.62	0.031	s001
−∞, −0.98	−∞	s000

characteristic for 4-bit DPCM voice frequency signals is shown in Table 2.2.

(6) The inverse quantizer, Q^{-1}, simply adds the scaling factor g_n to y_n and takes the inverse logarithm of the sum to obtain the quantized error $\tilde{e}_n$. This enables a reconstructed or quantized sample $\tilde{x}_n$ to be generated at the predictor input (Fig. 2.14).

2.2.2 DPCM quantizer design

There are several design philosophies here and we will concentrate first upon a classical statistical approach. This defines the optimum quantizer as one which minimizes the *mean-square quantizing error*, and so in terms of Fig. 2.15 we would minimize $E[(e - y_k)^2]$ where y_k represents one of a total of $N = 2^n$ output or *representative levels*. In this case the optimum values of y_k and input *decision levels* d_k are given by (Max, 1960)

$$\int_{d_{k-1}}^{d_k} (e - y_k)p(e)\,\mathrm{d}e = 0, \quad k = 1, 2, \ldots, \frac{N}{2} \tag{2.19}$$

$$d_k = \begin{cases} 0 & k = 0 \\ \dfrac{y_k + y_{k+1}}{2} & k = 1, 2, \ldots, \dfrac{N}{2} - 1 \\ \infty & k = \dfrac{N}{2} \end{cases} \tag{2.20}$$

where

$$d_{k-1} < y_k < d_k$$

Table 2.3. *Lloyd–Max quantizers assuming p(e) has gamma density (E[e] = 0, $\sigma_e^2 = 1$)(Paez and Glisson, © IEEE 1972).*

k	N 8 d_k	8 y_k	16 d_k	16 y_k	32 d_k	32 y_k
1	0.504	0.149	0.229	0.072	0.101	0.033
2	1.401	0.859	0.588	0.386	0.252	0.169
3	2.872	1.944	1.045	0.791	0.429	0.334
4	∞	3.799	1.623	1.300	0.630	0.523
5			2.372	1.945	0.857	0.737
6			3.407	2.798	1.111	0.976
7			5.050	4.015	1.397	1.245
8			∞	6.085	1.720	1.548
9					2.089	1.892
10					2.517	2.287
11					3.022	2.747
12					3.633	3.296
13					4.404	3.970
14					5.444	4.838
15					7.046	6.050
16					∞	8.043

Here, $p(e)$ is an even function representing the probability density of the quantizer input. Note in particular that the optimum decision levels are half-way between the values of adjacent representative levels. Equations (2.19) and (2.20) define the *Lloyd–Max* quantizer and solution of (2.19) is usually carried out numerically after assuming some appropriate form for $p(e)$. It is usual to assume that the effect of the quantizer on $p(e)$ is small (provided quantizing is not too coarse), and for speech $p(e)$ is well approximated by the gamma density function. Computed results for $N = 8$, 16 and 32 are given in Table 2.3 (Paez and Glisson, 1972). Note that levels y_k and d_k should be scaled by the actual standard deviation σ_e of error e, which implies that a fixed quantizer will be optimized for only one specific variance!

An alternative quantizer design technique for video systems has been described by O'Neal (1966). For video systems (and as a simpler but poorer approximation for speech) $p(e)$ is often assumed Laplacian, i.e.

$$p(e) = \frac{1}{\sqrt{2}\sigma_e} e^{-(\sqrt{2}/\sigma_e)|e|} \tag{2.21}$$

In this case, the mean-square quantizing error will be minimized by selecting the y_k levels according to the function

$$y(x) = -\frac{e_m}{r} \ln[1 - x(1 - e^{-r})]; \quad x \geqslant 0 \tag{2.22}$$

where

$$r = \frac{\sqrt{2}}{3} \frac{e_m}{\sigma_e} \tag{2.23}$$

Here, e_m is the maximum prediction error or, equivalently, the peak–peak value of the input signal, x_n. The input parameter x is a positive ratio, so (2.22) gives the positive part of the symmetrical characteristic in Fig. 2.15, and x takes on uniformly spaced values between zero and unity. For a quantizer with a total of 2^n levels there must be 2^{n-1} equally spaced input points in this range, so that

$$x = \frac{1}{2^n}, \frac{3}{2^n}, \ldots, \frac{2^n - 3}{2^n}, \frac{2^n - 1}{2^n} \tag{2.24}$$

Once the representative levels y_k have been found, the decision levels d_k can be determined using (2.20).

Example 2.12

A 5-bit DPCM video system might typically have an 8-bit PCM input (256 quanta) and the parameters (Devereux, 1975)

$$e_m = 130 \text{ quanta}, \quad \sigma_e = 12.3 \text{ quanta}$$

This gives

$$y(x) = -26.0923 \ln(1 - 0.993142x)$$

and the centre of the 32-level quantizing characteristic is therefore defined as follows:

Level, k	d_k	y_k
1	1.69 (1.81)	0.82 (0.89)
2	3.48 (3.71)	2.55 (2.73)
3	5.39 (5.74)	4.40 (4.70)
4	– –	6.39 (6.78)
.	.	.
.	.	.
.	.	.

For comparison purposes, the corresponding values for a scaled Lloyd–Max quantizer optimized to a Laplacian density are given in brackets.

A non-uniform law is a natural consequence of applying the minimum mean-square error (mmse) criterion to a non-uniform pdf, $p(e)$. Given $p(e)$ (or e_m and σ_e), mmse design is then fairly straightforward, as discussed. Clearly, this approach becomes more difficult if $p(e)$ varies, and in such cases the quantizing characteristic could be found by using a random search to minimize the mean-square error (Cohn and Melsa, 1975).

Subjectively optimized quantizers

For video signals it is often argued that the mmse criterion is too crude and that quantizers should be designed on *psychovisual* grounds (Netravali and Limb, 1980). At the outset we might argue that, for static pictures at least, the viewer can tolerate large quantizing errors on rapid transitions (edges), and so this itself suggests a non-uniform quantizing law.

One approach would be to determine the y_k values on a purely experimental basis (subjective testing to minimize noise *visibility*), and the d_k values could then be determined from (2.20). Using this technique it has been found that some optimal experimental values for y_k in Example 2.12 can differ from the mmse law by typically 40% (Devereux, 1975). More generally, subjective tests have shown that statistically optimized video quantizers tend to have too many levels for small prediction errors and too few levels for large prediction errors (Musmann *et al*, 1985). Also note that mmse design takes no account of any interaction between quantizer and predictor whereas, ideally, these designs should be combined.

An interesting systematic quantizer design technique is described by Sharma and Netravali (1977). They used psychovisual experiments to determine a *visibility threshold* and then employed this in several design strategies. In particular, a geometric design procedure is described whereby the minimum number of representative levels, the value of these levels and the value of the decision levels can all be found such that the quantizing errors are kept below the visibility threshold.

2.2.3 DPCM predictor design

For analysis purposes it is convenient to reduce the DPCM coder to the model in Fig. 2.16. Here, a signal x_n of variance σ_x^2 and zero mean is applied to the coder input. Invariably, prediction $\hat{x}_n$ is a *linear* combination of previous samples (as indicated), although it is widely accepted that a *non-linear* function will generally provide the best estimate. It is also convenient to assume that P can be designed independently of Q so that

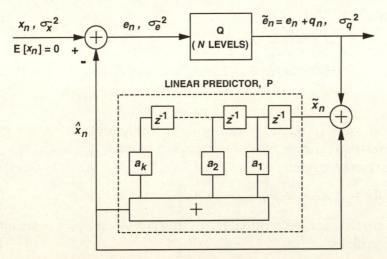

Fig. 2.16 Model for DPCM analysis assuming kth-order linear prediction.

any quantizing effects can be ignored, although, as stated above, the two designs should ideally be combined.

In the context of Fig. 2.16, a linear prediction of the next sample value x_0 given k previous sample values is defined as

$$\hat{x}_0 = \sum_{i=1}^{k} a_i \tilde{x}_i \qquad (2.25)$$

where $\tilde{x}_i$ is the quantized value of x_i and a_i is a predictor coefficient. Determination of the optimum coefficients is a 'least-squares' problem. If we regard x_n as a *random* signal, i.e. not a *deterministic* signal, then we minimize the mean-square error or error variance σ_e^2:

$$\text{minimize } E[e_0^2]; \quad e_0 = x_0 - \hat{x}_0 \qquad (2.26)$$

In a strict statistical sense, x_n will be regarded as a *random variable* belonging to a *random process*, and we will assume that this process is *wide-sense stationary* (essentially this means that the basic statistics of the process, such as its autocorrelation function, are independent of time). Differentiating and ignoring the effects of quantization

$$\frac{\partial E[e_0^2]}{\partial a_i} = E\left[\frac{\partial e_0^2}{\partial a_i}\right] = 2E\left[e_0 \frac{\partial e_0}{\partial a_i}\right] = -2E[e_0 x_i]$$

and so the optimum coefficients are given by the equations

$$E[x_i(x_0 - \hat{x}_0)] = 0, \qquad i = 1, \dots, k \qquad (2.27)$$

Expanding

$$E[x_0 x_i] = E[x_i(a_1 x_1 + a_2 x_2 + \ldots + a_k x_k)]$$

$$= a_1 E[x_1 x_i] + a_2 E[x_2 x_i] + \ldots + a_k E[x_k x_i] \quad i = 1, \ldots, k$$

$$(2.28)$$

If we let $\sigma_x^2 = 1$, terms $E[x_i x_j]$ will be recognized as values of the autocorrelation function, R_{ij}, of the random process, i.e. $E[x_i x_j] = R_{ij}$, and so (2.28) can be written

$$R_{0i} = a_1 R_{1i} + a_2 R_{2i} + \ldots + a_k R_{ki}, \quad i = 1, \ldots, k \qquad (2.29)$$

Noting that R_{ij} can be written $R_{|i-j|}$, this gives a set of k simultaneous linear equations

$$\left.\begin{aligned}
R_1 &= a_1 R_0 + a_2 R_1 + \ldots + a_k R_{k-1} \\
R_2 &= a_1 R_1 + a_2 R_0 + \ldots + a_k R_{k-2} \\
& \qquad\qquad \vdots \\
& \qquad\qquad \vdots \\
& \qquad\qquad \vdots \\
R_k &= a_1 R_{k-1} + a_2 R_{k-2} + \ldots + a_k R_0
\end{aligned}\right\} \qquad (2.30)$$

Expressing (2.30) in matrix form

$$\begin{bmatrix} R_1 \\ R_2 \\ \cdot \\ \cdot \\ \cdot \\ R_k \end{bmatrix} = \begin{bmatrix} R_0 & R_1 & \ldots & R_{k-1} \\ R_1 & R_0 & \ldots & R_{k-2} \\ & \cdot & & \\ & \cdot & & \\ & \cdot & & \\ R_{k-1} & R_{k-2} & \ldots & R_0 \end{bmatrix} \begin{bmatrix} a_1 \\ a_2 \\ \cdot \\ \cdot \\ \cdot \\ a_k \end{bmatrix} \qquad (2.31)$$

or

$$\mathbf{a} = \mathbf{R}_{xx}^{-1}\mathbf{R} \qquad (2.32)$$

Here $\mathbf{R}_{xx}$ is the $k \times k$ autocorrelation matrix for the input signal and it can always be inverted. Matrix representation of the problem suggests efficient recursive solution, as in the *Levinson–Durbin recursion* (Rabiner and Schafer, 1978). On the other hand, the major computational load often lies in the computation of the autocorrelation function rather than in the solution of (2.31). Given just N samples from a realization of a stationary zero-mean random process it is usual to compute R_i as a time average using the approximation

$$\hat{R}_i = \hat{R}_{-i} = \frac{1}{N} \sum_{n=0}^{N-i-1} x'_n x'_{n+i}, \quad i \geq 0 \tag{2.33}$$

Here, $x'_n = x_n w_n$ where w_n is some windowing function (often assumed rectangular) spanning N samples (see Section 4.4.3). Providing $N \gg i$ then (2.33) is a good estimate of the true autocorrelation function.

Equation (2.32) is a linear predictive form of the *Wiener–Hopf equations* and is sometimes referred to as *Weiner prediction*. A closely related problem is that of signal estimation in noise, or *Weiner filtering* (Section 4.6.1). In DPCM we seek the true value of the current input sample (x_n) given a linear combination of previous samples. In Wiener filtering we wish to estimate the true value of the current input sample given a linear combination of noisy samples, including the current sample.

Types of predictor

In practice the exact form of the linear predictor varies significantly between applications. If, for example, the input sequence $\{x_n\}$ can be modelled as an mth-order Markov (or AR(m)) sequence, then only m samples of x_n are needed to form the best estimate of x_0. In other words, the ideal predictor for an mth-order Markov source has order $k = m$, and it will fully decorrelate or 'whiten' the error sequence $\{e_n\}$. This is, after all, a main objective in DPCM. For example, the short-time spectral envelope for voiced speech can be well modelled by the spectrum of an AR(10) (or larger) process, and so predictors of this order should give good decorrelation for such a source (Flanagan *et al*, 1979). Simulation of 11th-order prediction for voiced speech is illustrated in Fig. 4.52 (note that this predictor is *adaptive* due to the nonstationary nature of speech signals).

Example 2.13

In the context of DPCM, a second-order linear prediction of sample x_0 is

$\hat{x}_0 = a_1 \tilde{x}_1 + a_2 \tilde{x}_2$

and the coefficients are found from

$R_1 = a_1 R_0 + a_2 R_1$

$R_2 = a_1 R_1 + a_2 R_0$

Assuming unit variance ($R_0 = 1$) we have

$$a_1 = R_1 \frac{1 - R_2}{1 - R_1^2} \tag{2.34}$$

$$a_2 = \frac{R_2 - R_1^2}{1 - R_1^2} \tag{2.35}$$

For a first-order Markov source (AR(1) model) it can be shown than $R_2 = R_1^2$, and so $a_2 = 0$. Therefore, as expected, only a first-order predictor is required for a first-order Markov source. The general form of the required coefficient will then be given by

$$a_1 = \frac{R_1}{R_0} = \frac{E[x_0 x_1]}{E[x_0^2]} = \rho_1 \tag{2.36}$$

where ρ_1 is the adjacent sample correlation coefficient. It is interesting to note that the autocorrelation function of one line of a monochrome video signal can be close to a Laplacian function, i.e. $R_\tau = e^{-\alpha|\tau|}$, τ being the spatial lag along a scan line, in samples. Therefore, again $R_2 = R_1^2$, and the optimum predictor uses only the most previous sample. Also, in such cases, ρ_1 may well be close to unity and so a_1 could be set to 1 to ease implementation.

The simplified approach in Example 2.13 suggests that, to a first approximation, second-order prediction for video signals is unnecessary if all samples are within the same scan line. In practice, second- and third-order prediction are found useful. Also, there is a worthwhile advantage in 2-D second-order prediction where x_2 is taken from the previous line of the same field. In the first place, 2-D prediction can greatly improve the subjective effect of transmission errors and, secondly, it improves prediction on large luminance transitions, especially when coding composite NTSC and PAL signals. Figure 2.17 illustrates two simple 2-D predictors for NTSC signals sampled at three-times colour sub-carrier frequency. The idea here is to select samples closest to the current input sample x_0 which have the same sub-carrier phase.

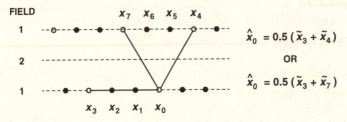

Fig. 2.17 Simple 2-D predictors for NTSC signals sampled at three-times colour sub-carrier frequency.

More complex 2-D fixed-coefficient predictors involving samples from several fields and third-order prediction along a scan line have been developed for YUV component-coded broadcast quality video signals (Knee and Wells, 1989). Here the prediction coefficients were chosen to minimize the mean-square prediction error measured over a wide range of picture material. A generalized fixed-coefficient predictor for component coded video signals is

$$\hat{x}_0 = \sum_a a(i, j, k)\tilde{x}(h - i, v - j, t - k) \tag{2.37}$$

where h, v and t are the horizontal, vertical and temporal coordinate values expressed in samples, picture (or frame) lines and frames respectively. Positive values of i, j or k correspond to sample locations earlier in time. For example, a two-field predictor utilizing third-order horizontal prediction would have coefficients $a(i, j, 0)$ $i = 0, 1, 2, 3$; $j = 0, 1, 2$, i.e. just *intraframe* prediction, whilst a simple *interframe* predictor using just the corresponding sample in the previous frame would have the single coefficient $a(0, 0, 1) = 1$. Clearly, this simple interframe predictor would be excellent for still pictures. On the other hand it is found that, as for speech, *adaptive* rather than fixed prediction tends to perform best for video signals (note that fixed prediction as designed by (2.32) assumes a *stationary* signal and so is often unsuitable for video signals). Adaptive techniques based upon the *LMS algorithm* (Section 4.6.2) are particularly attractive. For example, if P in Fig. 2.16 is considered to be an LMS predictor then, from (4.157), the predictor coefficients could be updated as

$$a_{i,n+1} = a_{i,n} + \mu\tilde{e}_n\tilde{x}_{n-i}, \quad i = 1, 2, \ldots, k \tag{2.38}$$

Using simple second-order 2-D prediction on video signals (i.e. $\tilde{x}_{n-1}$ and $\tilde{x}_{n-2}$ are not adjacent in time), and with $\mu \approx 10^{-5}$, Alexander and Rajala (1985) showed that this approach can give significant advantage compared to fixed prediction. More sophisticated LMS prediction for DPCM is described by Knee and Wells (1989) and Nishitani *et al* (1982).

As might be expected, interframe prediction is applicable to the quasi-static images found in videoconferencing. Essentially, interframe coding transmits the difference between successive frames and this can drastically reduce the bit rate; in fact, if there is no movement at all then, in principle, it is not necessary to transmit any information. Fig. 2.18 illustrates a popular form of interframe coding called *motion-compensated* coding. For still pictures the simple interframe prediction $\hat{x}_n = \tilde{x}_{n-1}$

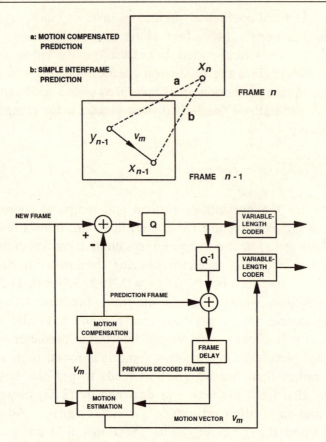

Fig. 2.18 Motion-compensated interframe coding.

performs well, but it is poor on moving pictures. Suppose that a point y_{n-1} on a moving object in frame $(n-1)$ appears in position x_n on frame n. In this case, a better prediction is $\hat{x}_n = \tilde{y}_{n-1}$ and it can be achieved by estimating the *motion vector* $\mathbf{v}_m$ and effectively using it to control a variable prediction delay of approximately one frame. Using this technique it is possible, for example, to transmit videoconference signals at the standard rate of 1.544 Mbit/s (see Table 1.3).

2.2.4 Adaptive DPCM (ADPCM)

As discussed in the preceding section, *adaptive* prediction tends to perform best, especially for speech. A fully adaptive DPCM system will, in fact, adapt both the predictor, P, and the quantizer, Q, as shown in Fig. 2.19. In general, adaptive information can be either separately transmitted

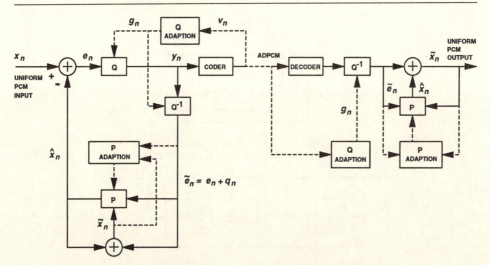

Fig. 2.19 General block diagram of ADPCM (backward adaption).

to the decoder (forward adaption), or it can be derived from recent quantized samples (backward adaption), and the latter is assumed in Fig. 2.19. In other words, the use of previous samples for adaption enables the decoder to operate without being sent any adaption information.

Predictor design for ADPCM can be considered in terms of the *decoder transfer function*, $D(z)$, Fig. 2.20. The predictor in Fig. 2.20(*a*) corresponds to (2.25), and using basic digital filter theory (Section 4.1) it is easily shown that

$$D(z) = \left(1 - \sum_{i=1}^{k} a_i z^{-i}\right)^{-1} \tag{2.39}$$

This is an 'all-pole' filter and, as shown in Section 4.1, its poles must fall within the unit circle for a stable decoder. However, in ADPCM the coefficients will adapt to minimize the mean-square error, and transmission errors may force the poles outside the unit circle (creating an unstable decoder). In fact, stability can be a crucial issue for high-order all-pole predictors.

A practical solution is to express $D(z)$ as a cascade of two filters as shown in Fig. 2.20(*b*), i.e.

$$D(z) = D_1(z)D_2(z)$$

$$= \left(1 + \sum_{i=1}^{m} b_i z^{-i}\right)\left(1 - \sum_{i=1}^{k} a_i z^{-i}\right)^{-1} \tag{2.40}$$

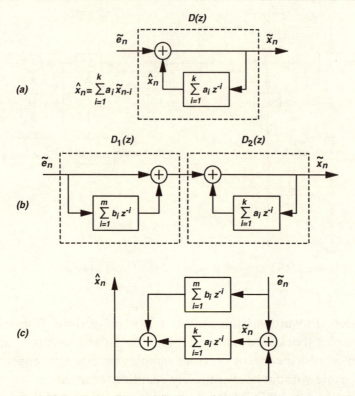

Fig. 2.20 Prediction filters: (*a*) all-pole (decoder); (*b*) pole-zero (decoder); (*c*) pole-zero (coder).

Filter $D_1(z)$ provides zeros in the transfer function and helps to stabilize the filter in the presence of transmission errors. It is easily seen from Fig. 2.20(*b*) that

$$\tilde{x}_n = \tilde{e}_n + \sum_{i=1}^{m} b_i \tilde{e}_{n-i} + \sum_{i=1}^{k} a_i \tilde{x}_{n-i} \qquad (2.41)$$

Therefore, since $\tilde{x}_n = \hat{x}_n + \tilde{e}_n$ (Fig. 2.14)

$$\hat{x}_n = \tilde{x}_n - \tilde{e}_n = \sum_{i=1}^{k} a_i \tilde{x}_{n-i} + \sum_{i=1}^{m} b_i \tilde{e}_{n-i} \qquad (2.42)$$

and the corresponding predictor at the coder is shown in Fig. 2.20(*c*). Usually, both a_i and b_i will be adaptive, although a_i could be fixed to ensure stability. The generalized ADPCM predictor defined by (2.40) could be described as an *adaptive filter* having m zeros and a relatively low number, k, of poles (ignoring poles at $z = 0$). It should be noted that coefficient optimization in the pole-zero approach involves *non-linear* equations (Makhoul, 1975).

Example 2.14

ADPCM is particularly important for the transmission of voice frequency signals. Under CCITT Rec. G.721 (1988), a 64 kbit/s A-law or μ-law PCM channel is first converted to a uniform PCM signal and the ADPCM coder transcodes this to a 4-bit ADPCM at 32 kbit/s. Both a_i and b_i are adaptive and so the actual prediction for G.721 can be written

$$\hat{x}_n = \sum_{i=1}^{2} a_{i,n-1}\tilde{x}_{n-i} + \sum_{i=1}^{6} b_{i,n-1}\tilde{e}_{n-i} \tag{2.43}$$

The coefficients are updated using a gradient algorithm.

There are various types of backward-adaptive quantizer (AQB) but the adaptive philosophy is essentially the same. Basically, a large quantizer output or codeword v_{n-1} at time $n-1$ will cause the quantizer effectively to increase its input dynamic range for time n. For the specific quantizing characteristic in Table 2.2, fast adaption for *speech* signals (not voiceband *data*) is achieved by adjusting the input scaling factor, g_n, according to

$$g_n = \alpha g_{n-1} + \beta W(v_{n-1}) \tag{2.44}$$

where W is some monotonically increasing function, i.e. it is large for a large quantizer output, and α and β are positive constants.

Effect of transmission errors
Since the all-pole decoder filter in Fig. 2.20(a) is a linear system we can consider its input to be a superposition of the signal and an error sequence, and treat the error sequence separately. Clearly, a single error sample at the filter input will generate an infinite sequence of decaying error samples at its output or, in other words, the all-pole decoder generates *error smearing* or *error propagation*. On the other hand, the all-zero filter $D_1(z)$ in Fig. 2.20(b) (an FIR filter) will generate only $(1 + m)$ error samples for a single input error, and so its inclusion helps to stabilize the decoder as previously noted.

The *subjective* performance of DPCM relative to PCM in the presence of channel errors depends upon the type of input signal. In the case of speech or music DPCM is subjectively more robust than PCM, typically by several orders of magnitude (Yan and Donaldson, 1972). This is because in speech or music a single error in PCM is subjectively more annoying (as in 'spiky clicks') than a relatively small error smeared over a long period.

Another way of looking at it is that a short noise burst will cause an all-pole filter to resonate at its *current* resonant frequency, thereby giving an 'acceptable' sound. Typically, DPCM speech and music systems can tolerate a channel bit error rate (BER) of 10^{-3} and 10^{-4} respectively, and FEC is often unnecessary.

In contrast, for image coding DPCM is subjectively *less* robust than PCM because error propagation is subjectively more annoying. The problem is worst for 1-D prediction since then an error generates a streak across the image, and for this case, the decoder loops could be reset during line blanking.

Sub-band coding (SBC)

ADPCM is sometimes used in conjunction with *sub-band coding* (SBC) in order to optimise compression. In a sub-band coder, the linearly quantized PCM signal is processed through a bank of filters and each sub-band is separately coded (quantized) according to the input signal energy. It is particularly applicable to audio signals and here the objective is to exploit the spectral redundancies within the audio spectrum. For example, the ear is particularly sensitive to low frequencies and so the lowest frequency band can be assigned the largest number of bits/sample. Conversely, quantization can be less accurate for bands to which the ear is more tolerant.

SBC of speech might typically be based on 11 bandpass filters (each 250 Hz wide) spanning the range 300 Hz–3 kHz. Heuval *et al* (1991) have shown that, out of these, the four most energetic bands generally contain some 95% of the energy 50% of the time, or 50% of the energy 88% of the time. They also showed by subjective testing that, essentially, it was only necessary to select the four most energetic bands for subsequent adaptive quantization and transmission in order to obtain acceptable speech quality. Similarly, compression of compact disc quality music typically uses four sub-bands, as shown in Fig. 2.21(*a*). This figure shows that excellent quality stereo music can be transmitted at 384 kbit/s using a combination of ADPCM and SBC, the corresponding compressed bit rate being just 4 bits/sample.

Example 2.15

SBC offers an opportunity to employ bandpass sampling. Suppose a bandwidth B is split into M equal sub-bands, as shown in Fig. 2.21(*b*). The bandwidth of each sub-band is therefore B/M and the maximum frequency of the nth band is nB/M. According to the bandpass sampling theorem (Equation (1.18)) we can therefore sample each band at a rate

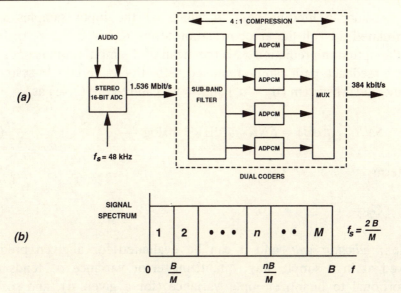

Fig. 2.21 SBC: (*a*) use of ADPCM and SBC for compact disc quality stereo music; (*b*) bandpass sampling for equal width sub-bands.

$$f_s = 2\frac{B}{M}\frac{n}{\lfloor n \rfloor} = \frac{2B}{M}$$

rather than at rate $2B$ prior to independent coding operations.

2.2.5 DPCM coding gain

It is important to establish formally the compression advantage of DPCM wrt linear PCM, and to do this we will refer to Figs. 2.14 and 2.16. First note that the decoded signal $\tilde{x}_n$ contains a normalized signal power σ_x^2 and a normalized quantizing noise power σ_q^2, and so we can define an output SNR. By making certain reasonable assumptions, such as a Laplacian pdf for $p(e)$, reasonably fine quantizing ($N \geq 8$) and small probability of slope overload, it can be shown that the maximum SNR at the decoder output is (O'Neal, 1966)

$$SNR_{DPCM} = 10\log \frac{\sigma_x^2}{\sigma_q^2|_{\min}}$$

$$= 10\log \frac{2N^2 \sigma_x^2}{9\sigma_e^2}$$

$$= 6n - 6.5 + 10\log \frac{\sigma_x^2}{\sigma_e^2} \text{ dB} \qquad (2.45)$$

Now consider a limiting case in which all the input samples x_n are uncorrelated, i.e. all the autocorrelation terms R_i, $i \neq 0$, are zero. In this case the optimum prediction $\hat{x}_n$ is the mean of the input sequence, or zero, and the feedback path can be removed. Effectively the DPCM system has become a PCM system ($\sigma_x^2 = \sigma_e^2$) and so we can rewrite (2.45) as

$$SNR_{DPCM}(\text{dB}) = SNR_{PCM}(\text{dB}) + 10 \log \frac{\sigma_x^2}{\sigma_e^2} \qquad (2.46)$$

The term

$$G_p = \frac{\sigma_x^2}{\sigma_e^2} \qquad (2.47)$$

is the *prediction gain* and it can be evaluated for a given predictor. Looked at in a simple way, quantizing error variance σ_q^2 tends to be proportional to quantizer input variance (for a given n), and so if we reduce the quantizer input variance by a factor G_p the output SNR must increase by G_p. Note that we can also express the output SNR as

$$SNR_{DPCM} = G_p \cdot SNR_Q \qquad (2.48)$$

where

$$SNR_Q = \frac{\sigma_e^2}{\sigma_q^2} \qquad (2.49)$$

The simple expression for G_p in (2.47) is approximately true for both voice frequency signals and video signals. Also, it is usual to take $G_p(\text{dB})$ as the approximate SNR improvement in going from PCM to DPCM assuming that both the PCM and DPCM systems code to n bits/sample. Alternatively, for a given SNR, the bit rate can be reduced for DPCM.

Example 2.16

Consider simple first-order prediction. According to (2.25), and ignoring quantization, we can define first-order prediction as $\hat{x}_0 = a_1 x_1$. Therefore, since $E[e_n] = 0$ in Fig. 2.16 we can write

$$\sigma_e^2 = E[(x_0 - a_1 x_1)^2]$$

$$= E[x_0^2] - 2a_1 E[x_0 x_1] + a_1^2 E[x_1^2]$$

Using (2.36) this reduces to

$$\sigma_e^2 = (1 - 2a_1 \rho_1 + a_1^2)\sigma_x^2$$

or

$$G_p = (1 - \rho_1^2)^{-1} \tag{2.50}$$

Measurements of the long-time autocorrelation function R_i for speech typically give $\rho_1 \approx 0.85$ (Flanagan *et al*, 1979), in which case $G_p \approx 5.6$ dB. Higher gain can be achieved by using higher-order prediction. Similarly, for monochrome video signals ρ_1 may be typically 0.96 for a 'head and shoulders' portrait and 0.75 for a 'crowd' scene, giving prediction gains of 11.1 and 3.6 dB respectively. A lower gain for high detail scenes is to be expected since there is less redundancy to remove.

So far we have *assumed* the SNR is a valid criterion for comparing PCM and DPCM systems and, for images at least, we might question this given that the visible (or subjective) effect of noise depends upon the noise spectrum. Fortunately, it turns out that quantizing noise in both PCM and DPCM video systems has a substantially flat spectrum, and so PCM and DPCM systems with the same SNR will have similar subjective impairment. Fig. 2.22 is therefore a reasonable comparison of colour video systems and indicates that substantial prediction gain is possible. For video systems it is generally accepted that DPCM gives approximately 12 dB SNR improvement for a given value of n, or, for a given SNR, DPCM requires about 2 bits/sample less than PCM (see (2.45)). Similar prediction gains are possible for speech using ADPCM and high-order prediction, e.g. $k = 10$.

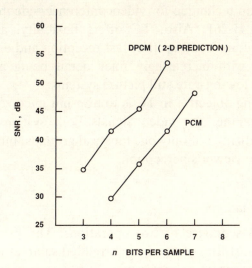

Fig. 2.22 Measured SNR for PCM and DPCM PAL colour signals (Devereux, 1975).

Example 2.17

Typical parameters for a monochrome video system might be

PCM input = 6 bits/sample

DPCM output = 4 bits/sample

σ_x = (peak–peak input signal)/3

$\rho_1 = 0.96$

Assuming first-order prediction, (2.36) gives $a_1 = 0.96$, and (2.50) gives

$$\sigma_e = \sqrt{(1 - \rho_1^2)}\sigma_x = 0.28\sigma_x$$

Taking the maximum prediction error, e_m as the peak–peak value of the PCM input (64 PCM quanta), then (2.23) gives $r = 5.05$. The quantizer representative levels can then be designed using (2.22), i.e.

$$y(x) = -12.67\ln(1 - 0.9936x)$$

2.3 Non-reversible coding – transform coding (TC)

TC is an alternative scheme to DPCM and it is useful for compressing both speech and video signals. Generally speaking it can be regarded as a non-reversible scheme if the transform coefficients are deliberately quantized for the purpose of achieving compression. For speech, (adaptive) TC has been shown to out-perform other frequency domain compression schemes, such as sub-band coding (Tribolet and Crochiere, 1979). For video signals, TC generally has greater potential for compression than DPCM although the signal processing tends to be more complex. In fact, TC is the selected compression technique for videoconference/videophone codecs under CCITT Rec. H.261. Also, TC offers flexibility, as for example when progressive picture build-up (i.e. fast reception and display of a low-resolution picture with increasingly finer detail being added subsequently) is employed in low-bit-rate still picture systems.

As in DPCM, the underlying objective in TC is to transmit uncorrelated samples and, generally speaking, for video signals both systems can achieve compression by quantizing not only on statistical grounds, but also in such a way as to exploit the viewer's perception.

2.3.1 Linear transformation

Transform coders perform a sequence of two operations. The first is a linear transformation that transforms a set of N correlated samples into a set of N 'more uncorrelated' samples or *coefficients*. In fact, a criterion for an optimal transform is that it yields uncorrelated coefficients, and so transformation is sometimes referred to as a 'whitening' process. After decorrelation, the coefficients can then be quantized *independently* in order to achieve data compression.

The underlying mathematics of discrete linear transforms is given in Section 5.1, and a general (1-D) TC system is shown in Fig. 2.23. Here, N PCM input samples are represented as a vector

$$\mathbf{X}^T = [x_0 x_1 \ldots x_{N-1}] \tag{2.51}$$

and this is transformed to N coefficients in the frequency (or more generally 'sequency') domain using the forward transformation

$$\mathbf{G} = \mathbf{AX} \tag{2.52}$$

where

$$\mathbf{G}^T = [g_0 g_1 \ldots g_{N-1}] \tag{2.53}$$

In (2.52), $\mathbf{A}$ is an $N \times N$ *transform matrix* whose elements are specified by the selected transform (in general they will be complex).

Suppose for the moment that we ignore the quantization process in Fig. 2.23 so that the input to the *inverse transform matrix* $\mathbf{B}$ is $\mathbf{G}$. Simply premultiplying $\mathbf{G}$ by $\mathbf{B} = \mathbf{A}^{-1}$ gives

$$\mathbf{Y} = \mathbf{A}^{-1}\mathbf{G} = \mathbf{A}^{-1}\mathbf{AX} = \mathbf{X} \tag{2.54}$$

and so, in this case, the TC process is completely reversible. In practice the inverse $\mathbf{A}^{-1}$ is not a problem since practical TC systems invariably use *unitary* transforms, i.e.

$$\mathbf{A}^{-1} = \mathbf{A}^{*T} \tag{2.55}$$

where T denotes transpose and * denotes complex conjugate. Practical sinusoidal transforms, such as the cosine, sine and Fourier transforms, and non-sinusoidal (sequency domain) transforms, such as the Walsh–Hadamard, Haar and Slant transforms, are all unitary. Therefore, ignoring quantization and assuming a unitary transform we can write

$$\mathbf{X} = \mathbf{A}^{*T}\mathbf{G} \tag{2.56}$$

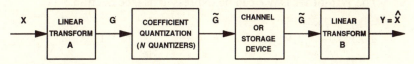

Fig. 2.23 Model of an Nth-order TC system.

which means that the inverse transformation is as easy to implement as the forward transform.

Example 2.18

One of the most useful transforms is the discrete cosine transform (DCT), and the 1-D forward transform (as defined in (5.77) and (5.78)) is

$$\mathbf{C} = \mathbf{WX} \tag{2.57}$$

Since this is a unitary transform, and since $\mathbf{W}$ has real coefficients (see (5.76)), then (2.56) gives

$$\mathbf{X} = \mathbf{W}^\mathsf{T}\mathbf{C} \tag{2.58}$$

Transposing the $\mathbf{W}$ matrix in (5.77) gives

$$
\begin{bmatrix} x_0 \\ x_1 \\ \cdot \\ \cdot \\ \cdot \\ x_{N-1} \end{bmatrix}
=
\begin{bmatrix}
w_{00} & w_{10} & \cdot & \cdot & \cdot & w_{N-1\,0} \\
w_{01} & w_{11} & & & & \\
\cdot & & & & & \\
\cdot & & & & & \\
\cdot & & & & & \\
w_{0\,N-1} & & \cdot & \cdot & \cdot & w_{N-1\,N-1}
\end{bmatrix}
\begin{bmatrix} C_0 \\ C_1 \\ \cdot \\ \cdot \\ \cdot \\ C_{N-1} \end{bmatrix}
\tag{2.59}
$$

i.e.

$$
x_n = w_{0n}C_0 + w_{1n}C_1 + \ldots + w_{N-1\,n}C_{N-1}
$$

$$
= \sum_{k=0}^{N-1} w_{kn}C_k, \qquad n = 0, 1, \ldots, N-1 \tag{2.60}
$$

Finally, substituting for w_{kn} from (5.76) gives the inverse transform in (5.11) (assuming the factor $2/N$ is evenly distributed between forward and inverse transforms).

As previously discussed, one objective of the transformation process is to transform a sequence of correlated samples into a sequence of uncorrelated coefficients. Although certainly not ideal, the simple discrete Walsh transform (DWT) is useful for demonstrating this concept.

Example 2.19

The Nth-order DWT can be expressed as (see Section 5.1)

$$
a(k) = \frac{1}{N}\sum_{n=0}^{N-1} x_n w(k, n), \qquad k = 0, 1, \ldots, N-1 \tag{2.61}
$$

where $w(k, n)$ is a sample value of the kth Walsh function in Fig. 5.1(b), i.e. it has value +1 or −1. Suppose $N = 8$ and that eight PCM input samples are represented on a 0–255 scale by the vector

$$\mathbf{X}^T = [x_0 \quad x_1 \quad \ldots \quad x_7]$$

$$= [202 \quad 231 \quad 190 \quad 174 \quad 137 \quad 110 \quad 115 \quad 130] \tag{2.62}$$

Inserting these values into (2.61) and using Fig. 5.1(b) gives

$$a(0) = \tfrac{1}{8}(202 + 231 + 190 + 174 + 137 + 110 + 115 + 130) = 161.125$$

$$a(1) = \tfrac{1}{8}(202 + 231 + 190 + 174 - 137 - 110 - 115 - 130) = 38.125$$

$$a(2) = \tfrac{1}{8}(202 + 231 - 190 - 174 - 137 - 110 + 115 + 130) = 8.375$$

$$\vdots$$

$$a(7) = \tfrac{1}{8}(202 - 231 + 190 - 174 + 137 - 110 + 115 - 130) = -0.125$$

The transformer output (or sequency spectrum) is therefore

$$\mathbf{a}^T = [161.125 \quad 38.125 \quad 8.375 \quad 8.875 \quad -0.375 \quad -10.875 \quad -3.125 \quad -0.125] \tag{2.63}$$

It is apparent from (2.62) and (2.63) that the transformer output is less correlated than its input.

Example 2.20

Suppose that a sine wave is sampled at exactly four times its frequency in the phase shown in Fig. 2.24. Applying the eight samples shown to the DWT in (2.61), and using Fig. 5.1(b) gives the transformer output

$$\mathbf{a}^T = [E00e0000]$$

In this carefully selected case, the transformer output samples are completely uncorrelated since E can be varied independently of e.

Like other practical transforms, the DWT would normally be evaluated

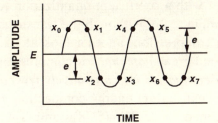

Fig. 2.24 Generation of an optimum data sequence for the DWT (Example 2.20).

using some form of fast algorithm, and in this case it is usual to express the forward transform in terms of an Nth-order *Walsh–Hadamard matrix*, $\mathbf{H}_N$ (giving the DWHT). For $N = 8$ the DWHT is then

$$\mathbf{a} = \tfrac{1}{8}\mathbf{H}_8\mathbf{X} \tag{2.64}$$

or

$$
\begin{bmatrix} a(0) \\ a(7) \\ a(3) \\ a(4) \\ a(1) \\ a(6) \\ a(2) \\ a(5) \end{bmatrix} = \tfrac{1}{8}
\begin{bmatrix}
+ & + & + & + & + & + & + & + \\
+ & - & + & - & + & - & + & - \\
+ & + & - & - & + & + & - & - \\
+ & - & - & + & + & - & - & + \\
+ & + & + & + & - & - & - & - \\
+ & - & + & - & - & + & - & + \\
+ & + & - & - & - & - & + & + \\
+ & - & - & + & - & + & + & -
\end{bmatrix}
\begin{bmatrix} x_0 \\ x_1 \\ x_2 \\ x_3 \\ x_4 \\ x_5 \\ x_6 \\ x_7 \end{bmatrix} \tag{2.65}
$$

where $\pm$ denote ± 1. A fast algorithm for (2.65) is shown in Table 2.4. Note that (2.65) requires $8(8 - 1) = 56$ additions whilst only $8 + 8 + 8 = 24$ additions are required for the fast Walsh–Hadamard transform (FWHT). For an input vector of order N the FWHT requires $N \log_2 N$ additions over $\log_2 N$ stages compared to $N(N - 1)$ additions by direct evaluation.

2.3.2 Energy compaction

Apart from computation efficiency, the FWHT in Table 2.4 highlights another practical point. Note that the addition in each stage requires an extra bit to represent the sum and so the wordlength increases through the transformer. To transform a block of N PCM samples each with a resolution of n bits, the transformer output samples must have a resolution of $n + \log_2 N$ bits if they are to be precisely represented. For example, if $N = 32$, then 13-bit coefficients are required for an 8-bit PCM input!

It should now be apparent that, in order to achieve compression, it is necessary to follow the transformer with a coefficient quantization (or 'bit allocation') process, as indicated in Fig. 2.23. Although the transformation itself invariably does not result in compression, another look at Example 2.19 will indicate that it has provided great *potential* for compression, i.e. *most of the signal energy has been compacted into a relatively few coefficients*. Similar (and generally somewhat better) energy compaction will be obtained using the DCT, as defined in (5.76)–(5.78), and Table 2.5 illustrates the performance of the DCT for several types of input vector.

Table 2.4. A FWHT algorithm for N = 8 (Walker, 1974).

	Stage 1	Stage 2	Stage 3
$x(0) = A$	$(A + E)$	$(A + E) + (C + D)$	$[(A + E) + (C + G) + [(B + F) + (D + H)] = a(0)$
$x(1) = B$	$(B + F)$	$(B + F) + (D + H)$	$[(A + E) + (C + G)] - [(B + F) + (D + H)] = a(7)$
$x(2) = C$	$(C + G)$	$(A + E) - (C + G)$	$[(A + E) - (C + G)] + [(B + F) - (D + H)] = a(3)$
$x(3) = D$	$(D + H)$	$(B + F) - (D + H)$	$[(A + E) - (C + G)] - [(B + F) - (D + H)] = a(4)$
$x(4) = E$	$(A - E)$	$(A - E) + (C - G)$	$[(A - E) + (C - G)] + [(B - F) + (D - H)] = a(1)$
$x(5) = F$	$(B - F)$	$(B - F) + (D - H)$	$[(A - E) + (C - G)] - [(B - F) + (D - H)] = a(6)$
$x(6) = G$	$(C - G)$	$(A - E) - (C - G)$	$[(A - E) - (C - G)] + [(B - F) - (D - H)] = a(2)$
$x(7) = H$	$(D - H)$	$(B - F) - (D - H)$	$[(A - E) - (C - G)] - [(B - F) - (D - H)] = a(5)$

Table 2.5. *DCT coefficients for two input vectors* (N = 16).

$\mathbf{X}^{\mathrm{T}} = [9139350126912048]$		$\mathbf{X}^{\mathrm{T}} = [5467677878978989]$	
k	C_k	k	C_k
0	63.0	0	115.0
1	5.3	1	−18.4
2	11.5	2	−5.1
3	4.5	3	−4.0
4	3.3	4	−0.8
5	−25.4	5	−3.0
6	26.5	6	−0.5
7	9.6	7	0.3
8	7.0	8	−1.0
9	13.3	9	4.5
10	17.6	10	5.5
11	3.8	11	−2.4
12	10.5	12	5.8
13	4.0	13	1.1
14	−18.0	14	−1.9
15	−2.4	15	−1.4

Again it is apparent that, for highly correlated inputs, the signal energy is compacted into a relatively few coefficients.

We can summarize these examples by stating that a good transform concentrates most of the energy from commonly occurring signals into as few coefficients as possible. The theoretically optimal transform is the *Hotelling* or *Karhunen–Loeve transform* (KLT) in the sense that it yields uncorrelated coefficients and packs the maximum average energy into the first M coefficients $(M \leq N)$. In practice, the complexity and data-dependent properties of the KLT mean that it is invariably replaced by a good sub-optimal transform, such as the DCT. In fact, the performance of the DCT (and other sinusoidal transforms) becomes equivalent to the KLT as the input vector size $N \Rightarrow \infty$. An efficient fast algorithm for the DCT is described in Section 5.7, and a 2-D DCT for images is discussed in Section 5.8.

2.3.3 TC of images and coefficient quantization

So far we have established that a good transform both decorrelates the signal and compacts the signal energy into a relatively few coefficients. It remains to quantize the coefficients in order to achieve compression and this is best discussed in the context of *image* compression, Fig. 2.25.

In practice an $L \times L$ image is usually partitioned into small $N \times N$ blocks and each block is transformed independently. Block coding is necessary in order to make the TC of images computationally efficient in terms of storage and speed, i.e. the DSP cost tends to decrease rapidly with decreasing block size, as indicated in Fig. 2.26. On the other hand, independently transform coding each block neglects the redundancies (pixel correlations) that exist between blocks, and on this basis the block size should be large. In fact, decreasing the block size leads to increased

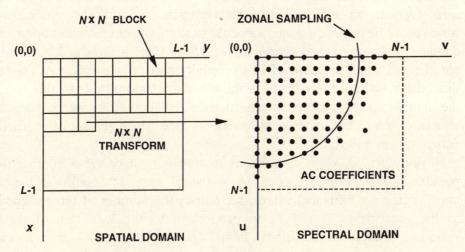

Fig. 2.25 TC: block coding for image compression (u and v are spatial frequencies).

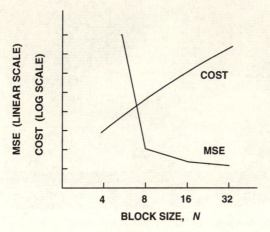

Fig. 2.26 TC: relative mean-square error (MSE)/cost vs block size.

mean-square error between the original image and the reconstructed image (Fig. 2.26), and so, in practice, the block size is a compromise (typically 8×8 or possibly 16×16). Fig. 2.27 illustrates typical reconstruction errors arising from block coding when TC is used to achieve excessive compression ratios.

Generally speaking, compression is achieved by a combination of coefficient selection, quantization and efficient coding. Fig. 2.25 illustrates a simple selection technique called *zonal sampling* and is based on the assumption that only coefficients within a certain fixed geometric zone (usually those concentrated around the DC coefficient) need be quantized and transmitted. At the decoder the discarded coefficients could be set to zero. In order to avoid the possible rejection of significant coefficients which could lie outside a zone, we could base coefficient selection upon an *adaptive* technique called *threshold sampling*. Here a threshold could be based upon some measure of block activity and only coefficients above this adaptively determined threshold would be transmitted. In this way, the number and location of the coefficients are adapted to the local block structure, but at the expense of having to code which sub-set of coefficients has been selected.

Having decided which coefficients to transmit, there are a number of possible quantization strategies. A statistical approach could be based upon coefficient variance, where, for mmse, the number of bits assigned to the coefficients should be in proportion to the logarithm of their variances (Netravali and Limb, 1980). On the other hand, in practice coefficient quantization for video transform coders is often done according

Fig. 2.27 TC: illustrating the nature of the reconstruction error when using block coding.

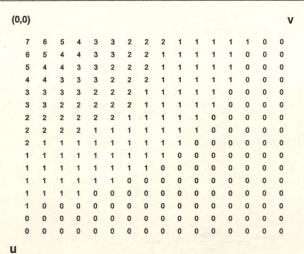

(0,0)   V

```
7  6  5  4  3  3  2  2  2  1  1  1  1  1  0  0
6  5  4  4  3  3  2  2  1  1  1  1  1  0  0  0
5  4  4  3  3  2  2  2  1  1  1  1  1  0  0  0
4  4  3  3  3  2  2  2  1  1  1  1  1  0  0  0
3  3  3  3  2  2  2  1  1  1  1  1  0  0  0  0
3  3  2  2  2  2  2  1  1  1  1  1  0  0  0  0
2  2  2  2  2  2  1  1  1  1  1  0  0  0  0  0
2  2  2  2  1  1  1  1  1  1  1  0  0  0  0  0
2  1  1  1  1  1  1  1  1  1  0  0  0  0  0  0
1  1  1  1  1  1  1  1  1  0  0  0  0  0  0  0
1  1  1  1  1  1  1  1  0  0  0  0  0  0  0  0
1  1  1  1  1  1  0  0  0  0  0  0  0  0  0  0
1  1  1  1  0  0  0  0  0  0  0  0  0  0  0  0
1  0  0  0  0  0  0  0  0  0  0  0  0  0  0  0
0  0  0  0  0  0  0  0  0  0  0  0  0  0  0  0
0  0  0  0  0  0  0  0  0  0  0  0  0  0  0  0
```

u

Fig. 2.28 Typical bit allocation after applying the DCT to a 16×16 image block. The compression rate corresponds to approximately 1 bit/pixel (Jain, 1981, © IEEE).

to psychovisual thresholds. Also, sometimes coefficient selection is implicit and not a deliberate first step, as in CCITT Rec. H.261. Here the coefficients are quantized to a fixed wordlength and then sequences of coefficients are efficiently coded using run-length/variable-length coding (see Example 2.2). High-frequency, zero-valued AC coefficients tend not to be selected due to the coding process. Similarly, Fig. 2.28 illustrates typical bit allocation after applying the DCT direct to a 16×16 image block, and again, only coefficients in a small zone around the origin will be transmitted since many coefficients are allocated zero bits. Here the average compressed rate is around 1 bit/pixel, although adaptive DCT techniques have been shown to yield usable photovideotex images at around 0.5 bit/pixel.

2.3.4 Hybrid DPCM–TC

Fig. 2.29 shows a hybrid coding scheme for video conferencing/video telephony and is specified for rates between 64 kbit/s and 2 Mbit/s (CCITT Rec. H.261). The input signal must first be converted to YUV format with a luminance and chrominance block size of 8×8 pixels (blocks are non-overlapping). In the intraframe mode pixel data in block form is passed into an 8×8 DCT, whilst in the interframe mode a DPCM loop is invoked and a prediction is read from the predictor framestore (as in Fig. 2.18).

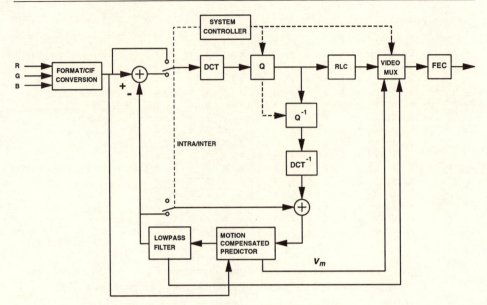

Fig. 2.29 Basic video coder specified in CCITT Rec.H.261.

This type of coder has to make two major decisions; namely, whether to use intraframe or interframe compression and, when using the latter, whether to use motion compensation, i.e. whether to suppress or not suppress the motion vector $\mathbf{v}_m$. Both decisions use computations made by the motion estimation processor (Fig. 2.18). In Rec. H.261 motion vector estimation is made by using 16×16 pixel luminance blocks and by searching in the previous frame. Consider a luminance block k in the current frame n having pixel intensity $b(k, n)$. We could compute the difference between this and the 'zero-displaced' block k in the previous frame as

$$d(k, n) = b(k, n) - b(k, n-1) \qquad (2.66)$$

where the actual computation finds some mean absolute error (MAE). Assuming lateral translation, we could also search for the 'best-match' block k' in the previous frame by invoking a vector shift – typically, the 16×16 pixel luminance block is shifted ± 7 pixels in both the horizontal and vertical directions. The new MAE value would then correspond to the 'displaced' block difference

$$d'(k, n) = b(k, n) - b(k', n-1) \qquad (2.67)$$

and both MAE values can then be used to make the motion compensation/no motion compensation decision. If motion compensation is selected, the corresponding vector is used for prediction, as indicated in Fig. 2.18. Finally, one of the MAE values together with a measure of the variance of the current block can also be used to make the interframe/intraframe decision.

Typically, the DCT output has 12-bit coefficients and is quantized using a midtread semi-uniform quantizer to 8 bits/coefficient. The quantizer output is then zig zag scanned and run-length coded (as discussed in Example 2.2) prior to video multiplexing. The lowpass filter in the DPCM loop performs a 3×3 convolution (see Section 4.8.2) on an 8×8 pixel block in order to reduce artifacts introduced by motion compensation.

2.4 Non-reversible coding – CELP coding

This section outlines a technique which has been selected as the US Federal Standard for 4.8 kbit/s speech coding (Campbell *et al*, 1989). The codebook-excited linear predictive (CELP) coder bridges the gap between waveform coders and conventional *analysis/synthesis* techniques for speech (vocoders), as indicated in Fig. 2.1. The concept originates from adaptive predictive coding (APC) (essentially a waveform coding technique) but it embraces the idea of analysis/synthesis common to vocoders. However, it has none of the usual vocoder noise problems and is, in fact, capable of providing high-quality synthetic speech (near toll quality).

The predictive coding system in Fig. 2.30 highlights the essential elements of a CELP coder. This differs from the conventional predictive coder (see Fig. 2.14) in several respects. First, the prediction $\hat{s}_n$ is derived from filtering a sequence v_n derived from a 'codebook'. This is called an *excitation* or *innovation* sequence, and very often it is a succession of

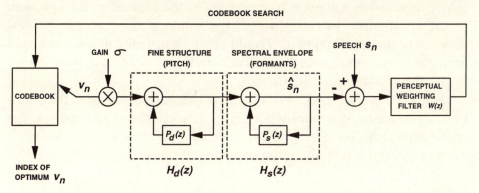

Fig. 2.30 Illustrating the prediction, codebook search and noise shaping concepts of the CELP coder.

independent Gaussian random numbers having zero mean and unit variance (a *stochastic* codebook). The samples of v_n are then scaled by a factor σ to produce an optimum match between the original signal s_n and the synthetic signal $\hat{s}_n$. To illustrate the basic idea, suppose speech sampled at 8 kHz is analysed over a 5 ms frame. The 40 samples s_n in this frame are compared to samples $\hat{s}_n$ derived from filtering a sequence v_n of 40 samples from the codebook. Sequence v_n is typically one of 1024 possible random number sequences or 'waveforms', and so the optimal sequence is identified by a 10-bit index. Transmission of this index will therefore correspond to a basic rate of 0.25 bit/sample or 2 kbit/s.

A second difference from conventional predictive coding arises from the use of a 'perceptual weighting filter', $W(z)$. The objective here is to minimise the subjective distortion in the synthetic signal by shaping the noise (speech error) spectrum, and this is of crucial importance for realizing low bit rates. In particular, noise components falling in frequency regions where speech energy is concentrated (formant regions) can have higher energy relative to other components since they will be masked by the speech itself. Filter $W(z)$ therefore attenuates those frequencies where the error is perceptually less important (i.e. the formant or high-energy frequencies) and amplifies those (lower-energy) frequencies where the error is perceptually more important.

In Fig. 2.30, determination of the optimum sequence v_n is done using an 'analysis by synthesis' approach. In other words, we might regard the predictors as filtering sequence v_n in such a way as to restore the essential spectral components of the speech. The reason for a cascade of two predictors is embedded in the classical assumptions behind speech synthesis. A basic assumption in speech synthesis is that the spectrum of the sound source (which is assumed to be either 'voiced' or 'unvoiced' sound) can be regarded as being spectrally shaped by the acoustic resonant system of the vocal tract. In other words, the sound source defines the *fine structure* of the spectrum (which could be line harmonics of a fundamental frequency, or a continuous 'noise-like' spectrum) and this is then shaped to give the actual *spectral envelope*.

Therefore, in Fig. 2.30, predictor $P_d(z)$ essentially restores the short-time spectral fine structure, whilst $P_s(z)$ restores the short-time spectral envelope. Prediction based on the spectral envelope involves relatively short delays and is typically 10th-order, i.e.

$$P_s(z) = \sum_{i=1}^{k} a_i z^{-i}, \quad k = 10 \tag{2.68}$$

where $P_s(z)$ is of the form shown in Fig. 2.16. Referring to filter $H_s(z)$ in Fig. 2.30, and denoting its input and output by the z-transforms $X(z)$ and $Y(z)$ respectively, we have

$$Y(z) = X(z) + P_s(z)Y(z)$$

$$H_s(z) = \frac{Y(z)}{X(z)} = [1 - P_s(z)]^{-1}$$

The significant point here is that, for a stable filter, the poles of transfer function $H_s(z)$ must lie inside the unit circle (Section 4.1.3).

Now consider the fine structure predictor. For voiced speech the fine structure is essentially a line spectrum resulting from harmonics of a fundamental frequency, and this suggests that the associated filter $H_d(z)$ should have regular peaks in its frequency response corresponding to the line harmonics. We therefore need a transfer function of the form

$$H_d(z) = [1 - P_d(z)]^{-1}$$

where

$$P_d(z) = \beta z^{-M} \tag{2.69}$$

and β and M are positive constants. In this case it is readily shown (see Example 4.6) that $H_d(z)$ has poles at

$$z = \beta^{1/M} e^{j2\pi m/M}, \qquad m = 0, 1, 2, \ldots, M - 1$$

corresponding to resonant peaks at frequencies m/MT where T is the sampling period. In other words, the fundamental or pitch period (typically 2–20 ms) will correspond to M sample periods. For unvoiced speech, which has a noise-like spectrum, β will be small and M would be random.

The basic CELP coder samples at 8 kHz, has a 30 ms frame (i.e. 240 samples/frame) and reserves 10 bits for a codebook size of 1024 sequences or waveforms. Essentially, CELP analysis requires:

(1) short delay 'spectral envelope' prediction once per frame for $P_s(z)$ (basically 10th-order LPC analysis, Section 2.5);
(2) Long delay 'pitch' search for $P_d(z)$, four times per frame;
(3) 1024 vector codebook search for the optimum v_n (essentially an exhaustive closed-loop search four times per frame to minimize the weighted mean-square error).

CELP synthesis (Fig. 2.31) consists of the same three functions performed in reverse order, with the addition of a postfilter to enhance the output speech. A number of fast algorithms for CELP coding are discussed by

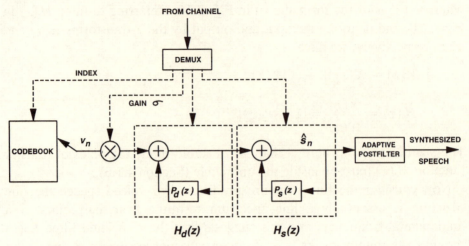

Fig. 2.31 Basic CELP decoder.

Kleijn *et al* (1990) and real-time implementations of the CELP algorithm are possible using single general purpose DSP devices.

2.5 Non-reversible coding – linear predictive coding (LPC)

In Section 2.4 we introduced the idea that linear predictors for restoring the short-time spectral envelope of speech will typically have order $k = 10$ or more. Put another way we could say that, over a short period of time, a speech sample s_n can be well estimated as a linear combination of k previous samples, as suggested by Fig. 2.32. Here, filter $A(z)$ is called a *short-term analysis filter*, or *inverse filter*, and due to the non-stationary nature of speech it could either be adaptive (see Fig. 4.52) or, as suggested by Fig. 2.32, the coefficients a_i, $i = 1, \ldots, k$ can be computed on a *frame-by-frame* basis. A short speech frame or segment is typically 20 ms.

Given good prediction (analysis) there will be a small *residual signal* given by

$$r_n = s_n - \sum_{i=1}^{k} a_i s_{n-i} \tag{2.70}$$

and taking the z-transform we have

$$R(z) = S(z) - \sum_{i=1}^{k} a_i z^{-i} S(z)$$

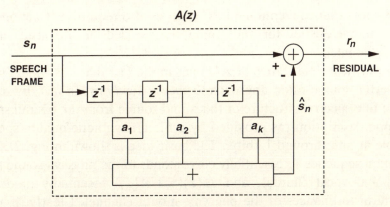

Fig. 2.32 Linear prediction (LP) for a speech frame.

giving

$$A(z) = \frac{R(z)}{S(z)} = 1 - \sum_{i=1}^{k} a_i z^{-i} \qquad (2.71)$$

In practice, coefficients a_i can be found for a short speech frame by using a least-square-error criterion, i.e.

$$\text{minimize} \sum_n r_n^2 \qquad (2.72)$$

where the summation is over the speech frame. As in DPCM predictor design, this minimization problem reduces to a set of k linear equations involving just the a_i terms and the signal autocorrelation function (Equation (2.30)). This is referred to as the *autocorrelation method* for determining the coefficients, and it requires the speech samples to be windowed to avoid transient effects arising from an abrupt start and finish. Conventional LPC analysis for the coefficients therefore involves:

(1) Segmenting the speech samples s_n into quasi-stationary frames by applying a window function w_n (see (2.33)). It is usual to use a Hamming window (Section 4.4.3) and to overlap adjacent windows.

(2) Using windowed samples to compute the *short-time* autocorrelation function, R_i, as in (2.33).

(3) Computing coefficients a_i via matrix inversion, Equation (2.32). As previously noted, this could be carried out using Levinson–Durbin recursion, or Schur recursion (ETSI, 1991).

Finally, note that conventional LPC analysis also requires extraction of the source characteristics for the frame, i.e. voiced or unvoiced, and, if voiced, the pitch.

The practical significance of $A(z)$ lies in the fact that it is widely used in a powerful 'source-filter' approach to *speech synthesis*. In fact, by separating the fine spectral structure of the sound source from the overall spectral envelope description (as provided by a filter), synthetic quality speech is possible at rates around 1 kbit/s. The basic idea is shown in Fig. 2.33. The excitation sequence v_n essentially corresponds to the physical sound source (lungs and vocal chords), and $H(z)(= 1/A(z))$ essentially models the short-term resonances of the pharynx–mouth channel. Clearly, if $v_n \approx r_n$ then simple inverse filtering gives

$$\hat{S}(z) = H(z)V(z) \approx H(z)A(z)S(z) = S(z)$$

and synthesis is very accurate. However, the simple synthesis model in Fig. 2.33 does have its limitations. For example, the 'all-pole' form of $H(z)$ can model physical resonances, or *formants*, but fails to model 'anti-resonance' (nasal) sounds. Also, some sounds cannot be clearly classified as either 'voiced' or 'invoiced' (noise-like) sounds. In addition, the periodic pulse generator in Fig. 2.33 gives a 'flat' amplitude spectrum for simplicity, but in practice the spectrum of a voiced source decreases by some 12 dB/octave (Hart *et al*, 1982). This implies that $H(z)$ should model not only the pharynx–mouth channel but also the spectral amplitude changes of the voiced source.

As far as communication systems are concerned, e.g. mobile radio, the main concern is to convey the filter coefficients and the excitation sequence v_n which gives the most accurate speech for a given filter $A(z)$.

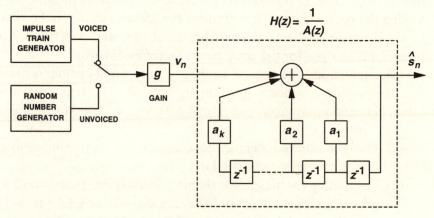

Fig. 2.33 Use of LPC for speech synthesis.

2.5.1 Regular-pulse excited linear predictive coding (RPE-LPC)

LPC is the basis of the European speech coding standard for digital mobile radio, or GSM (Natvig, 1988; ETSI, 1991), and a conceptual model for this system is shown in Fig. 2.34. As for LPC, the decoder generates synthesized speech by applying a sequence v_n to a time-varying filter $H(z) = 1/A(z)$ which models the short-time spectral envelope (formants or resonances) of the speech signal. Filter $A(z)$ is designed as before, i.e. some form of LPC analysis is performed on a speech frame and involves computation of the short-time autocorrelation function and some recursion rule to find the filter coefficients a_i. For GSM, each speech frame is 20 ms and the filter order is 8.

Having obtained $A(z)$, the remaining problem is to convey the prediction residual r_n that contains information describing the fine structure of the underlying spectrum. Essentially we need to transmit r_n in compressed form and for GSM this source coding process is performed on 5 ms sub-frames of length $L = 40$ samples (corresponding to an 8 kHz speech sampling rate). First note that in Fig. 2.34, r_n is modelled over a sub-frame by a sequence v_n, also of length L samples. The key point is that compression is achieved by constraining v_n to the form of Q $(Q < L)$ *equispaced* non-zero samples (of different amplitude) spaced by zero-valued samples. In other words, a pulse train or *regular-pulse excitation (RPE)* occurs at the input to $H(z)$ at the decoder.

The presence of zero-valued samples leads naturally to a set of N *candidate excitation vectors* $\mathbf{v}_k$, $k = 1, 2, \ldots, N$, where k denotes the position of the first non-zero sample in a particular sub-frame. Fig. 2.35 is a simple illustration for $L = 40$ and $N = 4$, i.e. there are four possible

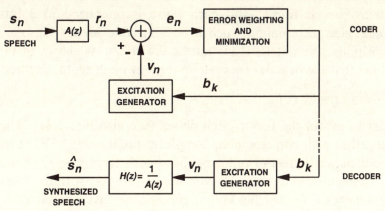

Fig. 2.34 A model for RPE-LPC for speech.

Fig. 2.35 Candidate regular-pulse excitation vectors $\mathbf{v}_k$ for $L = 40$, $N = 4$, $Q = 10$ (non-zero valued samples are represented by vertical lines).

phases or positions. Note that each candidate sequence has only $Q = 10$ non-zero sample values. The constraint on v_n enables it to be expressed as the excitation row vector

$$\mathbf{v}_k = \mathbf{b}_k \mathbf{M}_k, \qquad k = 1, \ldots, N \tag{2.73}$$

where $\mathbf{v}_k$ is of order L, $\mathbf{b}_k = [b_k(0)\ b_k(1)\ \ldots\ b_k(Q-1)]$ and $\mathbf{M}_k$ is a $Q \times L$ position matrix having entries of 1 or 0. In effect this means that $\mathbf{v}_k$ is entirely characterized by its position k and by $\mathbf{b}_k$ (a smaller vector). The coding objective is to select for each value of k the amplitudes $b_k(\cdot)$ which minimize some function of the error, e_n. The vector $\mathbf{b}_k$ that yields minimum error is then selected and transmitted, as indicated in Fig. 2.34. Given the optimum value of $\mathbf{b}_k$, the decoding procedure is then straight-forward.

Fig. 2.34 simply outlines the RPE-LPC concept and the actual compression scheme for (full-rate) GSM is shown in Fig. 2.36. This uses a $Q = 13$ pulse excitation 'grid' over a 5 ms sub-frame (52 RPE pulses over the 20 ms speech frame) and each pulse is quantized to 3 bits. With additional side information, such as a 2-bit code identifying one of four grid positions, the net RPE bit rate is 9.4 kbit/s, as shown in Fig. 2.36. Also note that the decoder uses a long-term or pitch synthesis filter

$$H_d(z) = (1 - \beta z^{-M})^{-1} \tag{2.74}$$

in order to model the major pitch pulses (see also Fig. 2.31). The use of this, together with corresponding long-term prediction (LTP) at the coder has the beneficial effect of reducing the speech error $s_n - \hat{s}_n$ (Kroon *et al*, 1986). Transmission of the LTP parameters together with the filter coefficients requires a further 3.6 kbit/s, giving a net average bit rate for full rate GSM of 13 kbit/s. In fact, the GSM coder can be defined as a

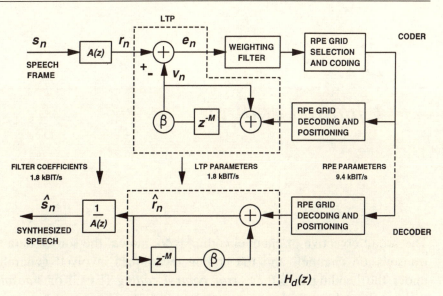

Fig. 2.36 Signal paths for a full rate GSM speech codec (RPE-LTP).

RPE-LTP transcoder between 13-bit uniform 8 kHz PCM (104 kbit/s) and 13 kbit/s. The system gives toll quality speech and a performance that degrades 'gracefully' as channel bit-error rate increases.

3

Channel coding

The broad objective of channel coding is to 'match' the source data to the transmission channel, and the coding techniques involved generally fall under the heading of either *error control coding* (ECC) or *transmission coding*. Scrambling techniques are often used to improve synchronization and can also be regarded as a form of channel coding.

3.1 Overview of ECC

Fig. 3.1 gives an overview of ECC techniques. Two main classes of error control are used to achieve an acceptable error rate at the receiver, namely, *automatic repeat request* (ARQ) and FEC. ARQ uses an error *detecting* code together with a feedback channel to initiate retransmission of any block received in error. It can be used where time delay is permissible, e.g. in data transmission. In contrast, FEC controls the received error rate via forward transmission only. In its simplest form, FEC amounts to error *concealment*, Fig. 3.2. Here, upon detection of a parity error the receiver could repeat the last most suitable word. Alternatively, after detection an isolated erroneous sample could be replaced by a linear interpolation between its two neighbours. The latter type of concealment is used on compact disc systems when an error *correction* code is overloaded, thereby avoiding sharp audio 'clicks'. Concealment techniques are simple to implement, but have restricted applications. For example, they are not applicable to computer systems, or to systems where sample-to-sample correlation has already been exploited, e.g. DPCM systems. Finally, it should be noted that ARQ (involving error detection) is sometimes combined with a moderate degree of error correction in order to improve the throughput efficiency. Such a hybrid ARQ–FEC scheme is indicated in Fig. 3.1.

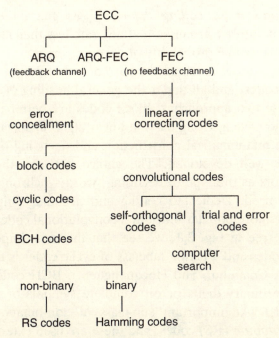

Fig. 3.1 Overview of ECC.

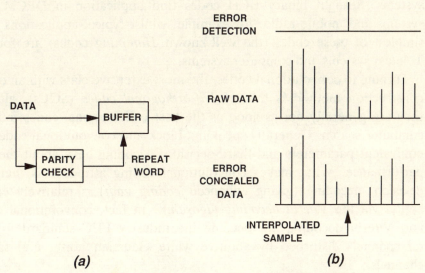

Fig. 3.2 Error concealment methods: (*a*) word repeat; (*b*) interpolation.

In this chapter we will concentrate on error correction coding. Most FEC systems use codes based on algebraic mathematics, and the objective is to achieve a particular decoded error rate given limited bandwidth and signal power in the channel. The fact that such codes are possible is underscored by *Shannon's noisy coding theorem*; this guarantees the existence of a code which allows transmission at any rate less than channel capacity and with arbitrarily small probability of decoding error.

Fig. 3.1 shows that it is usual to classify error control codes as either *block* or *convolutional* codes, and all under the general heading of *linear* or *group* codes. Of these two approaches, block codes are better understood mathematically and have a strong basis in linear algebra, specifically *Galois fields*, whilst the mathematical construction and decoding of convolutional codes is less well developed. The choice between the two depends upon such factors as the type of decoding, word synchronization and data format to be used. Dedicated coding and decoding chips are available for some of the more popular block and convolutional codes.

Continuing down the tree in Fig. 3.1, we see that the most important block codes are *cyclic* codes and a major subclass of cyclic codes is that of BCH codes (after Bose, Chaudhuri and Hocquenghem). BCH codes can operate on either single binary digits or on complete symbols or words (non-binary BCH codes). An important sub-class of non-binary BCH codes is that of *Reed–Solomon* (RS) codes, and these are used extensively in satellite communications systems and magnetic and optical recording systems. Straight binary BCH codes find application in DPCM video systems and mobile radio, for example, whilst typical applications of the simplest of these codes (the well-known *Hamming* codes) are found in Teletext systems and computer systems.

Turning to convolutional codes, the most extensive class with an explicit construction method is that of *self-orthogonal codes* (SOCs), although they are generally not as good as the best convolutional codes found by computer search. Generally speaking, block and convolutional codes with equivalent parameters and 'hard-decision' decoding have about the same performance. What makes convolutional coding attractive is that 'soft-decision' decoding (giving enhanced *coding gain*) is relatively easy to apply, via the *Viterbi decoding algorithm*. In fact, convolutional coding and Viterbi decoding has become the industry FEC standard for digital channels disturbed by additive white Gaussian noise, e.g. satellite channels.

Random and burst errors

Practical experience tells us that transmission errors can occur randomly or in bursts (when a single noise impulse affects a number of contiguous transmitted symbols) and codes have been devised specifically for both cases. For example, the Fire codes (Fire, 1959) and the RS codes are

Fig. 3.3 Concatenated code.

commonly used to combat *burst errors*, whilst the binary BCH codes are mainly used for correcting multiple random errors. In a *t*-error random error correcting code, any pattern of *t* errors will be corrected over a defined sequence length, and each symbol is assumed to be affected *independently* by noise. We will see that a *t*-error random error correcting code can also be used to correct burst errors when used in an *interleaved* encoding/decoding format. Interleaving is therefore particularly useful since it permits both random and burst errors to be corrected.

Other techniques for combatting mixtures of random and burst errors, or for combatting channels with high error rates may involve the use of a *concatenated* code. The usual approach (Fig. 3.3) is to use an *inner* code to correct most random errors and an *outer* code to correct those burst errors which overload the inner code. The inner code can be a binary or RS code, whilst the outer code is usually an RS code. Unfortunately, the overall *code rate* (Section 3.1.1) is the product of the two code rates and can be quite low. Also, a concatenated code is usually not as powerful as the best single-stage code with the same rate and block length. On the other hand, decoder complexity is reduced, and concatenated codes have been found useful in digital recorders, for example, where large data loss can occur due to 'dropout' (Gillard, 1986).

3.1.1 Distinctions between block and convolutional codes

At this point it is helpful to outline the essential differences (and some similarities) between block and convolutional error control codes. In general coding theory a block code always codes a particular message into the same fixed sequence of code symbols. In a 3-bit binary number code, for example, (represented by the cube vertices in Fig. 3.4(a)), message 'decimal 3' is always coded into the *codeword* or *vector* (011). Also, whatever the message, the codeword length is always fixed at the *block length*, *n*, of the code. In order to convert Fig. 3.4(*a*) to a simple error control block code, we could add an overall parity symbol (bit) as in Fig. 3.4(*b*). Here we have added an even parity symbol and the vertices of the cube represent the eight 'legal' code vectors. Odd parity vectors such as

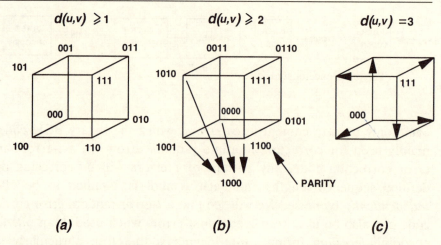

Fig. 3.4 Simple block codes: (*a*) no error control property; (*b*) SED; (*c*) SEC.

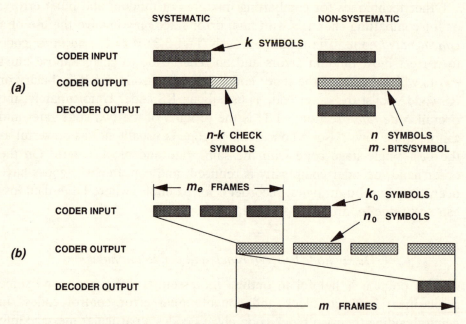

Fig. 3.5 Distinction between block and convolutional coding: (*a*) block coding; (*b*) convolutional coding.

(1000) are not part of the code and provide the redundancy which is so essential for an error control code.

In a general error control block code, k message symbols are coded into a block of n code symbols $(n > k)$ by adding $n - k$ check (or parity) symbols, as shown in Fig. 3.5(*a*). If the code vector contains the original k

symbol block unaltered, it is called a *systematic* (n, k) block code, otherwise it is *non-systematic*. Note that here we have referred to 'symbols' rather than to 'bits'. This is because, in practice, each of the k message symbols could be m bits deep (they could be ASCII characters), corresponding to 2^{km} different messages. Since each message has its own codeword, it follows that there are 2^{km} codewords, each codeword comprising n m-bit symbols.

In contrast to block coding, a convolutional coder will generally code a particular message block of k_0 symbols into different symbol sequences at different times. This is because, for every input block or *frame* of k_0 symbols, the resulting coded sequence or codeword frame of n_0 symbols depends upon a number of input blocks m_e of length k_0 (Fig. 3.5(b)). In a practical code to be discussed later, $k_0 = 1$, $n_0 = 2$, $m_e = 7$, and so for each input bit the coder generates two output bits from a span of 7 input bits. The *constraint length* K of a binary convolutional code can be defined as the span of input bits which influence the coder output at any one time, and so, in terms of Fig. 3.5(b)

$$K = m_e k_0 \tag{3.1}$$

The figure shows that the decoder for a convolutional code has a memory of m codeword frames (usually $m \geq m_e$), and it uses these to decode a k_0 symbol block for each input block of n_0 symbols.

Despite these fundamental differences between block and convolutional codes, there are some similarities. For example, in a systematic convolutional code the message symbols will be identifiable in the coded sequence, as in block coding. Also, block and convolutional codes have similar definitions of *coding efficiency* or *code rate*, R. For a block code

$$R = k/n \tag{3.2}$$

and for a fixed n this will decrease as more check symbols are added to achieve greater error control capability. For a convolutional code

$$R = k_0/n_0 \tag{3.3}$$

where R is usually of the form

$$R = 1/n \quad \text{or} \quad (n-1)/n$$

Typically,

$$R = \tfrac{1}{2}, \tfrac{1}{3}, \tfrac{1}{4}, \ldots \quad \text{or} \quad \tfrac{1}{2}, \tfrac{2}{3}, \tfrac{3}{4}, \tfrac{4}{5}, \ldots$$

As a general rule, both convolutional codes and short block codes are used for low and medium rates, and long block codes are used for high rates ($R > 0.95$). The minimum rate is determined by the available channel bandwidth since, for a fixed message rate and modulation scheme, coding will increase the bandwidth by a factor $1/R$ (Section 3.6). Finally, block and convolutional codes have a common link in their use of the concept of *code distance*.

3.1.2 Code distance

The concept of 'distance' is fundamental to both block and convolutional codes since it determines their error detection and correction capability. Distance is measured between two coded sequences or vectors, and the metric used is usually the *Hamming distance*. Bearing in mind that a coded sequence could be made up of *m*-bit symbols (Fig. 3.5(*a*)), we must define the Hamming distance between two coded sequences as the number of positions in which there are different *symbols*. It is immaterial how many corresponding bits differ from each other within corresponding symbols. For the binary sequences

$$\mathbf{u} = 10011010$$

$$\mathbf{v} = 01111000$$

the Hamming distance $d(\mathbf{u}, \mathbf{v})$ is four. The significant point is that the *minimum* Hamming distance, denoted d, can be simply related to the number of *correctable* errors, t, that can occur over a defined sequence length.

   Suppose that a code vector $\mathbf{v}_1$ is received as vector $\mathbf{r}$ with t errors, Fig. 3.6. If the nearest code vector $\mathbf{v}_2$ is at least $2t + 1$ positions away from $\mathbf{v}_1$, then $\mathbf{r}$ must differ from $\mathbf{v}_2$ by at least $t + 1$ positions. *Assuming that a maximum of t errors have occurred*, the receiver can then associate $\mathbf{r}$ with the legal vector t or less errors away, i.e. with $\mathbf{v}_1$. This general approach applies to both block and convolutional *t*-error correcting codes, and can

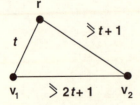

Fig. 3.6 Illustrating the code distance concept.

be formally expressed as

$$d \geq 2t + 1 \tag{3.4}$$

where d is the minimum Hamming distance between all possible pairs of coded sequences. We may, of course, simply wish to *detect* errors, in which case e errors can be detected provided $d \geq e + 1$. More generally, $d = e + t + 1$, where e denotes the number or errors that can be detected and t denotes the number of correctable errors ($t \leq e$).

For a block code, d is simply measured between n-bit codewords. For a convolutional code we could assume that the decoder has a memory of m frames, as in Fig. 3.5(b). In this case d is measured between coded sequences m frames long and which differ in their first frame (Section 3.7). Put another way, a convolutional decoder can correct any (isolated) pattern of t or less errors occurring in a contiguous sequence of mn_0 symbols providing (3.4) is satisfied.

Consider the simple block codes in Fig. 3.4. In Fig. 3.4(a) $d = 1$ and the code has no error control capability since a single error changes one code vector into another. Now consider Fig. 3.4(b) and assume that vector (1000) is received. An even parity check *detects* the error but, clearly, it cannot be corrected since any one of four code vectors could have been corrupted. In this case ($d = 2$) the code is only single error detecting (SED). In Fig. 3.4(c), $d = 3$ and so $t = 1$ and the code is single error correcting (SEC). A single error in the code vector (000), for example, changes it to a vector which is still a distance 2 from the other code vector (111), and so the received vector can be corrected on the assumption that only a single error has occurred. Similarly, if $d = 4$ the code will correct any single error and detect double errors (DED), and we could extend this concept as shown in Table 3.1.

Table 3.1. *Error control properties as a function of d.*

d	Property
1	nil
2	SED
3	SEC
4	SEC + DED
5	DEC
6	DEC + TED
⋮	⋮

3.1.3 Some algebraic concepts

The theory of error control codes (particularly block codes) is heavily based upon algebraic systems such as *groups* and *fields*. For example, a codeword of an (n, k) block code can be written

$$\mathbf{v} = (v_1, v_2, v_3, \ldots, v_n) \tag{3.5}$$

where, in general, symbol v_i could be drawn from a finite mathematical field of q elements. In this case the field is called a *Galois* field and is denoted $GF(q)$. For example, we might have $\mathbf{v} = (3410)$ where each symbol is drawn from the element range 0–4 and the relevant field is $GF(5)$. Any multiplication and addition would then be performed modulo 5. Similar *non-binary* $(q > 2)$ error control codes are discussed in Section 3.5.3. Much of this chapter concentrates on *binary* $(q = 2)$ codes. In this case we might have $\mathbf{v} = (1001101)$ and symbols 0 and 1 will correspond to the 0 and 1 elements of $GF(2)$. The significant point here is that $GF(2)$ defines the basic rules for any arithmetic operations required by the binary coder or decoder. The rules are simply those of modulo 2 arithmetic, i.e.

+	0	1		×	0	1
0	0	1		0	0	0
1	1	0		1	0	1

Note in particular that modulo 2 addition can be realized by the exclusive-OR operation, and that there is no difference between addition and subtraction over $GF(2)$. It is also worth noting that an *extension* of $GF(2)$, to $GF(2^m)$, m integer, is of practical significance when decoding binary codes (Section 3.5.2).

A linear binary block code is sometimes called a *group* code because its binary vectors $\mathbf{v}_1$, $\mathbf{v}_2$, $\mathbf{v}_3$ etc. can be regarded as elements of a mathematical group G. That is

$$\mathbf{v}_1, \mathbf{v}_2, \mathbf{v}_3, \ldots \in G \tag{3.6}$$

This is true if the group operation is defined as modulo 2 addition, i.e. any two elements $\mathbf{v}_i$, $\mathbf{v}_j$ could be added bit by bit over $GF(2)$. It then follows from the closure postulate of G under addition that

$$\mathbf{v}_i + \mathbf{v}_j = \mathbf{v}_k \in G \tag{3.7}$$

In words, the bit-by-bit modulo 2 addition of any two code vectors yields another code vector (the reader could check this for the simple parity check code in Fig. 3.4(b)). The number of 1s or non-zero elements of $\mathbf{v}_k$ is its *Hamming weight*, and, clearly, this will be identical to the Hamming distance between $\mathbf{v}_i$ and $\mathbf{v}_j$. It therefore follows that the *minimum* distance, d, between vectors must be defined by the (non-zero) vector of *minimum* weight:

Theorem 3.1 *The minimum distance d of a linear code equals the minimum Hamming weight of its non-zero vectors.*

This theorem is particularly useful for finding the minimum distance of convolutional codes.

3.2 Matrix description of block codes

This section discusses the matrix approach to block codes and then introduces the SEC Hamming codes. The reader interested in multiple random error correcting codes should also study the polynomial approach in Section 3.3.

3.2.1 Coding and decoding principles

A linear block code can be defined by a *check matrix* $\mathbf{H}$ as follows:

Definition 3.1 *A linear block code* V *is a set of vectors of length* n *where each vector* **v** *in* V *is orthogonal to every row of a matrix* **H**.

Using the conventional definition of orthogonality this statement can be expressed as

$$\mathbf{H}\mathbf{v}^{\mathrm{T}} = \mathbf{0} \tag{3.8}$$

where $\mathbf{0}$ is a column vector, T denotes transpose, $\mathbf{v} = (v_1, v_2, v_3, \ldots, v_n)$ and the equation is evaluated over $GF(2)$ for a binary code, i.e. we use modulo 2 arithmetic. For matrix multiplication, $\mathbf{H}$ must have n columns, and let $\mathbf{H}$ have $n - k$ rows, so that

$$
\begin{bmatrix}
h_{11} & h_{12} & \cdots & h_{1n} \\
\vdots & \vdots & \vdots & \vdots \\
h_{i1} & h_{i2} & \cdots & h_{in} \\
\vdots & \vdots & \vdots & \vdots \\
h_{(n-k)1} & h_{(n-k)2} & \cdots & h_{(n-k)n}
\end{bmatrix}
\begin{bmatrix}
v_1 \\
\vdots \\
v_i \\
\vdots \\
v_n
\end{bmatrix}
=
\begin{bmatrix}
0 \\
\vdots \\
0 \\
\vdots \\
0
\end{bmatrix}
\tag{3.9}
$$

This states that each code vector $\mathbf{v}$ must satisfy a set of $n - k$ independent equations each of the form

$$\sum_{j=1}^{n} h_{ij} v_j = 0 \quad \text{modulo } 2 \tag{3.10}$$

In practice, each equation is solved for a check (or parity) symbol and the resulting (n, k) linear block code will have vectors consisting of k information symbols and $n - k$ check symbols. The expression in (3.9) therefore represents the fundamental code generation mechanism (note that the equations check for even parity although they are trivially altered for odd parity). Since we can generate a complete binary code of 2^k code vectors from (3.9), i.e. from **H**, it is reasonable to assume that we can also deduce the minimum Hamming distance, d, of the code from **H**. In fact, we can, if we apply the following theorem:

> **Theorem 3.2** *There exists at least one subset of d columns of **H** that sum to **0** and no subsets of (d − 1) or fewer columns will sum to **0**.*

This theorem is somewhat more powerful than Theorem 3.1 since all we need is **H**, rather than the complete code in order to find d. Conversely, the theorem could be used to *construct* **H** in order to provide a code with a specific value of d.

A compact way of describing the 2^k vectors of a binary (n, k) block code is in terms of its *generator matrix* **G**, where

$$\mathbf{G} = \begin{bmatrix} a_{11} & a_{12} & \cdots & a_{1n} \\ a_{21} & a_{22} & \cdots & a_{2n} \\ \vdots & \vdots & & \vdots \\ a_{k1} & a_{k2} & \cdots & a_{kn} \end{bmatrix}$$

and terms a_{ij} are elements of $GF(2)$. All the row vectors in **G** must be linearly independent, and over $GF(2)$ this simply means that the bit-by-bit modulo-2 sum of two or more vectors must not be **0**. Vectors meeting this requirement form a *basis* for the code V. The significant point is that all linear combinations of the k row vectors (the *row space* of **G**) will generate a linear block code V of 2^k code vectors. As an example, the SED code in Fig. 3.4(b) could be described as the row space of, say,

$$\mathbf{G} = \begin{bmatrix} 1 & 0 & 0 & 1 \\ 1 & 1 & 0 & 0 \\ 1 & 0 & 1 & 0 \end{bmatrix}$$

where three linearly independent vectors have been selected from the eight code vectors (any three independent vectors could have been selected). For this example a linear combination of two or three vectors is simply the modulo-2 sum of the vectors taking them bit by bit, and it is easily seen that this leads to the code in Fig. 3.4(*b*).

Decoding mechanism

Now consider the use of **H** for decoding. Suppose the transmitted vector **v** is corrupted during transmission such that the received vector is

$$\mathbf{r} = \mathbf{v} + \mathbf{e} \tag{3.11}$$

where **e** is the error vector

$$\mathbf{e} = (e_1, e_2, \ldots, e_n) \quad e_i = \begin{cases} 0, \text{ no error} \\ 1, \text{ error} \end{cases}$$

The decoder performs the same checks on **r** as were carried out at the coder, so that

$$\mathbf{Hr}^{\mathrm{T}} = \mathbf{H}(\mathbf{v}^{\mathrm{T}} + \mathbf{e}^{\mathrm{T}})$$

$$= \mathbf{Hv}^{\mathrm{T}} + \mathbf{He}^{\mathrm{T}}$$

$$= \mathbf{He}^{\mathrm{T}} \quad \text{(even parity)} \tag{3.12}$$

$$= \mathbf{s}^{\mathrm{T}}$$

where

$$\mathbf{s} = (s_1, s_2, \ldots, s_{n-k})$$

The 'diagnostic' vector **s** is called the *syndrome*, and it is used to perform error correction. Note, however, that if more than t errors have occurred in the block then a t-error correcting code will breakdown. In terms of Fig. 3.6, if code vector $\mathbf{v}_1$ is received with $t + 1$ errors it will be 'corrected' to a word other than $\mathbf{v}_1$! On the other hand, if t (or less) randomly placed errors occur in a block they can be corrected provided they result in a unique (distinct) and non-zero syndrome. In order to achieve this, we need certain restrictions on the columns of **H**. Suppose $t = 2$ and two random errors occur at positions i and j, i.e. $e_i = e_j = 1$. Evaluating (3.12) gives

$$\mathbf{s}^{\mathrm{T}} = \begin{bmatrix} h_{1i}e_i & + & h_{1j}e_j \\ h_{2i}e_i & + & h_{2j}e_j \\ \vdots & & \vdots \\ h_{(n-k)i}e_i & + & h_{(n-k)j}e_j \end{bmatrix}$$

$$= \mathbf{h}_i + \mathbf{h}_j, \quad i, j = 1, \ldots, n; i \neq j \tag{3.13}$$

where $\mathbf{h}_i$ and $\mathbf{h}_j$ are vectors corresponding to columns i and j of $\mathbf{H}$. Clearly, in order to obtain a unique and non-zero syndrome, the sum of any two columns of $\mathbf{H}$ must also be unique and non-zero. Generalizing, for a t-error correcting code, the sum of any t or less columns of $\mathbf{H}$ must be unique and non-zero (note that this is simply a weaker restatement of Theorem 3.2).

3.2.2 Hamming codes

These are a class of SEC $(t = 1)$ codes and so, from (3.13), or Theorem 3.2, a sufficient condition is that all columns of $\mathbf{H}$ must be unique and non-zero. In addition, a good code will have the maximum number of information symbols, and therefore the maximum number of columns in $\mathbf{H}$ for a fixed number of check symbols or rows in $\mathbf{H}$. For $n - k$ rows, $\mathbf{H}$ can have up to $2^{n-k} - 1$ non-zero columns. In other words, the block length n of a SEC linear block code is given by

$$n \leqslant 2^{n-k} - 1 \tag{3.14}$$

and the code is optimum or *perfect* when

$$2^{n-k} = n + 1 \tag{3.15}$$

This equation defines a set of perfect SEC codes called the Hamming (n, k) codes (Hamming, 1950), where k is the number of information symbols and

$$(n, k) = (3, 1)(7, 4)(15, 11)(31, 26)(63, 57) \ldots$$

Note that the code rate R improves as n increases but this is balanced by an increased probability of decoding failure for a given random error rate, i.e. the longer the codeword the greater the chance of more than t errors occurring per block (a similar argument applies as the constraint length is increased in a convolutional code). The SEC property can be proved by finding the minimum distance d for the code (see Section 3.2.3), or by examining upper *coding bounds*. Either way, Section 3.3 shows that Hamming codes (in their *cyclic* form) are really only a form of t-error correcting binary BCH code, with $t = 1$.

Codeword format

Definition 3.2 *A linear code defined by a matrix* $\mathbf{H}$ *comprising* $2^r - 1$ *non-zero r-tuples arranged in any order is called a Hamming code.*

According to the above definition, a particular Hamming code can be constructed with an arbitrary ordering of the columns in **H**, although it is convenient to select the columns in ascending binary order. Taking the H(7, 4) code as an example, the usual form of **H** is

$$\mathbf{H} = \begin{bmatrix} 0 & 0 & 0 & 1 & 1 & 1 & 1 \\ 0 & 1 & 1 & 0 & 0 & 1 & 1 \\ 1 & 0 & 1 & 0 & 1 & 0 & 1 \end{bmatrix} \tag{3.16}$$

Now, according to (3.13) a single error at position i will give a syndrome

$$\mathbf{s}^T = \mathbf{h}_i, \qquad i = 1, \dots, 7$$

where $\mathbf{h}_i$ corresponds to the ith column of **H**. Therefore, if we use the binary ordered matrix in (3.16), it is clear that **s** will have a binary value which gives the position of the error, e.g. for $e_6 = 1$, $\mathbf{s} = (110)$ and position six would be complemented.

The code vector format follows from **H**. If the matrix in (3.16) is used for an even parity code the coder must obtain a check or parity symbol P from each of the following equations:

$$v_4 + v_5 + v_6 + v_7 = 0 \quad \text{giving } P_3$$

$$v_2 + v_3 + v_6 + v_7 = 0 \quad \text{giving } P_2$$

$$v_1 + v_3 + v_5 + v_7 = 0 \quad \text{giving } P_1$$

Note that these sums are equated to unity for odd parity and that all addition is modulo 2. Since each equation yields one parity symbol, it is apparent that these symbols should be in positions one, two and four, corresponding to a code vector format $P_1 P_2 M_1 P_3 M_2 M_3 M_4$. The message symbols M can be in any order and P_1 is in position one of the code vector.

Example 3.1

Using the **H** matrix in (3.16) the even parity H(7, 4) code vector for 'message' 13_{10} could be of the form

$$\mathbf{v} = (P_1 P_2 1 P_3 101)$$

where

$$P_3 + 1 + 0 + 1 = 0 \quad P_3 = 0$$

$$P_2 + 1 + 0 + 1 = 0 \quad P_2 = 0$$

$$P_1 + 1 + 1 + 1 = 0 \quad P_1 = 1$$

Therefore

$$\mathbf{v} = (1010101)$$

Example 3.2

Suppose that an odd parity H(7, 4) code defined by

$$\mathbf{H} = \begin{bmatrix} 0 & 0 & 1 & 1 & 0 & 1 & 1 \\ 0 & 1 & 1 & 0 & 1 & 0 & 1 \\ 1 & 1 & 0 & 0 & 0 & 1 & 1 \end{bmatrix}$$

is used to transmit a binary number in the range $0-15_{10}$. The code vector format is therefore $P_1 M_1 M_2 P_2 P_3 M_3 M_4$ where, as before, we assume that M_1 is the most significant bit of the number. If the received vector is $\mathbf{r} = (0111101)$, what is the probable value of the number? Since odd parity is used

$$\mathbf{Hr}^\mathsf{T} = \mathbf{Hv}^\mathsf{T} + \mathbf{He}^\mathsf{T}$$

$$= \mathbf{1} + \mathbf{s}^\mathsf{T}$$

which means that the check sums at the decoder should give unity unless there is an error. The check sums are as follows:

0	1	1	1	1	0	1		
		$\times$	$\times$		$\times$	$\times$	= 1;	$s_1 = 0$
	$\times$	$\times$		$\times$		$\times$	= 0;	$s_2 = 1$
$\times$	$\times$				$\times$	$\times$	= 0;	$s_3 = 1$

giving $\mathbf{s} = (011)$. This corresponds to the second column in $\mathbf{H}$, which means the error must be in position two ($e_2 = 1$) and the message in $\mathbf{r}$ would be decoded as (0101) or 5_{10}. This will be the correct number provided that only a single error has occurred in the 7-bit block. Following the discussion in Section 3.2.1, multiple errors in the block will cause decoding failure and, in general, the finite syndrome could lead to even more errors after 'correction'.

3.2.3 Modified Hamming codes

Basic codes are often modified to enhance their error control capability and/or to make them more applicable to a particular problem.

Extended codes

A simple technique is to *extend* the code by adding an overall parity check symbol. Fig. 3.7 shows the even parity H(7, 4) code generated from

POSITION

1 2 3 4 5 6 7

P P M P M M M   MESSAGE

0 0 0 0 0 0 0	0
1 1 0 1 0 0 1	1
0 1 0 1 0 1 0	2
1 0 0 0 0 1 1	3
1 0 0 1 1 0 0	4
0 1 0 0 1 0 1	5
1 1 0 0 1 1 0	6
0 0 0 1 1 1 1	7
1 1 1 0 0 0 0	8
0 0 1 1 0 0 1	9
1 0 1 1 0 1 0	10
0 1 1 0 0 1 1	11
0 1 1 1 1 0 0	12
1 0 1 0 1 0 1	13
0 0 1 0 1 1 0	14
1 1 1 1 1 1 1	15

$$\mathbf{H} = \begin{bmatrix} 0 0 0 1 1 1 1 \\ 0 1 1 0 0 1 1 \\ 1 0 1 0 1 0 1 \end{bmatrix}$$

Fig. 3.7 An even parity H(7, 4) code.

(3.16). Since the minimum Hamming weight is 3, then $d = 3$ (Theorem 3.1) and so the code is SEC. Clearly, the minimum Hamming weight, and therefore d, can be increased to 4 by adding an overall even parity symbol P_4 (obtained by checking all seven symbols). According to Table 3.1, this means that the extended code (the H(8, 4) code) could detect *double* errors over a block of eight symbols as well as correct any single error. The error correction/detection mechanism can be seen from the corresponding syndrome equations. Extending the matrix in (3.16) we have

$$\mathbf{s}^T = \mathbf{H}\mathbf{e}^T$$

$$\begin{bmatrix} s_1 \\ s_2 \\ s_3 \\ s_4 \end{bmatrix} = \begin{bmatrix} 0 & 0 & 0 & 0 & 1 & 1 & 1 & 1 \\ 0 & 0 & 1 & 1 & 0 & 0 & 1 & 1 \\ 0 & 1 & 0 & 1 & 0 & 1 & 0 & 1 \\ 1 & 1 & 1 & 1 & 1 & 1 & 1 & 1 \end{bmatrix} \begin{bmatrix} e_1 \\ e_2 \\ e_3 \\ e_4 \\ e_5 \\ e_6 \\ e_7 \\ e_8 \end{bmatrix} \qquad (3.17)$$

or

$$s_1 = 0 + 0 + 0 + 0 + e_5 + e_6 + e_7 + e_8$$

$$s_2 = 0 + 0 + e_3 + e_4 + 0 + 0 + e_7 + e_8$$

$$s_3 = 0 + e_2 + 0 + e_4 + 0 + e_6 + 0 + e_8$$

$$s_4 = e_1 + e_2 + e_3 + e_4 + e_5 + e_6 + e_7 + e_8$$

For single errors $s_4 = 1$ and the syndrome $\mathbf{s} = (s_1 s_2 s_3)$ indicates the position of the error. For double errors $s_4 = 0$ and $\mathbf{s}$ is still non-zero since at least one of the symbols s_1, s_2, s_3 is set. This means that the received vector having any pattern of two errors could be rejected by generating the 'reject' signal

$$R = \bar{s}_4 \cdot (s_1 + s_2 + s_3) \tag{3.18}$$

where, in this case, '+' denotes logical-OR. For a single error, $R = 0$ whilst two errors give $R = 1$. For three errors, $R = 0$ and, in general, the non-zero syndrome will result in even more errors, i.e. decoding fails. We could, of course, *detect* three errors since $d = 4$ (Section 3.1.2).

Example 3.3
An odd parity H(8, 4) code is useful for Teletext systems. Suppose such a code is derived using the arbitrary matrix

$$\mathbf{H} = \begin{bmatrix} 0 & 0 & 0 & 0 & 1 & 1 & 1 & 1 \\ 0 & 1 & 1 & 0 & 1 & 1 & 0 & 0 \\ 0 & 0 & 1 & 1 & 1 & 0 & 1 & 0 \\ 1 & 1 & 1 & 1 & 1 & 1 & 1 & 1 \end{bmatrix}$$

and that a Hamming decoder is required. First note that the (modulo-2) sum of the last four columns in $\mathbf{H}$ gives zero whilst no subset of three columns sum to zero. It therefore follows from Theorem 3.2 that $d = 4$, as expected for an (8, 4) code. As before we can write

$$\mathbf{Hr}^{\mathsf{T}} = \mathbf{Hv}^{\mathsf{T}} + \mathbf{He}^{\mathsf{T}} = \mathbf{1} + \mathbf{s}^{\mathsf{T}}$$

and over GF(2) this can be rewritten

$$\mathbf{s}^{\mathsf{T}} = \mathbf{1} + \mathbf{Hr}^{\mathsf{T}}$$

which gives the logic design equations for $\mathbf{s}$. For example,

$$s_1 = 1 + (r_5 + r_6 + r_7 + r_8)$$

In terms of error symbols, $\mathbf{s}$ is given by

$$s_1 = 0 + 0 + 0 + 0 + e_5 + e_6 + e_7 + e_8$$

$$s_2 = 0 + e_2 + e_3 + 0 + e_5 + e_6 + 0 + 0$$

$$s_3 = 0 + 0 + e_3 + e_4 + e_5 + 0 + e_7 + 0$$

$$s_4 = e_1 + e_2 + e_3 + e_4 + e_5 + e_6 + e_7 + e_8$$

A suitable format for each code vector is therefore $P_4 P_1 M_1 P_2 M_2 M_3 M_4 P_3$ where the overall parity symbol P_4 is in position one and an error in this symbol gives $\mathbf{s} = (s_1 s_2 s_3) = (000)$. For an error in message symbol M_1 ($e_3 = 1$) the syndrome will be $\mathbf{s} = (011)$ and the corrected message can be expressed over $GF(2)$ as

$$\hat{M}_1 = M_1 + (\bar{s}_1 s_2 s_3)$$

Similarly

$$\hat{M}_2 = M_2 + (s_1 s_2 s_3)$$

$$\hat{M}_3 = M_3 + (s_1 s_2 \bar{s}_3)$$

$$\hat{M}_4 = M_4 + (s_1 \bar{s}_2 s_3)$$

Finally, (3.18) could be used to reject double errors.

Shortened codes

The SEC + DED capability can also be achieved by *shortening* a basic $H(n, k)$ Hamming code. This code is defined by a matrix $\mathbf{H}$ having $2^{n-k} - 1$ unique and non-zero columns and $n - k$ rows (corresponding to $n - k$ check symbols). A shortened $H(n - q, k - q)$ Hamming code is generated by deleting q columns from the matrix, and we may wish to do this in order to match the number of message symbols to a particular application. Note that there are now only 2^{k-q} codewords (since we have deleted q message symbols) and so the minimum distance of the code may increase, i.e. $d \geqslant 3$. In fact, to ensure d is at least 4 (SEC + DED) it is necessary that each column has an odd number of 1s (Lin and Costello, 1983). Deleting the 'even 1s' columns in (3.16) for instance will give an $H(4, 1)$ code with $d = 4$.

Example 3.4

Consider the $H(63, 57)$ code. We could commence shortening this by deleting the 31 columns in $\mathbf{H}$ having an even number of 1s. This leaves a 6×32 matrix, where six of the 32 columns have a single 1 corresponding to the six check symbols, and the remaining 26 columns correspond to message symbols. If this code is required to protect 16-bit computer memory systems then a further ten 'message' columns must be deleted.

This can be done in an optimal way in order to minimise hardware (Hsiao, 1970), and in this respect one criterion is to select columns such that the total number of 1s in **H** is minimized. In this example, 20 of the remaining 26 'message' columns will have just three 1s, and 16 of these can be selected for the final code. For instance, an optimal matrix for the resulting H(22, 16) SEC + DED code is (Hsiao, 1970)

$$\mathbf{H} = \begin{bmatrix} 1 & 0 & 0 & 0 & 0 & 0 & 1 & 0 & 0 & 1 & 1 & 0 & 0 & 1 & 0 & 0 & 1 & 1 & 1 & 1 & 0 & 0 \\ 0 & 1 & 0 & 0 & 0 & 0 & 0 & 0 & 1 & 1 & 1 & 1 & 1 & 0 & 1 & 0 & 0 & 0 & 1 & 0 & 1 & 0 \\ 0 & 0 & 1 & 0 & 0 & 0 & 1 & 1 & 1 & 0 & 1 & 1 & 1 & 0 & 0 & 1 & 1 & 0 & 0 & 0 & 0 & 0 \\ 0 & 0 & 0 & 1 & 0 & 0 & 1 & 1 & 1 & 0 & 0 & 0 & 0 & 1 & 1 & 1 & 0 & 1 & 0 & 0 & 0 & 1 \\ 0 & 0 & 0 & 0 & 1 & 0 & 0 & 0 & 0 & 1 & 0 & 0 & 1 & 1 & 1 & 1 & 0 & 0 & 0 & 1 & 1 & 1 \\ 0 & 0 & 0 & 0 & 0 & 1 & 0 & 1 & 0 & 0 & 0 & 1 & 0 & 0 & 0 & 0 & 1 & 1 & 1 & 1 & 1 & 1 \end{bmatrix}$$

Fig. 3.8 illustrates how this code can be used to increase memory reliability. On a memory-write cycle, six check bits are generated and stored in a check memory. On a read cycle new check bits are generated from the data and these are compared with those in the check memory. The H(22, 16) device interrupts the CPU only when an error is detected,

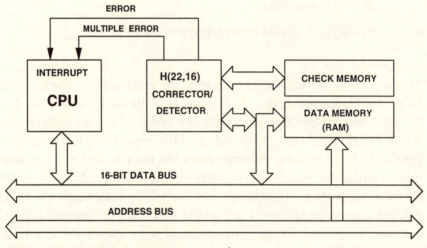

Fig. 3.8 Use of a Hamming error correction/detection chip to enhance the reliability of computer data memory.

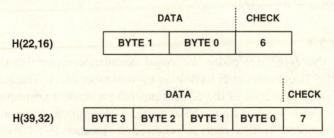

Fig. 3.9 Modified Hamming codes for 16- and 32-bit computer systems.

i.e. the system is high performance in that it achieves zero memory slow-down under zero error conditions. Note that a 32-bit data bus could be protected in a similar way using a H(39, 32) code, Fig. 3.9.

3.3 Polynomial description of block codes: cyclic codes

In this section we examine a polynomial rather than matrix description of block codes. This leads naturally to *cyclic* codes and to the multiple random error correcting binary BCH codes. The theory is also fundamental to the non-binary RS codes discussed in Section 3.5.3 (and so we will usually refer to 'symbols' rather than to 'bits'). As shown in Fig. 3.1, cyclic codes form the major class of error correcting block codes. This is because of their well-understood mathematical structure (which is polynomial based) and because they can be conveniently implemented using shift registers. The term 'cyclic' arises from the fact that a cyclic shift of a codeword of a cyclic code V generates a codeword in V.

In the polynomial description we represent a general n-symbol codeword by the code polynomial

$$v(x) = v_0 + v_1 x + v_2 x^2 + \ldots + v_{n-1} x^{n-1} \tag{3.19}$$

and a message or information vector $\mathbf{i} = (i_0, i_1, i_2, \ldots i_{k-1})$ as

$$i(x) = i_0 + i_1 x + i_2 x^2 + \ldots + i_{k-1} x^{k-1} \tag{3.20}$$

An advantage of this representation is that an (n, k) cyclic code can be represented in terms of a *generator polynomial* $g(x)$, where

$$g(x) = g_0 + g_1 x + g_2 x^2 + \ldots + g_{n-k} x^{n-k}; \qquad g_0 = g_{n-k} = 1 \tag{3.21}$$

Theorem 3.3 *An (n, k) cyclic code is completely specified by a generator polynomial $g(x)$ of degree $n - k$ that divides $x^n + 1$. Put another way, the basic block length, n, corresponds to the lowest degree polynomial $x^n + 1$ which is exactly divisible by $g(x)$.*

Example 3.5
Over $GF(2)$

$$x^7 + 1 = (1 + x)(1 + x^2 + x^3)(1 + x + x^3)$$

Factors $(1 + x^2 + x^3)$ and $(1 + x + x^3)$ *each* generate a $(7, 4)$ cyclic code. In general, for large n, $x^n + 1$ could have many factors of degree $n - k$ and only some of these will generate good codes. We shall see later that $g(x)$ can be selected from a table of polynomials.

It should be noted that, given the basic block length n as defined in Theorem 3.3, $g(x)$ will generate cyclic codes of block length $n' = in$, $i = 1, 2, 3, \ldots$. Also, since $g(x)$ completely specifies the code we might hope to relate it to the error control property of the code, i.e. to its minimum distance d. Unfortunately, in general, this turns out to be difficult, although we can state a few simple rules:

(a) If $g(x)$ has at least two terms, then $d \geqslant 2$.

(b) If the general block length of an error correcting code is $n' \leqslant n$, where n satisfies $g(x)$ as in Theorem 3.3, then $d \geqslant 3$. In other words, creating a shortened cyclic code (Section 3.3.2) sometimes leads to a more powerful code. Conversely, if $n' = in$, $i = 2, 3, 4, \ldots$ then in almost all cases $d = 2$, corresponding to an error detecting code.

(c) If $g(x) = (1 + x)g_1(x)$ and $n' \leqslant n_1$, where $g_1(x)$ 'belongs' to the polynomial $x^{n_1} + 1$, as in Theorem 3.3, then $d \geqslant 4$.

These rules can be useful when determining the error control property of practical codes.

Example 3.6

The ETSI-GSM channel coding specification for digital cellular radio (ETSI, 1991) protects speech bits by using a systematic cyclic code followed by a convolutional code. The cyclic code protects blocks of 50 speech bits using three check bits and is based on the generator

$$g(x) = 1 + x + x^3$$

According to Theorem 3.3, or Example 3.5, we see that $g(x)$ 'belongs' to the polynomial $x^7 + 1$, and so the basic block length is $n = 7$. On the other hand, $g(x)$ also defines the cyclic code (56, 53), corresponding to the block length $n' = 8n$, and since $n' > n$ this code will have $d = 2$. The GSM specification therefore simply uses the code for error *detection*, and in fact the code will reliably detect any burst error of length 3 bits or less (see Section 3.4.2) and 87.5% of all error patterns. Clearly, in order to protect the 50 speech bits, the code must be shortened to the (53, 50) code, which is the code defined in the specification.

3.3.1 *Polynomial multiplication and division*

The multiplication and division of polynomials is fundamental to the encoding and decoding of cyclic codes. Polynomial division tends to be of most practical significance, although sometimes both operations are necessary in a single system. Either way, hardware implementation is conveniently based upon shift register circuits.

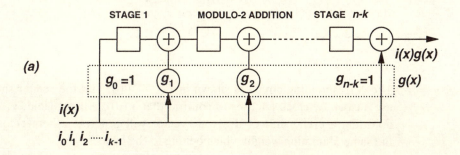

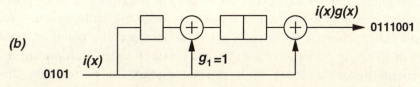

Fig. 3.10 Polynomial multiplication: (*a*) general shift register-based circuit; (*b*) multiplication of $i(x) = x + x^3$ by $g(x) = 1 + x + x^3$.

Consider the multiplication of a message $i(x)$ by a polynomial $g(x)$, where $i(x)$ and $g(x)$ are defined by (3.20) and (3.21) respectively. A general circuit for achieving this is shown in Fig. 3.10(*a*) and is based upon an $n - k$ stage shift register. The assumptions are that the message symbols enter high-order first (i_{k-1} first), followed by $n - k$ zeros, and that the register is initially cleared. The circuit output will then correspond to the product

$$i(x)g(x) = g_0 i_0 + (g_0 i_1 + g_1 i_0)x + \ldots + g_{n-k} i_{k-1} x^{n-1}$$

where the $g_{n-k} i_{k-1}$ term is generated first. Note that an equivalent 'transposed' form of Fig. 3.10(*a*) is possible, although this is less attractive in practice, especially when both multiplication and division are performed simultaneously by a single shift register (Example 3.9).

A primary application of polynomial multiplication is highlighted by the following theorem:

Theorem 3.4 *A binary codeword* $v(x)$ *in an* (n, k) *cyclic code can be derived as*

$$v(x) = i(x)g(x) \tag{3.22}$$

Example 3.7
Selecting $g(x) = 1 + x + x^3$ as the generator for a cyclic $(7, 4)$ code, a message vector $\mathbf{i} = (0101)$ is coded over $GF(2)$ as

$$v(x) = (x + x^3)(1 + x + x^3)$$

$$= x + x^2 + x^3 + x^6$$

$$\therefore \mathbf{v} = (0111001)$$

The corresponding circuit is shown in Fig. 3.10(b) and the reader should use a clock-by-clock analysis to confirm that a (0101) input yields $\mathbf{v}$. Also note that a cyclic shift of this code vector will generate other vectors of the same Hamming weight which belong to the code.

Example 3.7 shows that cyclic codes are easily generated using polynomial multiplication. On the other hand, the resulting code is non-systematic and, in practice, it is more usual to generate cyclic codes in systematic form. This requires *division* by $g(x)$ rather than multiplication and a general linear feedback shift register circuit for performing this over $GF(2)$ is shown in Fig. 3.11(a). Again, it is assumed that symbol i_{k-1} enters first and that the register is initially cleared. The circuit output is the quotient of the division and when all the quotient coefficients have appeared at the output the shift register contains the remainder.

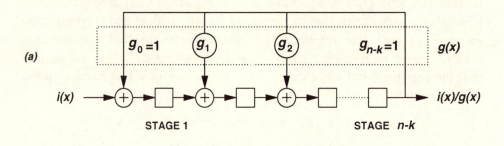

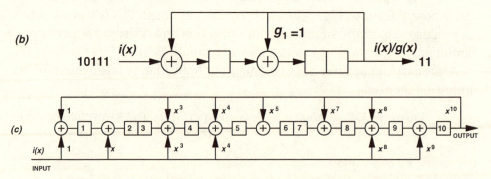

Fig. 3.11 Polynomial division: (a) division by a polynomial $g(x)$ of degree $n - k$; (b) division of $i(x) = 1 + x^2 + x^3 + x^4$ by $g(x) = 1 + x + x^3$; (c) simultaneous division by $1 + x^3 + x^4 + x^5 + x^7 + x^8 + x^{10}$ and multiplication by $1 + x + x^3 + x^4 + x^8 + x^9$.

Example 3.8

Assume the polynomial $i(x) = 1 + x^2 + x^3 + x^4$, i.e. $\mathbf{i} = (10111)$ is divided by $g(x) = 1 + x + x^3$. Noting that $+$ and $-$ are equivalent over $GF(2)$, we have

$$
\begin{array}{r}
x + 1 \\
x^3 + x + 1 \overline{)\ x^4 + x^3 + x^2 + 1} \\
x^4 + x^2 + x \\
\hline
x^3 + x + 1 \\
x^3 + x + 1 \\
\hline
\end{array}
$$

zero remainder

The corresponding circuit is shown in Fig. 3.11(b). The reader should show that the output will generate the sequence $\{11\}$ (corresponding to the quotient $x + 1$), and that immediately after this all register stages will go to zero (corresponding to zero remainder).

Example 3.9

As previously pointed out, sometimes it is necessary to perform both multiplication and division simultaneously, and a practical example is found in the European radio data system or RDS (CENELEC, 1990). This uses a $(26, 16)$ cyclic code to enable the receiver to detect and correct transmission errors. The decoder requires division by the generator polynomial

$$g(x) = 1 + x^3 + x^4 + x^5 + x^7 + x^8 + x^{10}$$

and simultaneous multiplication by the polynomial

$$g'(x) = 1 + x + x^3 + x^4 + x^8 + x^9$$

The basic shift register circuit at the decoder is therefore as shown in Fig. 3.11(c). Note that if $g'(x)$ had a higher degree than $g(x)$ by a factor m, then the register would need to be extended on the left by m stages.

3.3.2 Systematic cyclic codes

For a systematic (n, k) cyclic code we require a code vector of the form

$$\mathbf{v} = (p_0, p_1, \ldots, p_{n-k-1}, i_0, i_1, \ldots, i_{k-1}) \tag{3.23}$$

The coder will then generate k message symbols in unaltered form, followed by $n - k$ check symbols. The corresponding code polynomial is

$$v(x) = p_0 + p_1 x + \ldots + p_{n-k-1} x^{n-k-1} + i_0 x^{n-k} + i_1 x^{n-k+1}$$

$$+ \ldots i_{k-1} x^{n-1} \tag{3.24}$$

$$= p(x) + x^{n-k} i(x) \tag{3.25}$$

where $p(x)$ is a check or parity polynomial, and over $GF(2)$ this can be rearranged as

$$x^{n-k} i(x) = v(x) + p(x) \tag{3.26}$$

In order to derive a coding algorithm it will be helpful to restate Theorem 3.4 as follows:

Theorem 3.5 *A binary codeword polynomial $v(x)$ is divisible by $g(x)$.*

Applying this theorem to (3.26) gives

$$\frac{x^{n-k} i(x)}{g(x)} = a(x) + \frac{p(x)}{g(x)} \tag{3.27}$$

where $a(x)$ is the quotient (in this case of little interest) and $p(x)$ is the remainder. By definition, $p(x)$ must have degree $n - k - 1$ or less since $g(x)$ has degree $n - k$. Equation (3.27) fully defines the following coding algorithm for a systematic cyclic code:

(1) multiply $i(x)$ by x^{n-k};
(2) divide $x^{n-k} i(x)$ by $g(x)$. The remainder, $p(x)$, gives the required check symbols;
(3) combine information and check symbols.

Example 3.10
Find the $(7, 4)$ systematic code vector for message $\mathbf{i} = (1011)$ assuming $g(x) = 1 + x + x^3$. We have

$$x^3 i(x) = x^3 + x^5 + x^6$$

Dividing this by $g(x)$ gives remainder one, i.e.

$$p(x) = 1, \quad \mathbf{p} = (100)$$

$$\therefore v(x) = p(x) + x^{n-k} i(x)$$

$$= 1 + x^3 + x^5 + x^6$$

$$\mathbf{v} = (1001011)$$

A codeword for the $(15, 7)$ code is derived in Example 3.16.

Practical systematic coders

In practice it is not necessary to multiply by x^{n-k} separately. Referring to Fig. 3.11(c) we note that if we had simply multiplied by x^{10} then the input $i(x)$ would be applied at the output of stage 10, via a modulo-2 adder. The output of this adder would then be $x^{10}i(x)/g(x)$, where $g(x)$ is the divisor. Clearly then, multiplication by x^{n-k}, as required for systematic coding, can be achieved by applying $i(x)$ to one input of a modulo-2 adder, whilst the other input is fed from the output of stage $n - k$. The adder output is then fedback in order to perform division by $g(x)$.

Fig. 3.12(a) shows a practical systematic coder for a cyclic $(7, 4)$ code. Initially the gate is open (a short-circuit) and the switch is in position A. The k message symbols are then shifted into the register (i_{k-1} first) and simultaneously to the channel. As soon as the k message symbols have entered the register the $n - k$ symbols in the register are the remainder, $p(x)$. The gate is then closed (with its output forced to '0'), the switch is placed in position B, and the check symbols are clocked into the channel (symbol p_{n-k-1} first). Fig. 3.12(b) shows a $(26, 16)$ coder meeting the error protection requirements of the European radio data system (RDS). At the start of each 26-symbol block the shift register is cleared to the 'all 0s' state. After 16 message symbols have been clocked into the coder and simultaneously to the channel, gates A and B are closed and gate C is opened. The register is then clocked a further 10 times to shift the check symbols out to the channel.

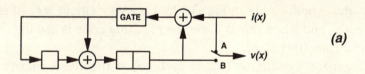

(a)

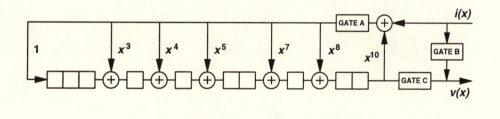

(b)

Fig. 3.12 Systematic coding: (a) $(7, 4)$ coder for $g(x) = 1 + x + x^3$; (b) $(26, 16)$ coder for $g(x) = 1 + x^3 + x^4 + x^5 + x^7 + x^8 + x^{10}$.

It is interesting to derive the general form of $p(x)$ for Fig. 3.12(a). Multiplying by x^{n-k}, we have

$$x^3 i(x) = i_0 x^3 + i_1 x^4 + i_2 x^5 + i_3 x^6 \tag{3.28}$$

and dividing (3.28) by $g(x) = 1 + x + x^3$ over $GF(2)$ gives the remainder

$$p(x) = (i_0 + i_2 + i_3) + (i_0 + i_1 + i_2)x + (i_1 + i_2 + i_3)x^2$$

$$= p_0 + p_1 x + p_2 x^2 \tag{3.29}$$

We now have three (even) parity check equations

$$p_0 + i_0 + i_2 + i_3 = 0$$

$$p_1 + i_0 + i_1 + i_2 = 0$$

$$p_2 + i_1 + i_2 + i_3 = 0$$

which can be expressed in matrix form

$$\begin{bmatrix} 1 & 0 & 0 & 1 & 0 & 1 & 1 \\ 0 & 1 & 0 & 1 & 1 & 1 & 0 \\ 0 & 0 & 1 & 0 & 1 & 1 & 1 \end{bmatrix} \begin{bmatrix} p_0 \\ p_1 \\ p_2 \\ i_0 \\ i_1 \\ i_2 \\ i_3 \end{bmatrix} = \begin{bmatrix} 0 \\ 0 \\ 0 \end{bmatrix} \tag{3.30}$$

The cyclic code is now represented by a parity check matrix and from Section 3.2 we can recognize this as a type of $(7, 4)$ Hamming code. Clearly, the simple circuit in Fig. 3.12(a) is a convenient way of generating such a code, and when this is done the resulting code is usually referred to as a *cyclic Hamming code*.

For a general systematic (n, k) cyclic code the check matrix in (3.30) becomes

$$H = \begin{bmatrix} 1 & 0 & 0 & \ldots & 0 & s_{00} & s_{10} & \ldots & s_{k-1,0} \\ 0 & 1 & 0 & \ldots & 0 & s_{01} & s_{11} & \ldots & s_{k-1,1} \\ \vdots & & & & & & & & \vdots \\ 0 & 0 & 0 & \ldots & 1 & s_{0,n-k-1} & & \ldots & s_{k-1,n-k-1} \end{bmatrix} \tag{3.31}$$

where

$$s_i(x) = \text{Rem}\left[\frac{x^{n-k+i}}{g(x)}\right]; \quad i = 0, 1, \ldots, k - 1$$

$$= s_{i0} + s_{i1}x + \ldots + s_{i,n-k-1}x^{n-k-1} \tag{3.32}$$

Example 3.11

Equation (3.31) suggests an alternative and flexible realization for a systematic coder (Bota, 1992). Consider the generation of a systematic $(7, 4)$ code (even parity). Using (3.8) we have

$$
\begin{bmatrix}
1 & 0 & 0 & s_{00} & s_{10} & s_{20} & s_{30} \\
0 & 1 & 0 & s_{01} & s_{11} & s_{21} & s_{31} \\
0 & 0 & 1 & s_{02} & s_{12} & s_{22} & s_{32}
\end{bmatrix}
\begin{bmatrix}
p_0 \\
\vdots \\
i_3
\end{bmatrix}
=
\begin{bmatrix}
0 \\
0 \\
0
\end{bmatrix}
$$

$$p_0 + i_0 s_{00} + i_1 s_{10} + i_2 s_{20} + i_3 s_{30} = 0$$

$$p_1 + i_0 s_{01} + i_1 s_{11} + i_2 s_{21} + i_3 s_{31} = 0$$

$$p_2 + i_0 s_{02} + i_1 s_{12} + i_2 s_{22} + i_3 s_{32} = 0$$

$$
\begin{bmatrix} p_0 \\ p_1 \\ p_2 \end{bmatrix}
= i_0 \begin{bmatrix} s_{00} \\ s_{01} \\ s_{02} \end{bmatrix}
+ i_1 \begin{bmatrix} s_{10} \\ s_{11} \\ s_{12} \end{bmatrix}
+ i_2 \begin{bmatrix} s_{20} \\ s_{21} \\ s_{22} \end{bmatrix}
+ i_3 \begin{bmatrix} s_{30} \\ s_{31} \\ s_{32} \end{bmatrix}
\tag{3.33}
$$

Taking the message $\mathbf{i} = (1011)$ in Example 3.10, and the corresponding $\mathbf{H}$ matrix in (3.30), gives

$$
\begin{bmatrix} p_0 \\ p_1 \\ p_2 \end{bmatrix}
= \begin{bmatrix} 1 \\ 1 \\ 0 \end{bmatrix}
+ \begin{bmatrix} 1 \\ 1 \\ 1 \end{bmatrix}
+ \begin{bmatrix} 1 \\ 0 \\ 1 \end{bmatrix}
= \begin{bmatrix} 1 \\ 0 \\ 0 \end{bmatrix}
$$

i.e. $\mathbf{p} = (100)$, as before. The significant point here is that the check-bit computation in (3.33) is attractive for *adaptive* FEC. For example, the 's' columns in $\mathbf{H}$ could be stored in EPROM and different sections of the EPROM used for different codes.

3.3.3 Binary BCH codes

We have used one particular generator polynomial $(g(x) = 1 + x + x^3)$ to explain the concept of cyclic code generation. For applications requiring a standard t-random error correction code we could refer to a list of generators, as in Table 3.2. This table lists some of the most popular cyclic codes, called BCH codes, after Bose and Chaudhuri (1960) and Hocquenghem (1959). Generators for n up to 1023 can be found in Lin and Costello (1983).

Each generator $g(x)$ is given in octal form for compact representation and is interpreted as in the following examples:

$$(7, 4) \text{ code: } g(x) = 13 \equiv 001011 \equiv 1 + x + x^3$$

$$(15, 7) \text{ code: } g(x) = 721 \equiv 111010001 \equiv 1 + x^4 + x^6 + x^7 + x^8$$

Table 3.2. *Generators of primitive BCH codes. (Reprinted with permission from IEEE Trans. Inf. Theory, Vol IT-10, No. 4, Oct. 1964, p. 391.)*

n	k	t	$g(x)$
7	4	1	13
15	11	1	23
	7	2	721
	5	3	2467
31	26	1	45
	21	2	3551
	16	3	107657
	11	5	5423325
	6	7	313365047
63	57	1	103
	51	2	12471
	45	3	1701317
	39	4	166623567
	36	5	1033500423
	30	6	157464165547
	24	7	17323260404441
	18	10	1363026512351725
	16	11	6331141367235453
	10	13	472622305527250155
	7	15	5231045543503271737
127	120	1	211
	113	2	41567
	106	3	11554743
	99	4	3447023271
	92	5	624730022327
	85	6	130704476322273
	78	7	26230002166130115
	71	9	6255010713253127753
	64	10	1206534025570773100045
	57	11	335265252505705053517721
	50	13	54446512523314012421501421
	43	14	17721772213651227521220574343
	36	15	31460746665220750447645774721735
	29	21	40311446136767060366753014117615 5
	22	23	12337607040472252243544562663764704 3
	15	27	22057042445604554770523013762217604353
	8	31	70472640527510306514762242715677331302 17
255	247	1	435
	239	2	267543
	231	3	156720665
	223	4	75626641375
	215	5	23157564726421
	207	6	16176560567636227
	199	7	7633031270420722341
	191	8	2663470176115333714567
	187	9	52755313540001322236351

Table 3.2. (*cont.*)

n	k	t	g(x)
255	179	10	22624710717340432416300455
	171	11	154162142123423560770616306357
	163	12	750041551007560255157472451601
	155	13	3757513005407665015722506464677633
	147	14	164213017353716552530416530544101711
	139	15	46140173206017556157072273024745356745
	131	18	2157133314715101512612502774421420241 65471
	123	19	1206140522420660037172103265161412262 72506267
	115	21	6052666557210024726363640460027635255 6313472737
	107	22	2220577232206625631241730023534742017 6574750154441
	99	23	1065666725347317422274141620157433225 2411076432303431
	91	25	6750265030327444172723631724732511075 550762720724344561
	87	26	1101367634147432364352316343071720462 06722545273311721317
	79	27	6670003563765750002027034420736617462 1015326711766541342355
	71	29	2402471052064432151555417211233116320 544425036255764322170 6035
	63	30	1075447505516354432531521735770700366 6111726455267613656702543301
	55	31	7315425203501100133015275306032054325 41432675501055704442 6035473617
	47	42	2533542017062646563033041377406233175 1233341454460450050660 24552543173
	45	43	1520205605523416113110134637642370156 367002447076237303320 2157025051541
	37	45	5136330255067007414177447245437530420 735706174323432347644 354737403044003
	29	47	3025715536673071465527064012361377115 342242324201174114060 25475741040356 5037
	21	55	1256215257060332656001773153607612103 22734140565307454252115312161446651 3473725
	13	59	4641732005052564544426573714250066004 3306774454765614031746772135702 6134 460500547
	9	63	1572602521747246320103104325535513461 4162367212044074545112766115547 7055 61677516057

Definition 3.3 *For any positive integer m (m ⩾ 3) and t (t < 2^{m-1}) there exists a binary BCH code having*

$$n = 2^m - 1$$

$$n - k \leqslant mt \qquad\qquad (3.34)$$

$$d \geqslant 2t + 1$$

Equation (3.34) states that these codes will correct any pattern of t or less errors occurring in the block length n. Note that BCH codes with $t = 1$ correspond to cyclic Hamming codes. The parameter m has important algebraic significance since it denotes the particular Galois field associated with a particular code (the reader unfamiliar with Galois fields is referred to Section 3.5.1). Here we are interested in *extension fields* of $GF(2)$, i.e. $GF(2^m)$ and it is shown in section 3.5.1 that

$$GF(2^m) = \{0, 1, \alpha, \alpha^2, \alpha^3, \ldots, \alpha^{2^m-2}\} \qquad\qquad (3.35)$$

In words, $GF(2^m)$ comprises a set of 2^m elements and α is the primitive element used to extend the basic field $GF(2)$. The significant point is that a class of (primitive) BCH codes is defined by specifying that $2t$ consecutive powers of α are roots of $g(x)$, as stated by the following theorem:

Theorem 3.6 *The generator polynomial $g(x)$ of a t-error correcting BCH code is the lowest degree polynomial over GF(2) which has α, α^2, α^3, . . . , α^{2t} as its roots, i.e. $g(\alpha^i) = 0$, $i = 1, 2, \ldots, 2t$.*

Consider the BCH(7, 4) code defined by

$$g(x) = 1 + x + x^3; \qquad m = 3, t = 1$$

This code is associated with $GF(2^3)$ and so any arithmetic must obey the rules of this field. In particular, for $GF(2^3)$ it is shown in Example 3.15 that

$$\alpha^3 = \alpha + 1; \qquad \alpha^6 = \alpha^2 + 1; \qquad \alpha^i + \alpha^i = 0$$

and we can use these relationships to demonstrate Theorem 3.6 for this code. Quite simply, according to the theorem, α and α^2 are roots of $g(x)$, so that

for α: $1 + \alpha + \alpha^3 = 1 + \alpha + (\alpha + 1) = 0$
for α^2: $1 + \alpha^2 + \alpha^6 = 1 + \alpha^2 + (\alpha^2 + 1) = 0$

Theorem 3.6 is fundamental not only to code generation, i.e. the derivation of $g(x)$ (see for example Lin and Costello (1983)), but also to the decoding of BCH codes. From Theorem 3.5 we can write $v(x) = a(x)g(x)$ so that

$$v(\alpha^i) = a(\alpha^i)g(\alpha^i) = 0; \quad i = 1, 2, \ldots, 2t \tag{3.36}$$

In other words, the elements α^i are also the roots of each code polynomial. We will return to (3.36) when discussing algebraic decoding of BCH codes.

In practice a BCH code is often shortened to give either improved error probability performance or a wordlength suitable for a particular application, and, as in Section 3.2.3, an (n, k) cyclic code can be shortened by deleting q message symbols from each codeword. Essentially, we select the 2^{k-q} codewords which have their q leading high-order message digits identical to zero, and then shorten them by removing these leading, zero-valued message digits. This gives a $(n - q, k - q)$ shortened cyclic code comprising 2^{k-q} codewords and with at least the error control capability of the original (n, k) code. The coding and decoding circuits are essentially the same as before.

Example 3.12

A particular application required a BCH code to protect television lines during 34 Mbit/s DPCM video transmission. For a 15.625 KHz line rate, a digitized television line comprised approximately 2199 bits, and so a block length of this order was required. The nearest BCH code corresponds to $m = 12$, i.e.

$n = 2^{12} - 1 = 4095$

and this had to be shortened to approximately 2199. The actual codeword length was restricted by counters to any length < 4095 which is divisible by 4, and so in practice the selected value was 2196.

In order to meet the required *decoded* or *residual* error rate (see Section 3.6) of 10^{-8} for a channel error rate of 10^{-4} it was found necessary to select a code with $t \geqslant 4$, and $t = 5$ was finally chosen. The corresponding generator polynomial then had a degree

$n - k = mt = 60$

and so a 60-stage linear feedback shift register was required at the coder. Apart from the larger register, the systematic coder was similar to those in Fig. 3.12. The final (2196, 2136) code was found by extensive search of possible codes and was defined by the polynomial

$g(x) = 101467071546246223445$ (octal)

Generally speaking, for any application, one objective is to maximise the coding efficiency (rate) R, or minimise the coding redundancy $1 - R$. In Example 3.12 the redundancy is an acceptable 2.7%. Note, however, that as the channel error rate increases, say to 10^{-2}, a BCH code becomes less efficient and other techniques should be used, such as convolutional codes or concatenated codes. Finally, note that most BCH codes can correct some error patterns which exceed the designed t-error capability.

3.3.4 Interleaving

Interleaving is an efficient way of using a t-random error correcting code to combat both random and burst errors. It can be implemented using either block or convolutional codes, and Fig. 3.13(a) illustrates the principle for an (n, k) block code. Effectively, a block of i codewords is formed into a $i \times n$ matrix and the matrix is transmitted *column by column* (as indicated) to give an (in, ik) *interleaved* code. At the receiver the data is fed into a similar matrix column by column and then de-interleaved by reading the matrix a row at a time, as indicated.

As in Section 3.2.1, we can model the errors by using an error vector $\mathbf{e} = (e_1, e_2, e_3, \ldots)$. This enables us to define a burst error of length b as a vector whose only non-zero symbols are among b consecutive symbols, the first and last of which are non-zero. For example, vector

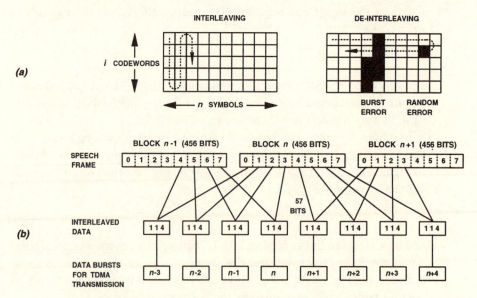

Fig. 3.13 Interleaving: (a) principle using an (n, k) block code; (b) interleaving for GSM cellular radio.

$\mathbf{e} = (00001001101000)$ corresponds to a burst of length 7. Clearly, if there are additional errors in the data stream we must also define a burst 'guard space' to separate the burst from these errors. For this example, $\mathbf{e}$ could then be regarded as containing a burst of length 7 *relative to a guard space of length 3*. Now consider the burst at the de-interleaver in Fig. 3.13(a). Clearly, the burst is distributed between codewords so if the basic code is double error correcting all errors can be corrected, including the random error. Generalizing this to a t-error correcting code we see that interleaving to a depth i will correct a burst of length

$$b \le it \tag{3.37}$$

and, in general, it will also correct some random errors. The essential practical requirements are a memory of ni symbols at both the coder and decoder, and the need for some form of frame synchronization to denote the start of each interleaved block. The latter could be achieved using a *unique word (UW)* with good autocorrelation properties. Note that long bursts require a large interleave depth, corresponding to a large interleave/de-interleave delay (in a basic system this is simply $2ni$ symbols). In some cases the required interleave depth may be excessive and an alternative scheme will have to be used. For example, assuming a return path, one strategy could be to identify burst periods quickly and then revert to a pure ARQ scheme (Ferebee, Tait and Taylor, 1989).

Burst errors are common in digital mobile radio systems since signal fades usually occur at a much slower rate than the transmission rate. Interleaving is therefore incorporated into the GSM digital cellular radio system for example (ETSI-GSM, 1991), and Fig. 3.13(b) illustrates the principle. Speech is coded into 456 bit blocks and each block is reordered and partitioned into 8 sub-blocks. The interleaver then pairs sub-blocks from adjacent blocks and these pairs are mapped to 'data bursts' for transmission in TDMA format. Clearly, since each 456 bit block contains ECC, interleaving will distribute a burst error between error protected blocks in a similar way to Fig. 3.13(a).

3.3.5 Cyclic redundancy check (CRC) codes

Binary BCH codes are the basis of a very common technique for checking the integrity of serial data transfers. The technique is used for example in data communications systems, computer networks and computer disk systems.

The basic error detection system has identical linear feedback shift registers at each end, as shown in Fig. 3.14. The message $i(x)$ is coded as in conventional systematic coding, i.e. $i(x)$ is transmitted first and then it is followed by the check polynomial $p(x)$. As $p(x)$ enters the receiver it is compared symbol by symbol with the check polynomial generated at the receiver and, clearly, if there is no transmission error the final contents of the receiver shift register will be zero. Looked at another way, for error-free transmission the receiver essentially divides the codeword by $g(x)$, and since all codewords are divisible by $g(x)$ the remainder will be zero. Conversely, any remainder indicates a transmission error, in which case an automatic request could be made for a repeat transmission (ARQ).

The transmitted check vector is called the *CRC character*, and CRC chips usually offer a range of standard polynomials (Table 3.3). The CRC-16 and CRC-CCITT standards, for example, operate with 8-bit or 16-bit characters and transmit a 16-bit CRC character. It is shown in Section 3.4 that an (n, k) BCH code can *reliably* detect a burst of errors of length $n - k$ or less, where $g(x)$ is of degree $n - k$. This means that the codes in Table 3.3 can reliably detect bursts up to and including 8, 12, or 16 bits.

Table 3.3. *Standard CRC polynomials.*

Standard	$g(x)$
CRC-8	$1 + x^8$
CRC-12	$1 + x + x^2 + x^3 + x^{11} + x^{12}$
CRC-16	$1 + x^2 + x^{15} + x^{16}$
CRC-CCITT	$1 + x^5 + x^{12} + x^{16}$

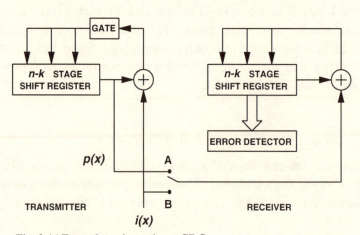

Fig. 3.14 Error detection using a CRC.

3.4 Non-algebraic decoding of cyclic codes

We now examine a simple technique for decoding binary (n, k) cyclic codes which avoids the detailed algebraic manipulations of the more complex methods that will be described in Section 3.5. It is suitable for relatively short length BCH codes and for modest values of t, say $t \leqslant 3$.

3.4.1 The Meggitt decoder

A general objective in the decoding of error control codes is that of *syndrome calculation*. This is true for the Hamming codes in Section 3.2, the algebraic techniques in Section 3.5, and the convolutional codes in Section 3.7. It is also true for Meggitt decoders. The received codeword in polynomial form is

$$r(x) = r_0 + r_1 x + r_2 x^2 + \ldots + r_{n-1} x^{n-1}$$

$$= v(x) + e(x) \tag{3.38}$$

where

$$v(x) = v_0 + v_1 x + v_2 x^2 + \ldots + v_{n-1} x^{n-1} \tag{3.39}$$

and

$$e(x) = e_0 + e_1 x + e_2 x^2 + \ldots + e_{n-1} x^{n-1} \tag{3.40}$$

Here, $e(x)$ is the *error polynomial* such that $e_i = 1$ for a received error in symbol r_i. The syndrome polynomial $s(x)$ is generated by dividing $r(x)$ by $g(x)$, and since all codewords are divisible by $g(x)$ we can write

$$\frac{r(x)}{g(x)} = a(x) + \frac{e(x)}{g(x)} \tag{3.41}$$

The required syndrome $s(x)$ is then the remainder resulting from the division $e(x)/g(x)$ and, clearly, $s(x)$ will be zero if $e(x)$ is zero. As in the encoding of cyclic codes, we use a linear feedback $n - k$ stage shift register to perform the division by $g(x)$ and Fig. 3.15 shows the basic circuit.

The syndrome register (SR) is initially reset and $r(x)$ is applied to the input (symbol r_{n-1} first). After the entire polynomial $r(x)$ has been clocked in, the SR contains the syndrome $s(x)$ of $r(x)$. The received codeword will have up to t correctable errors and one might reasonably ask how a single syndrome can correct them? In fact, error correction is

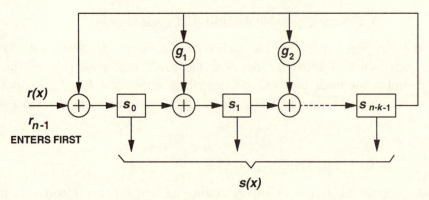

Fig. 3.15 Syndrome calculation for the Meggitt decoder.

performed over a number of clock cycles using a number of different syndromes, and is based upon the following cyclic decoding theorem:

Theorem 3.7 *If $s(x)$ is the syndrome of $r(x)$, then the remainder $s'(x)$ resulting from dividing $xs(x)$ by $g(x)$ is the syndrome of $r'(x)$, which is a cyclic shift of $r(x)$ (the division $xs(x)$ by $g(x)$ is achieved by shifting the SR once with $s(x)$ as the initial contents).*

This theorem states that, given the initial syndrome $s(x)$ of $r(x)$, then repeated clocking of the SR with the input open circuit will generate syndromes corresponding to cyclically shifted versions of $r(x)$. If then $r(x)$ is held in an n-symbol (circulating) buffer, we can decode those syndromes corresponding to an error at the end of the buffer and so perform *symbol-by-symbol* decoding. The basic Meggitt decoder for a binary BCH code therefore requires an n-stage buffer, an $n - k$-stage SR, and syndrome detection logic, as shown in Fig. 3.16. Note that it is not essential to shift the buffer *cyclically*.

Decoder operation

Step 1 With gate 1 open (a short-circuit) and gate 2 closed the entire $r(x)$ is clocked into the buffer and simultaneously into the SR. When symbol r_{n-1} occupies the last stage of the buffer (as shown), the SR contains $s(x)$ corresponding to $r(x)$.

Step 2 Symbol-by-symbol error correction is now performed with gate 1 closed and gate 2 open. If r_{n-1} is not in error, then both the SR and the buffer are shifted once simultaneously giving $s'(x)$ corresponding to $r'(x)$.

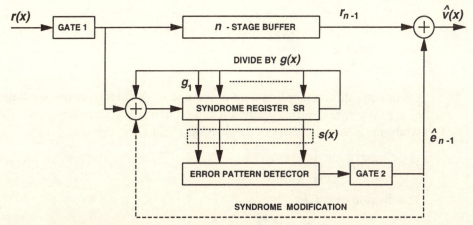

Fig. 3.16 A Meggitt decoder for cyclic codes.

If $s(x)$ corresponds to a correctable error pattern, i.e. t or less errors in $r(x)$, with an error in symbol r_{n-1}, then the pattern detector generates $\hat{e}_{n-1} = 1$ to complement or correct r_{n-1}. The corrected polynomial is

$$r_1(x) = r_0 + r_1 x + \ldots + r_{n-2} x^{n-2} + (r_{n-1} + \hat{e}_{n-1}) x^{n-1}$$

and a cyclic shift right would give

$$r_1'(x) = (r_{n-1} + \hat{e}_{n-1}) + r_0 x + \ldots + r_{n-2} x^{n-1}$$

Theoretically, the SR should also be modified by adding a 1 to its input. This removes the effect of the error from $s(x)$, giving $s_1(x)$, and subsequent shifting gives $s_1'(x)$ corresponding to $r_1'(x)$.

Step 3 Symbol r_{n-2} is now at the end of the buffer and can be corrected if necessary using the new $s(x)$, i.e. $s'(x)$ or $s_1'(x)$. Decoding stops when $r(x)$ has been clocked out of the buffer.

In practice syndrome modification is not essential and is therefore drawn dotted in Fig. 3.16. When it is used, the SR will be zero at the end of decoding, and when it is omitted the SR will generally be non-zero after decoding, although decoding will still be correct. Also note that the basic circuit in Fig. 3.16 is non-real-time since only alternate words are decoded. A real-time decoder could be realized using two n-stage buffers, or dual decoders.

Example 3.13

Suppose we require a Meggitt decoder for the $(15, 11)$ cyclic Hamming code defined by

$g(x) = 1 + x + x^4$

The single error can occur in any one of 15 bits, i.e. the possible error polynomials are

$e(x) = 1, x, x^2, \ldots, x^{14}$

Eventually, the error will end up at the end of the buffer, corresponding to $e(x) = x^{14}$, and in this position it can be corrected. The corresponding syndrome is the remainder of the division $x^{14}/g(x)$:

$$s(x) = \text{Rem}\left[\frac{x^{14}}{g(x)}\right] \tag{3.42}$$

The division over $GF(2)$ is

$$
\begin{array}{r}
x^{10} + x^7 + x^6 + x^4 + x^2 + x + 1 \\
x^4 + x + 1\overline{\smash{\big)}\ x^{14}} \\
\underline{x^{14} + x^{11} + x^{10}} \\
x^{11} + x^{10} \\
\underline{x^{11} + x^8 + x^7} \\
x^{10} + x^8 + x^7 \\
\underline{x^{10} + x^7 + x^6} \\
x^8 + x^6 \\
\vdots \\
x^3 + 1 = s(x)
\end{array}
$$

$\therefore s = (s_0, s_1, s_2, s_3)$

$ = (1001)$

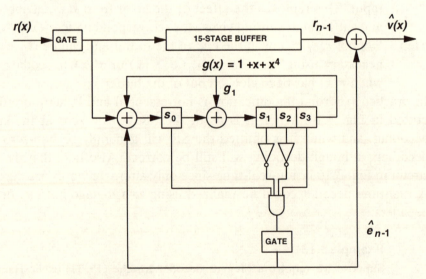

Fig. 3.17 Meggitt decoder for the BCH(15, 11) code.

The pattern detector must therefore generate $\hat{e}_{n-1} = 1$ for this particular vector, and the basic circuit is shown in Fig. 3.17. Note that the cyclic structure of the code results in a simpler decoder circuit when compared to a (15, 11) decoder derived using the approach in Example 3.3. On the other hand, decoding now takes longer since it is carried out in a serial rather than a parallel manner.

Example 3.14

The decoder for the DEC BCH(15, 7) code is more complex since there are significantly more correctable error patterns. The simplest case corresponds to a single error, and so $s(x)$ is given by (3.42) with

$$g(x) = 1 + x^4 + x^6 + x^7 + x^8$$

However, 14 other correctable error patterns are possible, namely,

$$x^{13} + x^{14} \quad \text{(adjacent errors)}$$

$$x^{12} + x^{14}$$

$$e(x) = \;\; \vdots$$

$$1 + x^{14}$$

For example, when adjacent errors occur with one error in symbol r_{n-1}, the syndrome is

$$s(x) = \text{Rem}\left[\frac{x^{13} + x^{14}}{g(x)}\right]$$

$$= x^2 + x^3 + x^4 + x^7$$

$$\mathbf{s} = (00111001)$$

This is one of the 15 possible patterns which must generate $\hat{e}_{n-1} = 1$, and so a small PROM could be used for pattern detection, as shown in Fig. 3.18. Clearly, the complexity of the pattern detector will rapidly increase with t and this is probably the main limitation of the Meggitt decoder. The C-routine in Appendix E could be used to illustrate the performance of the Meggitt decoder under random and burst channel errors.

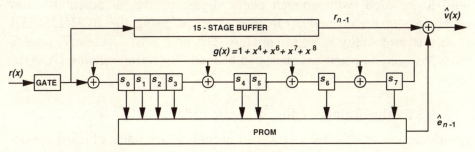

Fig. 3.18 Meggitt decoder for the BCH(15, 7) code.

3.4.2 Burst detection

Besides random error correction, an (n, k) cyclic code can also detect *burst errors*, as defined in Section 3.3.4. We know from (3.41) that the syndrome $s(x)$ is guaranteed to be non-zero if the degree of $e(x)$ is less than that of $g(x)$. In other words, since $g(x)$ has degree $n - k$, the division of $r(x)$ by $g(x)$ will yield a non-zero syndrome provided $e(x)$ has degree $n - k - 1$ or less (corresponding to a maximum burst length of $n - k$). A burst of length $n - k$ or less will therefore be detected by simply examining the syndrome register (via an OR-gate) for at least one non-zero bit after division. As an example, the BCH$(15, 11)$ code defined by

$$g(x) = 1 + x + x^4$$

could detect the burst

$$e(x) = 1 + x + x^2 + x^3$$

$$\mathbf{e} = (1111000 \ldots)$$

Similarly, a shifted version of $e(x)$ is also non-divisible by $g(x)$ and so is detectable. In fact, many bursts of length *greater* than $n - k$ will be detected whilst only a relatively few will go undetected. Clearly, detection will fail when the error pattern is of the form $e(x) = a(x)g(x)$, i.e. when it is actually a codeword. For example,

$$e(x) = (x + x^3)(1 + x + x^4) = x + x^2 + x^3 + x^4 + x^5 + x^7$$

$$\mathbf{e} = (01111101000 \ldots)$$

will go undetected by the BCH$(15, 11)$ code since $e(x)$ is divisible by $g(x)$. On this basis, since there are 2^k codewords and very nearly 2^n possible n-bit error sequences, the probability of an undetected error is very nearly $2^k/2^n = 2^{-(n-k)}$. In practice, efficient high-rate, long block length codes tend to be used, with enough check digits to give an acceptably low probability of undetected error. For example, using the BCH$(255, 231)$ code this probability is approximately 2^{-24} or 6×10^{-8}. Clearly, once a burst has been detected a request can be made for retransmission (ARQ).

3.5 Algebraic decoding of cyclic codes

In this section we discuss a powerful decoding technique which is applicable to word orientated codes (RS codes) as well as to binary cyclic codes.

Table 3.4. *Postulates of a mathematical field, F (a* and a⁻¹ denote inverse elements).*

	Under addition	Under multiplication
Closure	$a + b \in F$	$a \cdot b \in F$
Associative	$(a + b) + c = a + (b + c)$	$(a \cdot b) \cdot c = a \cdot (b \cdot c)$
Commutivity	$a + b = b + a$	$a \cdot b = b \cdot a$
Identity	$a + 0 = a$	$a \cdot 1 = a$
Inverse	$a + a^* = 0$	$a \cdot a^{-1} = 1, a \neq 0$
Distributivity		$a(b + c) = ab + ac$

It relies heavily upon Galois field theory to find *algebraic solutions* for the error *locations* and also, in the case of RS codes, the error *values*. We therefore first review aspects of Galois field theory which are relevant to algebraic decoding.

3.5.1 Galois field theory

A mathematical field, F, is a system having two operations, both with inverses. Usually these are addition (with its inverse, subtraction) and multiplication (with its inverse, division). If a, b and c are elements of F, i.e. a, b, $c \in F$, then the postulates or laws of a field are defined as in Table 3.4.

A field with a finite number of elements, q, is called a finite field or *Galois* field and is denoted $GF(q)$. In general, finite fields only exist when q is prime, or when $q = p^m$ where p is prime and m is integer. The prime field $GF(q)$, $(q > 1$ and q prime), will have elements $0, 1, 2, \ldots, q - 1$, i.e. the set of all integers modulo q, and the operations will be addition and multiplication modulo q. The simplest prime field uses modulo 2 arithmetic and is defined as

$$GF(2) = \{0, 1\} \tag{3.43}$$

where 0 and 1 are the additive and multiplicative identity elements respectively. These elements must satisfy the field postulates, which leads to the simple arithmetic rules for binary error control codes discussed in Section 3.1.3.

Extension fields

The field $GF(2^m)$, $m > 1$, is called an *extension* field of $GF(2)$ and these fields are particularly relevant to cyclic codes. To generate a field $GF(2^m)$ we extend the elements 0, 1 using a primitive element α. If

$$\alpha \in GF(2^m)$$

then the closure postulate under multiplication means that $\alpha \cdot \alpha = \alpha^2$, $\alpha \cdot \alpha^2 = \alpha^3$ etc. are also elements of $GF(2^m)$. Therefore,

$$GF(2^m) = \{0, 1, \alpha, \alpha^2, \alpha^3, \ldots\} \qquad (3.44)$$

By definition, $GF(2^m)$ must be finite and this can be achieved by associating it with a polynomial $p(x)$ of degree m which is *primitive* over $GF(2)$ (see Definition 3.6). Essentially, $p(x)$ limits the field to a finite number of elements and defines the addition and multiplication rules which must be satisfied by these elements. This can be achieved if we state that, for element α

$$p(\alpha) = 0 \qquad (3.45)$$

A list of primitive polynomials is given in Table 3.15, Section 3.9.

Example 3.15

Consider $GF(2^3)$. From Table 3.15 we select a primitive polynomial of degree 3, e.g.

$$p(x) = 1 + x + x^3 \qquad (3.46)$$

and according to (3.45) this gives the identity

$$\alpha^3 + \alpha + 1 = 0 \qquad (3.47)$$

Equation (3.47) can be used to define rules for addition and multiplication and to limit the field to a finite number of elements. First note that for $GF(2^m)$ each element is its own additive inverse, so that from Table 3.4 we can state $\alpha^i + \alpha^i = 0$. Put another way, addition and subtraction are identical over $GF(2^m)$. Therefore, using (3.47) we can write

$$\alpha^3 = \alpha + 1$$

$$\alpha^4 = \alpha\alpha^3 = \alpha^2 + \alpha$$

$$\alpha^5 = \alpha\alpha^4 = \alpha^3 + \alpha^2 = \alpha^2 + \alpha + 1$$

$$\alpha^6 = \alpha\alpha^5 = \alpha^3 + \alpha^2 + \alpha = \alpha^2 + 1$$

Table 3.5. *Arithmetic rules for $GF(2^3)$ generated by $p(x) = 1 + x + x^3$.*

+	0	1	α	α^2	α^3	α^4	α^5	α^6
0	0	1	α	α^2	α^3	α^4	α^5	α^6
1	1	0	α^3	α^6	α	α^5	α^4	α^2
α	α	α^3	0	α^4	1	α^2	α^6	α^5
α^2	α^2	α^6	α^4	0	α^5	α	α^3	1
α^3	α^3	α	1	α^5	0	α^6	α^2	α^4
α^4	α^4	α^5	α^2	α	α^6	0	1	α^3
α^5	α^5	α^4	α^6	α^3	α^2	1	0	α
α^6	α^6	α^2	α^5	1	α^4	α^3	α	0

$\times$	0	1	α	α^2	α^3	α^4	α^5	α^6
0	0	0	0	0	0	0	0	0
1	0	1	α	α^2	α^3	α^4	α^5	α^6
α	0	α	α^2	α^3	α^4	α^5	α^6	1
α^2	0	α^2	α^3	α^4	α^5	α^6	1	α
α^3	0	α^3	α^4	α^5	α^6	1	α	α^2
α^4	0	α^4	α^5	α^6	1	α	α^2	α^3
α^5	0	α^5	α^6	1	α	α^2	α^3	α^4
α^6	0	α^6	1	α	α^2	α^3	α^4	α^5

Note in particular that

$$\alpha^7 = \alpha\alpha^6 = \alpha^3 + \alpha = 1$$

This means that $\alpha^5\alpha^4 = \alpha^2$ etc., and that the elements of $GF(2^3)$ are restricted to

$$GF(2^3) = \{0, 1, \alpha, \alpha^2, \alpha^3, \alpha^4, \alpha^5, \alpha^6\}$$

Generalizing (see also (3.34)),

$$\alpha^{2^m - 1} = \alpha^n = 1 \tag{3.48}$$

$$GF(2^m) = \{0, 1, \alpha, \alpha^2, \ldots, \alpha^{2^m - 2}\} \tag{3.49}$$

The full addition and multiplication rules for $GF(2^3)$ are given in Table 3.5. Note that a polynomial $r(\alpha)$ can be reduced modulo $p(\alpha)$ by saving the remainder after division by $p(\alpha)$, i.e.

$$r(\alpha) \bmod p(\alpha) = \text{Rem}\left[\frac{r(\alpha)}{p(\alpha)}\right] \tag{3.50}$$

For example, if $r(\alpha) = \alpha^6$ the remainder is easily shown to be $\alpha^2 + 1$.

Table 3.6. *Galois field* $GF(2^4)$ *generated by* $p(x) = 1 + x + x^4$.

Power representation	Polynomial representation	4-tuple representation
0	0	(0 0 0 0)
1	1	(1 0 0 0)
α	α	(0 1 0 0)
α^2	α^2	(0 0 1 0)
α^3	α^3	(0 0 0 1)
α^4	$1 + \alpha$	(1 1 0 0)
α^5	$\alpha + \alpha^2$	(0 1 1 0)
α^6	$\alpha^2 + \alpha^3$	(0 0 1 1)
α^7	$1 + \alpha \quad + \alpha^3$	(1 1 0 1)
α^8	$1 \quad + \alpha^2$	(1 0 1 0)
α^9	$\alpha \quad + \alpha^3$	(0 1 0 1)
α^{10}	$1 + \alpha + \alpha^2$	(1 1 1 0)
α^{11}	$\alpha + \alpha^2 + \alpha^3$	(0 1 1 1)
α^{12}	$1 + \alpha + \alpha^2 + \alpha^3$	(1 1 1 1)
α^{13}	$1 \quad + \alpha^2 + \alpha^3$	(1 0 1 1)
α^{14}	$1 \quad + \alpha^3$	(1 0 0 1)

Example 3.16

Using Table 3.2, the BCH(15, 7) code is defined by

$$g(x) = 1 + x^4 + x^6 + x^7 + x^8$$

Suppose we apply the systematic coding procedure in Section 3.3.2 to the message $i(x) = 1 + x^3 + x^4 + x^5$. We have

$$x^8 i(x) = x^8 + x^{11} + x^{12} + x^{13}$$

and dividing this by $g(x)$ gives the remainder or check polynomial $x + x^7$. Therefore the code polynomial is

$$v(x) = x + x^7 + x^8 + x^{11} + x^{12} + x^{13} \tag{3.51}$$

The code is defined on $GF(2^4)$ (i.e. from (3.34) $15 = 2^4 - 1$) which can be constructed from primitive polynomial $p(x) = 1 + x + x^4$ (see Table 3.15). Various forms of this Galois field are shown in Table 3.6. Note that polynomials in α can be represented as 4-tuples, and since $\alpha^i + \alpha^i = 0$, these can be added modulo 2 in order to reduce a larger polynomial modulo $p(\alpha)$. Consider, from (3.51),

$$v(\alpha) = \alpha + \alpha^7 + \alpha^8 + \alpha^{11} + \alpha^{12} + \alpha^{13}$$

Expressing this in terms of 4-tuples from Table 3.6 we have

0	1	0	0
1	1	0	1
1	0	1	0
0	1	1	1
1	1	1	1
1	0	1	1

0	0	0	0	modulo 2

i.e. $v(\alpha) = 0$, as expected, since α is a root of $v(x)$ (Equation (3.36)). The reader should show that the same result is obtained if we use (3.50) to reduce $v(\alpha)$ modulo $p(\alpha)$, i.e. the remainder will be zero.

3.5.2 Decoding binary BCH codes

Much of the following discussion also applies to the decoding of RS codes. Consider a t-error correcting binary BCH(n, k) code defined on $GF(2^m)$, i.e. $n = 2^m - 1$. Following the approach in Section 3.4.1, the received polynomial is

$$r(x) = v(x) + e(x)$$

and we require the syndrome of $r(x)$. Using (3.36) we can write

$$r(\alpha^i) = v(\alpha^i) + e(\alpha^i)$$

$$= e(\alpha^i)$$

$$= s_i \tag{3.52}$$

where s_i is a syndrome symbol. Restating this result, the syndrome symbols can be found from

$$s_i = r(\alpha^i), \qquad i = 1, 2, \ldots, 2t \tag{3.53}$$

As we shall see, there will be a need to reduce $r(\alpha^i)$ modulo $p(\alpha)$ and we could use one of the techniques discussed in Examples 3.15 and 3.16, or we could use a Galois field table.

Error location polynomial

In Section 3.2 we used syndrome symbols to determine the error location for a Hamming code. Here, symbols s_i perform the same role but in a less direct way. Essentially, they will be used to determine the coefficients of an *error location polynomial*, the roots of which define the error locations. Suppose that for a particular received polynomial

$$e(x) = x^8 + x^{11}$$

so that

$$e(\alpha^i) = (\alpha^8)^i + (\alpha^{11})^i$$

$$= X_2^i + X_1^i \tag{3.54}$$

Here, $X_1 = \alpha^{11}$ is an *error locator*, denoting an error in received symbol r_{11}, and $X_2 = \alpha^8$ is an error locator denoting an error in symbol r_8. The kth error locator will be of the form

$$X_k = \alpha^j, \quad \alpha^j \in GF(2^m) \tag{3.55}$$

and denotes an error in symbol r_j, $j = 0, 1, \ldots, n-1$. Since we are considering binary codes, r_j will then be complemented to perform error correction. As just intimated, (3.55) is obtained from symbols s_i in a somewhat indirect way. Let us generalize (3.54) to

$$s_i = e(\alpha^i) = \sum_{k=1}^{t} X_k^i, \quad i = 1, 2, \ldots, 2t \tag{3.56}$$

In principle we could solve (3.56) for the values of X_k, but since it involves *non-linear* equations the usual approach is to use an *error location polynomial*, $\sigma(x)$. The significance of $\sigma(x)$ is that it translates the problem into a set of *linear* equations, and that its roots turn out to be the error locators, X_k.

There are a number of algorithms for finding $\sigma(x)$, and here we use *Peterson's direct method* (Peterson, 1960). This approach is suitable for decoding many binary BCH codes, but becomes inefficient for large values of t, say $t > 5$. In such cases, *Euclid's algorithm for polynomials* (McEliece, 1977), or *Berlekamp's iterative algorithm* (Berlekamp, 1968) could be applied. Considering Peterson's method, one form for $\sigma(x)$ is

$$\sigma(x) = x^t + \sigma_1 x^{t-1} + \ldots + \sigma_t \tag{3.57}$$

and since the roots are the error locators $X_1, X_2, \ldots, X_t$, we can write

$$X_k^t + \sigma_1 X_k^{t-1} + \ldots + \sigma_t = 0, \quad k = 1, 2, \ldots t \tag{3.58}$$

Multiplying by X_k^j, and then expanding

$$X_k^{t+j} + \sigma_1 X_k^{t+j-1} + \ldots + \sigma_t X_k^j = 0 \quad k = 1, 2, \ldots, t$$

$$\therefore X_1^{t+j} + \sigma_1 X_1^{t+j-1} + \ldots + \sigma_t X_1^j = 0$$

Table 3.7. *Coefficients of the error location polynomial*
$\sigma(x)$ *for some t-error correcting binary BCH codes.*

t	Coefficients σ_i
1	$\sigma_1 = s_1$
2	$\sigma_1 = s_1$
	$\sigma_2 = \dfrac{s_3 + s_1^3}{s_1}$
3	$\sigma_1 = s_1$
	$\sigma_2 = \dfrac{s_1^2 s_3 + s_5}{s_1^3 + s_3}$
	$\sigma_3 = (s_1^3 + s_3) + s_1\sigma_2$

$$X_2^{t+j} + \sigma_1 X_2^{t+j-1} + \ldots + \sigma_t X_2^j = 0$$

$$\vdots$$

$$X_t^{t+j} + \sigma_1 X_t^{t+j-1} + \ldots + \sigma_t X_t^j = 0$$

$$\therefore \sum_{k=1}^{t} X_k^{t+j} + \sigma_1 \sum_{k=1}^{t} X_k^{t+j-1} + \ldots + \sigma_t \sum_{k=1}^{t} X_k^j = 0$$

Finally, substituting for the summation terms from (3.56) gives

$$s_{t+j} + \sigma_1 s_{t+j-1} + \ldots + \sigma_t s_j = 0, \quad j = 1, 2, \ldots, t \tag{3.59}$$

Equation (3.59) represents a set of t *linear* equations which can be solved for the coefficients σ_i, $i = 1, 2, \ldots, t$, hence giving $\sigma(x)$. In addition, for binary codes

$$s_{2k} = s_k^2, \quad k = 1, 2, 3, \ldots \tag{3.60}$$

and, in fact, (3.53) need only be computed for 'odd' values of i. Using (3.59) and (3.60), a little straightforward manipulation gives the results in Table 3.7.

Chien search

So far we have used the syndrome symbols to determine the error location polynomial $\sigma(x)$. It is now necessary to determine the roots of $\sigma(x)$ in order to determine the error locators X_k, $k = 1, 2, \ldots, t$. The most obvious approach is to substitute the n values of X_k given by (3.55) into (3.58) and note those values which satisfy (3.58). The substitution is usually carried out in the following way. First rewrite (3.57) as

$$\frac{\sigma(x)}{x^t} = 1 + \sigma_1 x^{-1} + \ldots + \sigma_t x^{-t} \tag{3.61}$$

Over $GF(2^m)$, equating $\sigma(x)$ to zero is equivalent to finding the roots of

$$\sigma_1 x^{-1} + \sigma_2 x^{-2} + \ldots + \sigma_t x^{-t} = 1$$

$$\sum_{i=1}^{t} \sigma_i x^{-i} = 1 \tag{3.62}$$

To reiterate, the roots of $\sigma(x)$, or (3.62), are the error locators and these are of the form α^j, $j = 0, 1, \ldots, n-1$. In other words, an error in symbol r_{n-1} will correspond to an error locator or root of the form α^{n-1}. Suppose we first test for an error in symbol r_{n-1} by substituting α^{n-1} for x. Term x^{-i} then reduces to (see (3.48))

$$\alpha^{-i(n-1)} = \alpha^{-ni}\alpha^i = \left(\frac{1}{\alpha^n}\right)^i \alpha^i = \alpha^i$$

Similarly, x^{-i} reduces to α^{2i} when substituting α^{n-2}, and to α^{ji} when substituting α^{n-j}. Equation (3.62) can now be written as the 'search' equation

$$\sum_{i=1}^{t} \sigma_i \alpha^{ji} \overset{?}{=} 1, \quad j = 1, 2, \ldots, n \tag{3.63}$$

A circuit for implementing (3.63) is commonly called the *Chien search* (Chien, 1964). This circuit sets a correction bit to 1 whenever (3.63) is satisfied, and this complements the appropriate bit in the received word. For example, if (3.63) is satisfied for $j = 3$, then symbol r_{n-3} is complemented. Error correction starts on symbol r_{n-1} and proceeds bit by bit just as in the Meggitt decoder (Fig. 3.16). Note that the Chien search is not required for $t = 1$ since solution of (3.58) is then trivial, i.e.

$$X_1 + \sigma_1 = 0$$

$$\therefore X_1 = \sigma_1 = s_1$$

In words, if $s_1 = \alpha^j$, $j = 0, 1, \ldots, n-1$, then symbol r_j must be complemented.

Example 3.17

Suppose the BCH(15, 7) codeword $v(x)$ in (3.51) is received as

$$r(x) = x + x^7 + x^{12} + x^{13} \tag{3.64}$$

Using (3.53) and reducing polynomials in α modulo $p(\alpha)$ via Table 3.6 gives

$$s_1 = r(\alpha)$$

$$= \alpha + \alpha^7 + \alpha^{12} + \alpha^{13}$$

$$= \alpha^7 \text{ modulo } p(\alpha)$$

$$s_3 = r(\alpha^3)$$

$$= \alpha^3 + \alpha^{21} + \alpha^{36} + \alpha^{39}$$

$$= \alpha^3 + \alpha^6 + \alpha^6 + \alpha^9 \quad (\text{note } \alpha^{15} = 1)$$

$$= \alpha \text{ modulo } p(\alpha)$$

The full syndrome vector is, in fact, $\mathbf{s} = (\alpha^7, \alpha^{14}, \alpha, \alpha^{13})$. Using Table 3.7 gives

$$\sigma_1 = \alpha^7$$

$$\sigma_2 = (\alpha + \alpha^{21})(\alpha^{-7})$$

Here, α^{-7} is the multiplicative inverse of α^7 (Table 3.4), i.e.

$$\alpha^7 \cdot \alpha^{-7} = 1$$

Clearly, since $\alpha^{15} = 1$, then $\alpha^{-7} = \alpha^8$ and

$$\sigma_2 = (\alpha + \alpha^6)\alpha^8$$

$$= \alpha^4 \text{ modulo } p(\alpha)$$

The error location polynomial is therefore

$$\sigma(x) = x^2 + \alpha^7 x + \alpha^4 \tag{3.65}$$

Applying the Chien search in (3.63) for the roots of (3.65), and reducing using Table 3.6, we have

$$j = 1: \sigma_1\alpha + \sigma_2\alpha^2 = \alpha^7\alpha + \alpha^4\alpha^2 = \alpha^{14}$$
$$j = 2: \sigma_1\alpha^2 + \sigma_2\alpha^4 = \alpha^7\alpha^2 + \alpha^4\alpha^4 = \alpha^{12}$$
$$j = 3: \sigma_1\alpha^3 + \sigma_2\alpha^6 = \alpha^7\alpha^3 + \alpha^4\alpha^6 = 0$$
$$j = 4: \sigma_1\alpha^4 + \sigma_2\alpha^8 = \alpha^7\alpha^4 + \alpha^4\alpha^8 = 1$$

Therefore, $\alpha^{n-4} = \alpha^{11}$ is a root of $\sigma(x)$, i.e. $X_1 = \alpha^{11}$ satisfies (3.58), and symbol r_{11} should be complemented. Similarly, the reader should show that (3.63) is also satisfied for $j = 7$, i.e. $X_2 = \alpha^8$, and that r_8 should be complemented. Error correction effectively adds $x^8 + x^{11}$ to $r(x)$ thereby regenerating the original code polynomial $v(x)$.

Summary

We can summarize the algebraic decoding procedure for binary BCH codes as follows:

(1) Compute the syndrome $\mathbf{s} = (s_1, s_2, \ldots, s_{2t})$ from $r(x)$. For software implementation $\mathbf{s}$ is best computed using

$$s_i = r(\alpha^i), \quad i = 1, 2, \ldots, 2t, \text{ or } i \text{ odd}$$

(2) If $\mathbf{s}$ is non-zero, find the error location polynomial $\sigma(x)$. If $t \leqslant 5$, say, use Peterson's direct solution for roots $\sigma_1, \sigma_2, \ldots, \sigma_t$ by solving

$$s_{t+j} + \sigma_1 s_{t+j-1} + \ldots + \sigma_t s_j = 0, \quad j = 1, 2, \ldots, t$$

Otherwise, use Euclid's algorithm, or Berlekamp's algorithm.

(3) Determine the locators (error location numbers) X_k, $k = 1, 2, \ldots, t$ and perform the error correction. If $t = 1$ then $X_1 = s_1 = \alpha^j$, $j = 0, 1, \ldots, n - 1$ and symbol r_j must be complemented. Otherwise, use the Chien search (trial substitution) for the roots of $\sigma(x)$, starting with the substitution α^{n-1}, i.e. commence testing for an error at symbol r_{n-1}. Do this using the search equation

$$\sum_{i=1}^{t} \sigma_i \alpha^{ji} \stackrel{?}{=} 1, \quad j = 1, 2, \ldots, n$$

starting with $j = 1$.

3.5.3 Reed–Solomon (RS) codes (Reed and Solomon, 1960)

These non-binary BCH codes are *symbol* (m-bit binary byte) orientated, rather than bit orientated, and since data is often processed in byte form, this makes the codes extremely attractive. The codes are also attractive where burst errors will span a whole symbol or a number of adjacent symbols.

Definition 3.4 *A t-error correcting RS code with symbols drawn from GF(2^m) has*

$$\left.\begin{array}{l} n = 2^m - 1 \\ n - k = 2t \\ d = 2t + 1 \end{array}\right\} \tag{3.66}$$

The 2^m different elements of $GF(2^m)$ can be represented by binary m-bit words and so each symbol in the n-symbol codeword will be m-bits deep, as suggested in Fig. 3.5(a). Put another way, the number of bits in a symbol determines the size of the Galois field, and hence the number of symbols in the codeword. Also note that the minimum distance d between codewords is the number of positions in which there are different *symbols*, and it does not matter how many corresponding bits differ from each other within the corresponding symbols. Equation (3.66) also shows the high efficiency of these codes, i.e. to correct t errors each codeword only requires $2t$ check symbols. The DEC RS(15, 11) code for example, needs only four check symbols compared with eight symbols for the DEC binary BCH(15, 7) code (see (3.34)).

Several practical illustrations are useful at this point to establish the 'symbol' concept.

(1) A particular satellite paging system used an RS(30, 18) code to protect text (ASCII characters). This is a shortened RS(31, 19) code and so is capable of correcting up to 6 *symbol* errors over a block of 30 symbols. Each symbol represents an element from $GF(2^5)$ and so is 32 levels or 5-bits deep. It was therefore necessary to shorten each ASCII character to 5 bits. Encoding was achieved by dividing the message into groups of 18 5-bit characters and each group was then encoded into a RS(30, 18) codeword. The paging system also used the double error correcting RS(7, 3) code to protect 'unique word' transmissions (which were used for synchronization).

(2) The DEC RS(63, 59) code has been used to protect 68-Mbits/digital PAL television transmissions (Stott, 1984). The code used 6-bit symbols (corresponding to $GF(2^6)$) and each symbol was actually a 6-bit DPCM sample. An interleave depth of 6 generated a matrix of 378 symbols (see Fig. 3.13) and provided significant burst correction capability. The reader should show that this arrangement can correct a single burst of 67 bits, or less, or two bursts each of length 31 bits, or less. It is interesting to note that in magnetic recording interleave depths up to 16 have been used to combat long bursts, i.e. a single error correcting RS code using m-bit symbols would then be able up to correct burst lengths up to $15m + 1$ bits.

Encoding RS codes

RS codewords are found in a similar way to codewords for binary BCH codes, and a coding algorithm for a systematic RS code will follow the same form as (3.27). First we need the generator polynomial $g(x)$ of the RS code. For binary BCH codes, Theorem 3.6 implies that $g(x)$ will be a product of the *minimal polynomials* of the elements α^i $i = 1, 2, \ldots, 2t$. A similar argument applies for non-binary BCH codes and for RS codes the minimal polynomial of α^i is simply $(x - \alpha^i)$. Therefore

$$g(x) = (x - \alpha)(x - \alpha^2) \ldots (x - \alpha^{2t})$$

$$= (x + \alpha)(x + \alpha^2) \ldots (x + \alpha^{2t}) \tag{3.67}$$

As before, α is a primitive element of $GF(2^m)$ and, clearly, $\alpha, \alpha^2, \ldots, \alpha^{2t}$ are all roots of $g(x)$.

As an example, consider the DEC RS(7, 3) code defined over $GF(2^3)$. In this case

$$g(x) = (x + \alpha)(x + \alpha^2)(x + \alpha^3)(x + \alpha^4)$$

and using Table 3.5 this reduces to

$$g(x) = x^4 + \alpha^3 x^3 + x^2 + \alpha x + \alpha^3$$

Note that the coefficients of $g(x)$ are now drawn from $GF(2^3)$, rather than from $GF(2)$. We now assume the systematic form for a $(7, 3)$ codeword, i.e.

$$\mathbf{v} = (p_0, p_1, p_2, p_3, i_0, i_1, i_2)$$

$$v(x) = p(x) + x^{2t} i(x)$$

where $i(x) = i_0 + i_1 x + i_2 x^2$ and $p(x)$ is the check polynomial or remainder from the division $x^{2t} i(x)/g(x)$. For the RS(7, 3) code we therefore divide $x^6 i_2 + x^5 i_1 + x^4 i_0$ by $g(x)$, and after some manipulation (using Table 3.5) we obtain the following check equations:

$$p_0 = \alpha^3 i_0 + \alpha^6 i_1 + \alpha^5 i_2$$

$$p_1 = \alpha i_0 + \alpha^6 i_1 + \alpha^4 i_2$$

$$p_2 = i_0 + i_1 + i_2$$

$$p_3 = \alpha^3 i_0 + \alpha^2 i_1 + \alpha^4 i_2$$

Since message symbols i_0, i_1 and $i_2 \in GF(2^3)$ we could assume the arbitrary message

$$i(x) = \alpha^3 + \alpha x + \alpha^6 x^2$$

Use of the above check equations then gives the complete RS(7, 3) codeword

$$\mathbf{v} = (\alpha, \alpha^2, \alpha^2, \alpha^6, \alpha^3, \alpha, \alpha^6)$$

This is, in fact, the codeword we will use in Example 3.19. Similarly, the reader should show that message $i(x) = \alpha^4 + x + \alpha^5 x^2$ yields the codeword $\mathbf{v} = (\alpha^5, \alpha^4, 0, 1, \alpha^4, 1, \alpha^5)$.

Decoding RS codes

For binary BCH codes the essential task is to solve (3.56) for the error locators X_k, $k = 1, 2, \ldots, t$. For RS codes it is necessary to compute not only the error locators (positions) but also the error *values*, Y_k, $k = 1, 2, \ldots, t$ (since each symbol is m-bits deep). In this case, for t errors in the received codeword, (3.56) can be rewritten as

$$s_i = \sum_{k=1}^{t} Y_k X_k^i, \quad i = 1, 2, \ldots, 2t \tag{3.68}$$

The RS decoding algorithm therefore includes all the steps for decoding binary BCH codes, plus a further step to compute the Y_k from (3.68). Note that, once the X_k have been determined, e.g. via the Chien search, then (3.68) reduces to a set of *linear* simultaneous equations which are readily solved for the Y_k. Solving (3.68) for single and double errors in the received word, gives

$$t = 1: Y_1 = \frac{s_1^2}{s_2} \tag{3.69}$$

$$t = 2: Y_1 = \frac{s_1 X_2 + s_2}{X_1 X_2 + X_1^2} \tag{3.70}$$

$$Y_2 = \frac{s_1 X_1 + s_2}{X_1 X_2 + X_2^2} \tag{3.71}$$

Table 3.8. *Coefficients of the error location polynomial for single and double errors.*

t	Coefficients σ_i
1	$\sigma_1 = \dfrac{s_2}{s_1}$
2	$\sigma_1 = \dfrac{s_1 s_4 + s_3 s_2}{s_2^2 + s_1 s_3}$
	$\sigma_2 = \dfrac{s_2 s_4 + s_3^2}{s_2^2 + s_1 s_3}$

Error correction is then performed by adding Y_k to the symbol identified by X_k, the addition being carried out using the rules of $GF(2^m)$.

Finally, note that (3.60) does not hold for RS codes. This means that *all* syndrome symbols will be required and that the coefficients σ_i of the error location polynomial $\sigma(x)$ will, in general, be given by more complex expressions than those in Table 3.7. These are given in Table 3.8 for single and double errors.

A decoding algorithm for the RS(7, 3) code

The following discussion is for the systematic DEC RS(7, 3) code, although it is easily generalized. As discussed, the codeword format is

$$\mathbf{v} = (\underbrace{v_0, v_1, v_2, v_3}_{\text{check}}, \underbrace{v_4, v_5, v_6}_{\text{message}})$$

where each symbol v_i is drawn from $GF(2^3)$, i.e.

$$v_i \in \begin{Bmatrix} 0, & 1, & \alpha, & \alpha^2, & \alpha^3, & \alpha^4, & \alpha^5, & \alpha^6 \\ 0, & 1, & 2, & 3, & 4, & 5, & 6, & 7 \end{Bmatrix} \tag{3.72}$$

The decimal numbering of the field elements could be used in a practical realization using, say, a DSP chip. Each symbol is therefore 3 bits deep. Similarly, the received vector is $\mathbf{r} = (r_0, r_1, r_2, \ldots, r_6)$ corresponding to the polynomial

$$r(x) = r_0 + r_1 x + \ldots + r_6 x^6 \tag{3.73}$$

where the r_i are also elements from $GF(2^3)$.

Step 1 The four syndrome symbols are computed using (3.53) and noting that $\alpha^7 = 1$;

$$s_1 = r_0 + r_1\alpha + r_2\alpha^2 + r_3\alpha^3 + r_4\alpha^4 + r_5\alpha^5 + r_6\alpha^6$$

$$s_2 = r_0 + r_1\alpha^2 + r_2\alpha^4 + r_3\alpha^6 + r_4\alpha + r_5\alpha^3 + r_6\alpha^5$$

$$s_3 = r_0 + r_1\alpha^3 + r_2\alpha^6 + r_3\alpha^2 + r_4\alpha^5 + r_5\alpha + r_6\alpha^4$$

$$s_4 = r_0 + r_1\alpha^4 + r_2\alpha + r_3\alpha^5 + r_4\alpha^2 + r_5\alpha^6 + r_6\alpha^3$$

For a DSP realization these syndromes are easily computed in a recursive loop, and multiplication and addition over $GF(2^3)$ can be implemented using look-up tables.

Step 2 Determine the number of errors present. Table 3.8 suggests the following rules:

if $s_1 = s_2 = s_3 = s_4 = 0$ no errors have occurred.

if $\quad s_2^2 + s_1 s_3 = 0$ and $s_1 \neq 0 \quad$ assume one error

if $\quad s_2^2 + s_1 s_3 \neq 0 \quad$ assume two errors

Step 3 Compute the coefficients σ_i of the error location polynomial using Table 3.8. Assuming a single error, (3.57) gives

$$\sigma(x) = x + \sigma_1$$

$$\therefore X_1 = \sigma_1 = \frac{s_2}{s_1} \tag{3.74}$$

Step 4 For two errors, use the Chien search (either in hardware or as a software equivalent) to find the error locators X_1 and X_2. Here, $\sigma(x)$ is assumed to have degree two and if the Chien search fails to find two distinct roots, more than two errors have probably occurred.

Step 5 Determine the error value(s) using (3.69)–(3.71) and correct the appropriate symbol(s) by adding the value(s) to the symbol(s) over $GF(2^3)$.

Step 6 Recompute the syndromes for the corrected vector. If these are non-zero then indicate a decoding failure.

When implemented in assembly language on a TMS320C25 processor, the above algorithm took 2440 machine cycles (0.2 ms) to correct two errors and 1350 cycles to correct a single error.

Example 3.18

An RS(7, 3) coder generates the codeword $\mathbf{v} = (1226316)$ for a particular message. Here we have used the practical notation in (3.72). Suppose this vector is received with an error in symbol r_0, e.g. $\mathbf{r} = (5226316)$. In order to solve for the syndrome symbols (step 1) it is helpful first to extract the following identities from Example 3.15:

$\alpha^3 = \alpha + 1$

$\alpha^4 = \alpha^2 + \alpha$

$\alpha^5 = \alpha^2 + \alpha + 1$

$\alpha^6 = \alpha^2 + 1$

$\alpha^7 = 1$

$\therefore s_1 = \alpha^4 + \alpha^2 + \alpha^3 + \alpha^8 + \alpha^6 + \alpha^5 + \alpha^{11}$

$\quad = \alpha^2 + \alpha + 1$

$\quad = \alpha^5 \equiv 6$

Solving for all symbols we find

$$\mathbf{s} = (s_1 s_2 s_3 s_4) = (6666)$$

Step 2 indicates a single error, and so the error location is simply

$$X_1 = \sigma_1 = \frac{s_2}{s_1} = 1$$

Using (3.55) we deduce that $j = 0$ and so the error is in symbol r_0. Finally, since we are assuming a single error the error value is obtained from (3.69) as

$$Y_1 = \frac{s_1^2}{s_2} = \alpha^5$$

Symbol r_0 is therefore corrected by adding its value (α^4) to Y_1 giving (see Table 3.5)

$$\alpha^5 + \alpha^4 = 1$$

Example 3.19

Let the RS(7, 3) codeword $\mathbf{v} = (2337427)$ be received as $\mathbf{r} = (2737426)$. Step 1 gives $\mathbf{s} = (\alpha^3 1 \alpha^2 0) \equiv (4130)$ and step 2 gives

$$s_2^2 + s_1 s_3 = 1 + \alpha^3 \alpha^2 = \alpha^4$$

This indicates two errors, and so the equations for σ_1 and σ_2 in Table 3.8 (corresponding to $t = 2$) must be evaluated. The denominator in these equations is α^4 and requires the use of the multiplicative inverse defined in Table 3.4. Since $\alpha^4 \alpha^{-4} = 1$ and $\alpha^7 = 1$ it follows that $\alpha^{-4} = \alpha^3$. Using this and the identities in Example 3.18 then gives

$$\sigma_1 = \alpha^5, \quad \sigma_2 = 1$$

Now apply the Chien search. This starts by looking for an error in symbol $r_{n-1} = r_6 (\equiv 6)$ by evaluating (3.63) starting from $j = 1$:

$$j = 1: \quad \sigma_1 \alpha + \sigma_2 \alpha^2 = \alpha^5 \alpha + \alpha^2 = 1$$

This means $\alpha^{n-1} = \alpha^6$ is a root and that r_6 is in error. Proceeding with the search subsequently locates the second error, i.e.

$$j = 6: \quad \sigma_1 \alpha^6 + \sigma_2 \alpha^{12} = \alpha^4 + \alpha^5 = 1$$

This means that $\alpha^{n-6} = \alpha$ is a root and that r_1 is in error. The error locators are therefore

$$X_1 = \alpha^6, \quad X_2 = \alpha$$

Finally, using (3.70) and (3.71) for the error values gives

$$Y_1 = \frac{\alpha^3 \alpha + 1}{\alpha^6 \alpha + \alpha^{12}}$$

which reduces to $Y_1 = \alpha$. Similarly, $Y_2 = 1$. Finally

correcting position r_6: $\alpha + \alpha^5 = \alpha^6 \equiv 7$

correcting position r_1: $1 + \alpha^6 = \alpha^2 \equiv 3$

3.6 Performance of hard decoders

A FEC system can use either hard decision or soft decision decoding.

Consider the model of hard decoding given in Fig. 3.19. For many communications links, such as satellite and deep-space links, we can assume an additive white Gaussian noise (AWGN) channel, as indicated. Figure 3.19 also indicates that in hard decoding the samples in the demodulator are applied to a decision device or threshold circuit which makes a hard 1–0 decision. Clearly then, the 1–0 levels at the threshold circuit will be disturbed by Gaussian noise and there will be a finite probability of error, p, at each decision. This error probability can also be regarded as a *channel (bit) error rate (BER)*, as shown.

The discrete-time channel comprising modulator, AWGN channel and demodulator is conveniently modelled as a *binary symmetric channel (BSC)* with transition probability p. That is, for any transmitted bit (1 or 0) there is a probability $1 - p$ that the bit is correctly detected and a probability p that it is detected in error. Moreover, bit errors are assumed

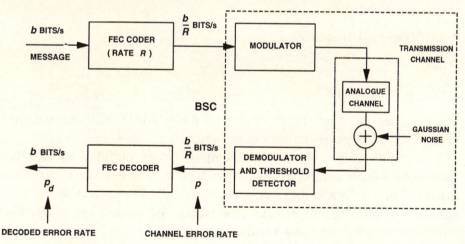

Fig. 3.19 Bit rates and error rates in a channel disturbed by AWGN (hard decision decoding).

to be independent of one another. We can determine p by noting that $p = P(1|0) = P(0|1)$, where $P(1|0)$ is the probability that a transmitted 0 is detected as a 1, i.e. $P(1|0)$ is the area under the noise pdf above the binary threshold. For Gaussian noise this can be evaluated for a typical modulation scheme, such as binary PSK, by using tables of the complementary error function (see (3.87)). Finally, Fig. 3.19 shows that the demodulator output is applied to the FEC decoder which then attempts to correct the errors.

In contrast to hard decoding, soft decoding eliminates the thresholding circuit and so avoids the information loss associated with hard decoding. Soft decoding therefore gives superior performance but at the expense of increased complexity. In fact, it is an integral part of the demodulation process and, for block codes, is considerably more difficult than hard decision to implement. For this reason, we will restrict the following discussion to hard decoding.

3.6.1 Decoded error rate

As discussed, an FEC system with hard decision decoding could be modelled as in Fig. 3.19. Unfortunately, practical error control systems can be overloaded and so there is always a finite bit error probability, p_d, or BER at the FEC decoder output. System design therefore usually specifies a maximum decoded BER. For example, a particular satellite communications system required an output (decoded) 40 character message to be completely error-free for at least 99% of the time. Using 6-bit characters we have

$$((1 - p_d)^6)^{40} \geqslant 0.99$$

giving

$$p_d \leqslant 4.2 \times 10^{-5}$$

The FEC decoder must therefore convert the relatively high channel error rate, p, to this target specification, or better.

A word of caution before we compute some decoded error rates. Consider a t-error correcting binary block code. The decoder is certain to correct t or less errors but, in general, it will also correct some error patterns of *more* than t errors! This makes the calculation of decoded error rate complex, and so we will *assume* that only t or less errors can be corrected. Our calculations for decoded error rate will then simply be an upper bound for most t-error correcting block codes, although they will be exactly correct for the so-called 'perfect codes', such as the Hamming codes and the Golay (23, 12) code.

Decoded block error rate

Suppose an n-bit codeword of an (n, k) block code is received at the FEC decoder input with a single error in a particular position. The probability of this particular word (n-symbol block) occurring is simply

$$P(\text{specific word}) = p(1 - p)^{n-1} \tag{3.75}$$

where p is the BER of the assumed BSC feeding the decoder. Since there are n possible error locations the probability of a codeword with a single error *somewhere* is the sum of the probabilities of n independent events

$$P(1 \text{ error}) = np(1 - p)^{n-1}$$

$$= \binom{n}{1} p(1 - p)^{n-1} \tag{3.76}$$

where

$$\binom{n}{r} = \frac{n!}{r!(n - r)!}$$

Similarly, the probability of a specific 2-error codeword is

$$P(\text{specific word}) = p^2(1 - p)^{n-2} \tag{3.77}$$

and the number of different codewords having exactly two errors is given by

$$\binom{n}{2} = \frac{n(n - 1)}{2}$$

$$\therefore P(2 \text{ errors}) = \binom{n}{2} p^2(1 - p)^{n-2} \tag{3.78}$$

Finally, generalizing (3.78) gives

$$P(m \text{ errors}) = \binom{n}{m} p^m(1 - p)^{n-m} \tag{3.79}$$

Assuming that a maximum of t errors can be corrected for each n-symbol block, then a block decoding error is sure to occur if $m > t$. Therefore,

$$P(\text{block error}) = \sum_{m=t+1}^{n} P(m \text{ errors}) = 1 - \sum_{m=0}^{t} P(m \text{ errors}) \tag{3.80}$$

Equivalently,

$$P(\text{correct block}) = \sum_{m=0}^{t} \binom{n}{m} p^m (1-p)^{n-m} \tag{3.81}$$

As previously noted, most *t*-error correcting codes will perform better than this, in which case (3.80) is simply an upper bound. It is interesting to observe that, with correct interpretation, (3.80) and (3.81) also apply to non-binary (RS) coding. In this case we assume the channel makes independent *symbol* errors with probability *p*.

Example 3.20

Consider the decoded block error rate for Hamming codes. From (3.79) and (3.80) we have

$$P(\text{block error}) = 1 - (1-p)^n - np(1-p)^{n-1} \tag{3.82}$$

We could expand this using the binomial theorem to give an approximate expression:

$$P(\text{block error}) \approx \frac{n}{2}(n-1)p^2, \quad p \ll 1 \tag{3.83}$$

Therefore, if $p = 10^{-3}$, the probability of a H(7, 4) codeword being in error at the decoder output is approximately 2×10^{-5}.

Example 3.21

A similar approximate expression can be found for DEC block codes. We have

$$P(\text{correct decoding}) \geqslant \sum_{m=0}^{2} P(m \text{ errors})$$

$$\geqslant (1-p)^n + np(1-p)^{n-1}$$

$$+ \frac{n}{2}(n-1)p^2(1-p)^{n-2}$$

which reduces to

$$P(\text{correct decoding}) \geqslant 1 - \frac{n}{6}(n-1)(n-2)p^3, \quad p \ll 1$$

$$\therefore P(\text{decoding error}) = 1 - P(\text{correct decoding})$$

$$\approx \frac{1}{6}n^3 p^3, \quad p \ll 1 \tag{3.84}$$

Decoded BER

Let the probability of a block error in the decoder output be x^{-1}. This means that, on average, one in x n-bit words is in error, or that there is a decoding failure every nx bits. The decoded BER, p_d, is simply the proportion of erroneous bits in the decoder output. Suppose that, on average, γ bits are in error at the decoder output for each decoding failure. Then

$$p_d = \frac{\gamma}{nx} = \frac{\gamma}{n}P(\text{block error})$$

For SEC codes (3.83) gives

$$p_d \approx \frac{\gamma}{2}(n-1)p^2$$

where, in general, $2 < \gamma < 3$. Therefore, when $p \ll 1$,

$$p_d \approx (n-1)p^2, \quad t = 1 \tag{3.85}$$

For DEC codes, γ is somewhat larger (typically 4).

3.6.2 Coding gain

In the last section we examined the improvement in BER provided by the FEC decoder. In practice the improvement achieved by FEC is usually measured in terms of *coding gain*, as defined in Fig. 3.20. Here, E_b is the mean received energy per *information* (or message) bit and N_o is the

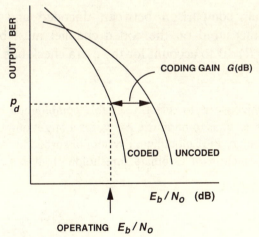

Fig. 3.20 Defining coding gain.

single-sided received noise power spectral density (it is usually assumed that the channel is disturbed by AWGN). The ratio E_b/N_o is therefore a signal-to-noise measure and it acts as a figure of merit for different combinations of coding and modulation schemes.

Fig. 3.20 shows that coding gain, G, is measured relative to the channel error rate curve for a particular, uncoded modulation scheme, e.g. PSK. Clearly, the gain varies and so must be quoted at a particular output error rate, p_d (typically 10^{-5}). Also note that G can become negative for high error rates (typically around 10^{-2} or 10^{-1}), in which case there is no advantage in using FEC.

It is important to realize that the use of ECC results in an increased channel rate, increased channel bandwidth, and increased channel error rate, p, compared with an uncoded system! Suppose the information rate is fixed at b bits/s (Fig. 3.19). The n bits of a rate R block code must then be transmitted in the time taken to transmit k message bits, which means that the *coded* bit rate will be increased to b/R bits/s. Inevitably this increases the required channel bandwidth (by a factor $1/R$). Also, since the $n - k$ check bits require some energy to transmit them, the mean energy per *coded* bit will be only $E_c = RE_b$. In turn, this means that for a fixed noise density, N_o, more bits will be received in error at the detector output compared with an uncoded system. Consequently, the code must have enough error control capability to more than compensate for the increased channel errors and so provide useful gain. Finally, since $E_b = E_c/R$, we have

$$\frac{E_b}{N_o} = \frac{E_c}{N_o} + 10\log\frac{1}{R}\,\text{dB} \qquad (3.86)$$

In other words, for a fair comparison between uncoded and coded systems, the ratio E_c/N_o measured on the coded channel must be corrected by a factor $10\log(1/R)$ dB to account for the extra check bits.

Example 3.22

A common design objective is to select a suitable combination of FEC and modulation schemes so as to minimize E_b/N_o for a target output error rate, p_d. Consider a binary PSK modulation system disturbed by AWGN. Coherent detection (which involves binary thresholding) gives a channel BER of

$$p = \text{erfc}\,\sqrt{\left(2\frac{E_b}{N_o}\right)} \qquad (3.87)$$

where, here, the complementary error function is defined as

$$\mathrm{erfc}(x) = \frac{1}{\sqrt{(2\pi)}} \int_x^\infty \mathrm{e}^{-y^2/2} \, \mathrm{d}y$$

This provides a realistic reference curve for the uncoded system, as in Fig. 3.20. Assuming a target output BER $= 10^{-5}$, Equation (3.87) shows that the uncoded system requires $E_b/N_o = 9.6$ dB.

Using the results of Section 3.6.1, and accounting for the code rate R, it is readily shown that the DEC BCH$(15, 7)$ code will reduce the operating E_b/N_o to 8.7 dB – a gain of 0.9 dB, whilst the higher rate BCH$(127, 113)$ code will give some 2 dB gain at $p_d = 10^{-5}$. On the other hand, a high-performance system using convolutional coding and soft decision Viterbi decoding (Section 3.7.6) can yield over 5 dB gain. A gain of 5 dB means that the transmitter power can be reduced by 5 dB for the same information rate as the PSK system. Alternatively, a 5 dB gain allows the information rate to be increased by a factor $10^{0.5} \approx 3.2$ compared with the uncoded information rate.

Finally, it is interesting to note that the use of FEC simply illustrates a fundamental principle in communication theory – the exchange of bandwidth for SNR (Section 1.1). Referring to Fig. 3.20, for a fixed information rate a particular output error rate p_d can be achieved by either

(*a*) increasing the modulated carrier power, i.e. E_b/N_o, or

(*b*) applying FEC, which requires increased bandwidth. Generally speaking, FEC is the more economical approach.

3.7 Convolutional codes

As shown in Fig. 3.1, convolutional codes are the main alternative to block codes when FEC is required. They are particularly attractive when used with soft decision decoding and probabilistic decoding algorithms, such as Viterbi decoding.

A convolutional coder can be modelled as shown in Figs. 3.5 and 3.21. The message is split into frames of k_0 symbols and is stored in a register of

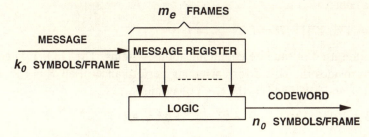

Fig. 3.21 Structure of a convolutional coder.

length m_e frames, or $m_e k_0$ symbols. The contents of the register are examined by a logic circuit, which essentially involves modulo 2 addition, and the coder generates a frame of n_0 symbols for every input frame of k_0 symbols. The code rate is therefore $R = k_0/n_0$ as discussed in section 3.1. The constraint length K of a binary convolutional code can be defined as the total span of symbols (bits) which influence the coder output at any one time, and in terms of Fig. 3.21 or Fig. 3.5

$$K = m_e k_0 \tag{3.88}$$

Therefore, if we define the coder as in Fig. 3.21, K is simply the length of the shift register.

3.7.1 Polynomial representation

A block code can be defined by a single generator polynomial, $g(x)$, whilst a convolutional code can be defined by $k_0 n_0$ polynomials. Very often $k_0 = 1$ and so this will be assumed from now on. In this case, a convolutional code can be compactly defined by the generator matrix

$$\mathbf{G}(x) = [g_1(x), \quad g_2(x), \quad \cdots \quad g_{n_0}(x)] \tag{3.89}$$

where

$$g_j(x) = g_{j0} + g_{j1}x + g_{j2}x^2 + \ldots + g_{jK-1}x^{K-1} \tag{3.90}$$

As for block codes, the problem is to find generator polynomials which have good error control properties, and most convolutional codes are found by computer search. Optimal codes (codes with maximum *free distance*, d_∞) are described in the literature (Larsen, 1973; Paaske, 1974) and an abbreviated list is given in Table 3.9. Here the polynomials are in octal format and are interpreted as in the following example.

Example 3.23

The code

$$\mathbf{G} = [133, 171] \quad R = 1/2 \quad K = 7$$

is often used in satellite communications systems. The [133, 171] generator provides the best error correcting performance of all $R = 1/2$, $K = 7$ codes, and is defined by the two polynomials

$$133 \equiv 1011011 \equiv 1 + x^2 + x^3 + x^5 + x^6 = g_1(x)$$

$$171 \equiv 1111001 \equiv 1 + x + x^2 + x^3 + x^6 = g_2(x)$$

Table 3.9. *Optimal convolutional codes*.

	K	d_∞	$g_1(x)$	$g_2(x)$	$g_3(x)$	$g_4(x)$
$R = 1/2$	3	5	5	7		
	4	6	15	17		
	5	7	23	35		
	6	8	53	75		
	7	10	133	171		
	8	10	247	371		
	9	12	561	753		
	10	12	1167	1545		
$R = 1/3$	3	8	5	7	7	
	4	10	13	15	17	
	5	12	25	33	37	
	6	13	47	53	75	
	7	15	133	145	175	
	8	16	225	331	367	
	9	18	557	663	711	
	10	20	1117	1365	1633	
$R = 1/4$	3	10	5	7	7	7
	4	13	13	15	15	17
	5	16	25	33	35	37
	6	18	53	67	71	75
	7	20	135	135	147	163
	8	22	235	275	313	357
	9	24	463	535	733	745
	10	27	1117	1365	1633	1653

The corresponding coder is shown in Fig. 3.22. This example highlights several inportant points. First, neither $g_1(x)i(x)$ nor $g_2(x)i(x)$ are identical to $i(x)$ and so the message is not identifiable in codeword $v(x)$. In other words, the code is *non-systematic*. Non-systematic codes are usually preferred when Viterbi decoding is used since, in general, they offer the maximum possible free distance for a given rate and constraint length (Viterbi and Omura, 1979). Second, polynomial representation enables the output sequence to be easily computed by simply multiplying polynomials.

Example 3.24
Consider the simple code

$$G = [5, 7] \quad R = 1/2 \quad K = 3$$

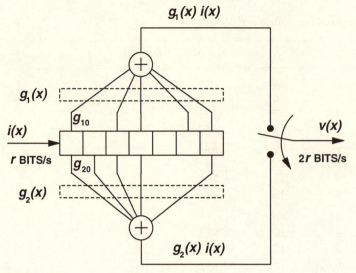

$g_1(x) i(x)$

$g_1(x)$

g_{10}

$i(x)$

r BITS/s

g_{20}

$g_2(x)$

$g_2(x) i(x)$

$v(x)$

$2r$ BITS/s

Fig. 3.22 [133, 171], $R = 1/2$, $K = 7$ coder.

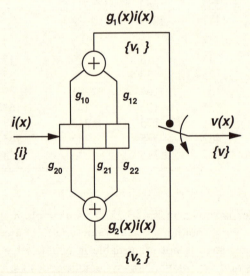

$g_1(x)i(x)$

$\{v_1\}$

g_{10} g_{12}

$i(x)$

$\{i\}$

g_{20} g_{21} g_{22}

$g_2(x)i(x)$

$\{v_2\}$

$v(x)$

$\{v\}$

Fig. 3.23 [5, 7], $R = 1/2$, $K = 3$ coder.

The two polynomials are

$$101 \equiv 1 + x^2 = g_1(x)$$

$$111 \equiv 1 + x + x^2 = g_2(x)$$

and these give the circuit in Fig. 3.23. The polynomials behave as *sequence transforms* and so we can use them in the same way as we use Fourier or z transforms. For example, the sequence $\{v_1\}$ can be written

$$\{v_1\} \Leftrightarrow g_1(x)i(x) \tag{3.91}$$

where multiplication is carried out over $GF(2)$. Sequence $\{v_1\}$ can therefore be regarded as the result of a filtering or *convolution* operation upon input $i(x)$ – hence the term convolutional coding. To illustrate, let the input sequence be

$$\{i\} = 011010 \dots$$

Therefore

$$g_1(x)i(x) = (1 + x^2)(x + x^2 + x^4)$$

$$= x + x^2 + x^3 + x^6$$

$$\{v_1\} = 0111001 \dots$$

Similarly

$$g_2(x)i(x) = (1 + x + x^2)(x + x^2 + x^4)$$

$$\{v_2\} = 0100011 \dots$$

giving the semi-infinite code sequence

$$\{v\} = 00111010000111 \dots$$

3.7.2 Trellis representation

Convolutional codes fall under the general heading of *tree codes* since their coding operation can be visualized by tracing through a *coding tree* (Viterbi, 1971). The tree comprises nodes and branches and it directly gives the coded sequence for an arbitrary input sequence. However, it also contains a lot of redundant information and so, for codes with small constraint length, a more compact visualization is the *trellis diagram*. Consider the coder in Fig. 3.23. Its *state* is defined as the contents of the first $K - 1$ shift register stages, or, equivalently, by the most recent $K - 1$ input symbols. Fig. 3.23 therefore has $2^{K-1} = 4$ states, and these can be used to define a repetitive trellis structure for the coder as shown in Fig. 3.24. Assume that the register is initially cleared (coder state $= 00$, corresponding to the two left-most stages in Fig. 3.23) and that the coder input is the sequence 10110001 . . . . The first 1 input will change the state to 10 and reference to Fig. 3.23 shows that the coder output will be 11. The following 0 will change the coder state to 01 and the coder output will also be 01. The third input bit, a 1, will change the state back to 10 and, since the contents of the register are now 101, Fig. 3.23 shows that the

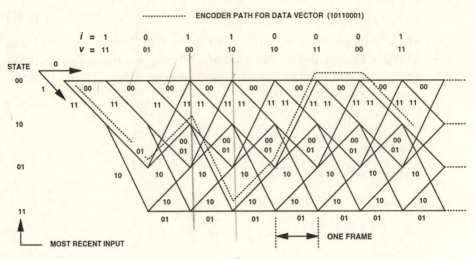

Fig. 3.24 Trellis for the [5, 7], $R = 1/2$, $K = 3$ code.

coder output will be 00. Continuing in this way gives the full coding path (dotted) and codeword **v** for the above sequence. Note that, in general, each branch will be labelled with n_0 output symbols for rate $R = 1/n_0$, and that a *message tail* of $K - 1$ zeros could be used to return the coder (decoder) to the all zeros state. For example, adding two zeros to the message in Fig. 3.24 will always return the coder to this state. Similarly, the ETSI-GSM specification for cellular radio uses a $K = 5$ convolutional code and adds a tail of four zeros to force the coder to the all-zeros state. In this way, the decoder has prior knowledge of the message tail, which helps to reduce the decoded error rate for the last few bits of the message.

The trellis concept is fundamental to Viterbi decoding (Section 3.7.6), although the diagram itself is clearly limited to very low values of K due to the exponential growth in the number of states (2^{K-1} states). More significantly, this exponential growth also means that Viterbi decoding is limited to low values of K; usually $K < 10$ and typically $K = 5$ or 7.

Example 3.25

The trellis for the [5, 7, 7] $R = 1/3$ $K = 3$ code is easily constructed. By definition

$$G(x) = [1 + x^2, \quad 1 + x + x^2, \quad 1 + x + x^2]$$

and so the coder will be as shown in Fig. 3.25(a). The reader should now use this circuit to deduce the trellis in Fig. 3.25(b).

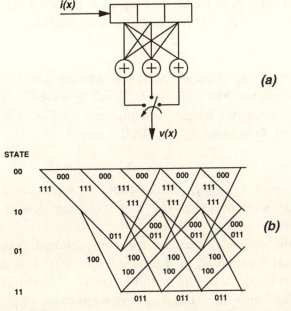

Fig. 3.25 (*a*) Coder and (*b*) trellis for the [5, 7, 7], $R = 1/3$, $K = 3$ code.

3.7.3 Distance measures

The minimum distance of a convolutional code is a fundamental para-meter since it determines its error control capability. However, compared with the distance measure for block codes, the measure for convolutional codes is somewhat more complex since it depends upon the type of decoding used. More specifically, it depends upon the number of frames m used by the decoder, see Fig. 3.5(*b*).

> **Definition 3.5** *The mth-order minimum distance d_m of a convolu-tional code is the minimum Hamming distance between all possible pairs of coded sequences or codewords m frames (branches) long which differ in their initial frame. Mathematically*
>
> $$d_m = \min d_H(\mathbf{u}_m, \mathbf{v}_m), \quad \mathbf{u}_1 \neq \mathbf{v}_1 \tag{3.92}$$
>
> *where the first m frames of two codewords are denoted $\mathbf{u}_m$ and $\mathbf{v}_m$.*

Since either $\mathbf{u}_m$ or $\mathbf{v}_m$ could be the all-zeros word, then, without loss of generality, (3.92) reduces to finding the minimum Hamming distance from this specific word to all other words. This means we can look for the minimum weight codeword m branches long that is non-zero in the first branch, as implied by Theorem 3.1. Once d_m is determined, we can use the same error control concept as for block codes, i.e. the code can correct any pattern of t errors or less in any m adjacent frames provided

$$d_m \geqslant 2t + 1 \tag{3.93}$$

As previously mentioned, d_m depends upon the type of decoding used. Simple decoding techniques like *threshold decoding* have a decoding memory of only one constraint length, and in terms of Fig. 3.5(b) this means that $m = K = m_e$ (assuming $k_0 = 1$). In this case

$$d_m = d_K = d_{\min} \tag{3.94}$$

where $d_{\min}$ is the *minimum distance* of the code. In words, $d_{\min}$ is the minimum weight code sequence over the first constraint length whose initial frame is non-zero.

The most important distance measure for convolutional codes corresponds to $m \Rightarrow \infty$ and is called d_∞ or d_{free} (see Table 3.9). This distance is appropriate to probabilistic decoding methods, such as Viterbi decoding and *sequential decoding*, since here the decoding memory is, in principle, unlimited. Since d_m is monotonic in m then clearly $d_\infty \geqslant d_{\min}$.

Example 3.26

The trellis diagram can be used to determine both $d_{\min}$ and d_∞. Consider the $[5, 7, 7]$, $R = 1/3$, $K = 3$ code represented by Fig. 3.25(b). For $d_{\min}(\equiv d_K)$ we determine the Hamming weights for all the three-branch paths starting with a non-zero frame, and select the path(s) with minimum weight. It is easily seen that $d_{\min} = 5$, corresponding to the codewords $(111, 011, 000)$ and $(111, 100, 100)$. To find d_∞ we examine all paths leaving the zero state and returning (or 'remerging') to that state, and the path with minimum Hamming weight gives d_∞. For most codes the path corresponding to d_∞ is only a few constraint lengths long, and often is just one constraint length long. For example, in Fig. 3.25(b), $d_\infty = 8$, corresponding to codewords $(111, 011, 111)$ and $(111, 100, 100, 111)$. This confirms the corresponding entry in Table 3.9.

Similarly, d_∞ for the $[5, 7]$, $R = 1/2$, $K = 3$ code represented by Fig. 3.24 is easily seen to be 5, corresponding to the codeword $(11, 01, 11)$.

3.7.4 Threshold decoding

Decoding techniques for convolutional codes tend to fall under the headings of sequential decoding, Viterbi decoding and threshold (majority) decoding. The last is relatively simple to implement and is conceptually closest to block coding in that it computes syndrome digits (in fact, threshold decoding is also applicable to some block codes). The price paid for simplicity is that only moderate coding gain can be achieved relative to sequential and Viterbi decoding techniques.

Threshold decoders usually use a systematic code. This can be generated by setting $g_1(x) = 1$ in (3.89), so for a rate $1/2$ code the check bits will be generated by a single polynomial $g(x)$, as shown in Fig. 3.26. Here the message and check sequences are shown as parallel streams for ease of discussion, but in practice they would be multiplexed for serial transmission and demultiplexed at the decoder. The figure also shows the corresponding syndrome decoder and models channel errors by error polynomials $e^i(x)$ and $e^p(x)$. The received sequences are therefore

$$i'(x) = i(x) + e^i(x) \tag{3.95}$$

$$p'(x) = p(x) + e^p(x) \tag{3.96}$$

where the general error polynomial is

$$e(x) = e_0 + e_1 x + e_2 x^2 + \ldots + e_n x^n + \ldots \tag{3.97}$$

The decoder generates a check sequence in the same way as the coder, and the generated and received sequences are compared by modulo 2 addition. The resulting syndrome polynomial is

$$s(x) = g(x)i'(x) + p'(x)$$

$$= g(x)[i(x) + e^i(x)] + g(x)i(x) + e^p(x)$$

$$= g(x)e^i(x) + e^p(x) \tag{3.98}$$

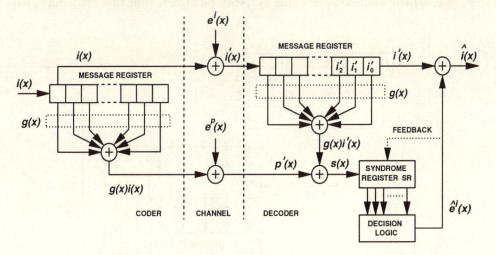

Fig. 3.26 General parallel coder and syndrome decoder for a systematic code ($R = 1/2$, $k_0 = 1$).

where $s(x)$ corresponds to a syndrome sequence $\{s\}$. Clearly, for no transmission error, symbols e_j are all zero and so $e^i(x) = e^p(x) = s(x) = 0$ and zero will be stored in the SR. When the syndrome is finite, the decision logic makes a decision from the contents of the SR and attempts to correct the error in the last symbol held in the message register. The resulting decoder output is

$$\hat{i}(x) = i'(x) + \hat{e}^i(x)$$

$$= i(x) + e^i(x) + \hat{e}^i(x) \tag{3.99}$$

where $\hat{e}^i(x)$ is an estimate of the actual transform $e^i(x)$. If the decision is correct, then the error is correctly removed. In Fig. 3.26 the decision attempts to correct received information symbol i_0' by subtracting the estimate $\hat{e}_0^i$ of the error symbol e_0^i. If this estimate is correct, the error in i_0' is removed and symbol i_0 is output from the decoder. Clearly, errors in the check sequence must also be taken into account (as will be shown) but they do not need to be corrected.

Use of feedback
Feedback to the SR is not always essential but it often gives improved decoding. Consider the systematic code defined by

$$R = 1/2 \quad K = 7 \quad g(x) = 1 + x^2 + x^5 + x^6$$

The corresponding syndrome decoder is shown in Fig. 3.27. When an error occurs in symbol i_0' the contents of the SR are decoded to give $\hat{e}_0^i = 1$, which then complements i_0'. Note however, that this error has also

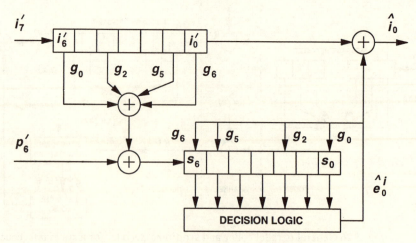

Fig. 3.27 Syndrome decoder for the code, $R = 1/2$, $K = 7$, $g(x) = 1 + x^2 + x^5 + x^6$.

Table 3.10. *Syndrome equations for feedback decoding of code*

$R = 1/2$ $K = 7$ $g(x) = 1 + x^2 + x^5 + x^6$

$$s_0 = e_0^i + e_0^p$$
$$s_1 = e_1^i + e_1^p$$
$$s_2 = e_0^i + e_2^i + e_2^p$$
$$s_3 = e_1^i + e_3^i + e_3^p$$
$$s_4 = e_2^i + e_4^i + e_4^p$$
$$s_5 = e_0^i + e_3^i + e_5^i + e_5^p$$
$$s_6 = e_0^i + e_1^i + e_4^i + e_6^i + e_6^p$$

affected parts of the SR as it passed down the message register and that information about the error will be passed on down the SR when decoding symbols i_1', i_2' etc. Ideally, i_1' should be decoded independently of any errors in preceding message symbols and this can be achieved by applying feedback to the appropriate stages of the SR, as shown in Fig. 3.27. When correction is performed, feedback will then complement the appropriate stages prior to clocking the SR.

Analytically, using (3.98)

$$s(x) = g(x)e^i(x) + e^p(x)$$
$$= (1 + x^2 + x^5 + x^6)e^i(x) + e^p(x) \tag{3.100}$$

Also, since

$$s(x) = s_0 + s_1 x + s_2 x^2 + \ldots + s_n x^n + \ldots$$

and

$$xe(x) = e_0 x + e_1 x^2 + e_2 x^3 + \ldots + e_{n-1} x^n + \ldots$$

then the nth syndrome symbol, i.e. the symbol associated with x^n, is

$$s_n = e_n^i + e_{n-2}^i + e_{n-5}^i + e_{n-6}^i + e_n^p \tag{3.101}$$

The significant point here is that, when i_0' is being corrected, the use of feedback should make s_n, $n = 0, 1, 2, \ldots$ independent of any symbols preceding i_0'. This being the case, (3.101) then leads to the set of syndrome equations in Table 3.10.

We can see from Table 3.10 that the effective memory span of this decoder is just K frames or $n_0 K$ coded symbols (corresponding to $i_0 - i_6$, $p_0 - p_6$) and so d_{min} rather than d_∞ is appropriate in this case (Section 3.7.3).

Majority logic
The decision logic is usually implemented by using a majority vote technique on the SR outputs and this requires a set of parity check equations which are *orthogonal* on some symbol. For example, the following equations are orthogonal on symbol e_0^i since only e_0^i is checked by each equation and no other symbols are checked by more than one equation:

$$A_1 = e_0^i + e_1^i + e_1^p$$
$$A_2 = e_0^i + e_2^i + e_2^p$$
$$A_3 = e_0^i + e_3^i + e_4^i + e_3^p$$
$$A_4 = e_0^i + e_5^i + e_4^p$$

Suppose we now form the arithmetic sum

$$A = \sum_{j=1}^{4} A_j \qquad\qquad (3.102)$$

Any single error will give only $A = 1$, unless $e_0^i = 1$ when $A = 4$, which means that it is possible to detect an error in position zero ($e_0^i = 1$) by taking a majority vote on A. As symbols are clocked through the decoder, the erroneous symbol will inevitably fall into position zero and so it can be corrected. The A_j are called *estimates* of the error symbol e_0^i and for J estimates the majority decoding rule is

$$\hat{e}_0^i = 1 \quad \text{iff} \quad \left(\sum_{j=1}^{J} A_j > \frac{J}{2} \right) \quad \text{else} \quad \hat{e}_0^i = 0 \qquad\qquad (3.103)$$

This rule gives a correct estimate $\hat{e}_0^i$ and therefore correct decoding provided that there are $\lfloor J/2 \rfloor$ or less errors in the symbols checked by the orthogonal set (as we have just seen, for the rate $R = 1/n_0$ code in Table 3.10 this corresponds to a span of $n_0 K$ coded symbols). According to (3.93) and (3.94) this correction capability can also be related to the d_{min} of the code. In fact, if $d_{min} - 1$ estimates orthogonal on e_0^i can be formed

Table 3.11. *SOCs for R = 1/2.*

J	K	First row of triangle	$g(x)$
2	2	1	$1 + x$
4	7	2, 3, 1	$1 + x^2 + x^5 + x^6$
6	18	2, 5, 6, 3, 1	$1 + x^2 + x^7 + x^{13} + x^{16} + x^{17}$
8	36	7, 3, 6, 2, 12, 1, 4	$1 + x^7 + x^{10} + x^{16} + x^{18} + x^{30} + x^{31} + x^{35}$

then i_0 will be correctly decoded for any pattern of $\lfloor (d_{min} - 1)/2) \rfloor$ or fewer errors (Massey, 1963).

SOCs

These codes directly generate a set of orthogonal check equations and so are very applicable to majority decoding. Code generators for SOCs are easily constructed from difference triangles. If the first row of this triangle is of the form $d_{11}, d_{12}, d_{13}, \ldots$ then the generator polynomial is

$$g(x) = 1 + x^{d_{11}} + x^{d_{11}+d_{12}} + x^{d_{11}+d_{12}+d_{13}} + \ldots \qquad (3.104)$$

Difference triangles can be found in the literature (Robinson and Bernstein, 1967; Klieber, 1970) and a few of the simpler SOCs are given in Table 3.11.

Example 3.27

Consider the SOC defined by

$$R = 1/2 \quad K = 7 \quad J = 4 \quad g(x) = 1 + x^2 + x^5 + x^6$$

The syndrome equations for this particular code have already been given in Table 3.10, and from this table we can select the orthogonal set

$$A_1 = s_0 = e_0^i + e_0^p$$

$$A_2 = s_2 = e_0^i + e_2^i + e_2^p$$

$$A_3 = s_5 = e_0^i + e_3^i + e_5^i + e_5^p$$

$$A_4 = s_6 = e_0^i + e_1^i + e_4^i + e_6^i + e_6^p$$

These estimates lead directly to the decoder in Fig. 3.28. Note that up to two errors can be tolerated over the symbols checked by the orthogonal set – a span of $n_0 K = 14$ coded symbols. For example, if $i_0' = i_0$, and i_3' and i_6' are received incorrectly then $\sum A_j = 2$ and i_0' will still be decoded correctly if the check symbols are correctly received.

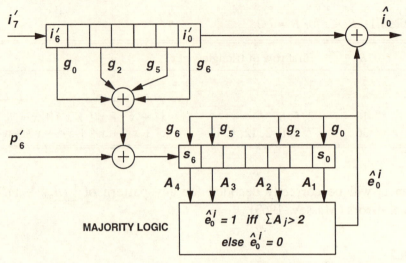

Fig. 3.28 Majority logic feedback decoder for SOC code $R = 1/2$, $K = 7$, $J = 4$.

Trial-and-error codes

These codes are more efficient than SOCs in the sense that they achieve a given value of J, and therefore error correction capability, with a shorter constraint length (compare Tables 3.11 and 3.12). Their construction is poorly understood and they are derived by a trial-and-error search (Massey, 1963). Feedback decoding is invariably used. They are also sometimes called *orthogonizable* codes since they can yield an orthogonal set of syndrome equations by taking linear combinations of syndrome symbols.

Example 3.28

Consider the trial-and-error code in Table 3.12 defined by

$$R = 1/2 \quad K = 6 \quad J = 4 \quad g(x) = 1 + x^3 + x^4 + x^5$$

From this table we see that

$$A_1 = s_0$$

$$A_2 = s_3$$

$$A_3 = s_4$$

$$A_4 = s_1 + s_5$$

To check that this is an orthogonal set, recall that

$$s(x) = g(x)e^i(x) + e^p(x)$$

$$= (1 + x^3 + x^4 + x^5)e^i(x) + e^p(x)$$

Table 3.12. *Trial-and-error codes for R = 1/2.*

J	K	Orthogonal estimates	$g(x)$
2	2	s_0, s_1	$1 + x$
4	6	$s_0, s_3, s_4, s_1 + s_5$	$1 + x^3 + x^4 + x^5$
6	12	$s_0, s_6, s_7, s_9, s_1 + s_3 + s_{10}, s_4 + s_8 + s_{11}$	$1 + x^6 + x^7 + x^9 + x^{10} + x^{11}$
8	22	$s_0, s_{11}, s_{13}, s_{16}, s_{17},$	$1 + x^{11} + x^{13} + x^{16} + x^{17} + x^{19} + x^{20} + x^{21}$
		$s_2 + s_3 + s_6 + s_{19}, s_4 + s_{14} + s_{20},$	
		$s_1 + s_5 + s_8 + s_{15} + s_{21}$	

giving

$$s_n = e_n^i + e_{n-3}^i + e_{n-4}^i + e_{n-5}^i + e_n^p$$

Therefore, for feedback decoding,

$$A_1 = s_0 = e_0^i + e_0^p$$

$$A_2 = s_3 = e_0^i + e_3^i + e_3^p$$

$$A_3 = s_4 = e_0^i + e_1^i + e_4^i + e_4^p$$

$$A_4 = s_1 + s_5 = e_0^i + e_2^i + e_5^i + e_1^p + e_5^p$$

and so the set is orthogonal on e_0^i. The corresponding decoder is shown in Fig. 3.29.

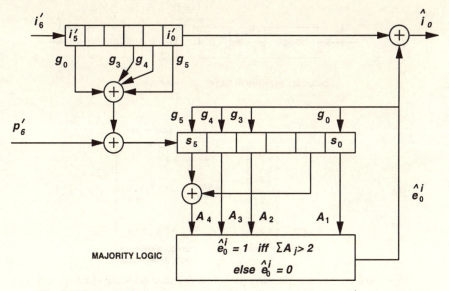

Fig. 3.29 Majority logic decoder for trial-and-error code $R = 1/2$, $K = 6$, $J = 4$.

3.7.5 Maximum likelihood decoding

An error control decoder can use either a hard decision input (single bit, 1 or 0 input) or a soft decision input (typically a 3-bit PCM input). As discussed in Section 3.6, a hard decision input can be regarded as the output of a BSC having a channel BER (transition probability) equal to p. The BSC compactly models the modulator, noisy analogue channel and hard decision demodulator of Fig. 3.19, as shown in Fig. 3.30(a). We will use this hard decision model to introduce maximum likelihood decoding.

Assume a message k generates a sequence or codeword $\mathbf{v}_k$ and that this is received as codeword $\mathbf{r}$. If all message sequences are equally likely then the (sequence) error probability is minimized if the decoder decides on a transmitted sequence $\mathbf{v}_j$ such that

$$P(\mathbf{r}|\mathbf{v}_j) \geqslant P(\mathbf{r}|\mathbf{v}_k), \quad \forall k \neq j \tag{3.105}$$

Here the conditional probability $P(\mathbf{r}|\mathbf{v}_j)$ is called a *likelihood function* and represents the 'likelihood' that $\mathbf{r}$ comes from $\mathbf{v}_j$. Clearly, decoding error occurs if the decoded message $j \neq k$. This optimum decoder, based on likelihood functions, is called a *maximum likelihood decoder (MLD)*. If the input (or code) sequences are not equally likely, the MLD is not necessarily optimum, although it is often the best feasible decoding rule.

Fortunately, the MLD rule in (3.105) can be reduced to a simple form for implementation. For a BSC it can be shown (Viterbi, 1971) that

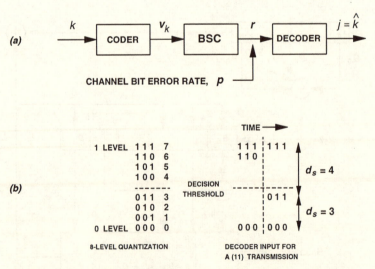

Fig. 3.30 Hard and soft decision decoding: (*a*) hard decision; (*b*) 3-bit soft decision.

maximum likelihood decoding can be realized by selecting the code sequence $\mathbf{v}_j$ which has minimum Hamming distance from the received sequence $\mathbf{r}$. Put another way, for a BSC the MLD reduces to a *minimum distance decoder*, and a coded path must be found through the trellis which has minimum Hamming distance from $\mathbf{r}$. This concept is used by the Viterbi decoding algorithm.

3.7.6 *Viterbi decoding algorithm (Viterbi, 1967)*

This is an optimal decoding technique since it observes the maximum likelihood criterion. It is also a very practical technique (at least for low-constraint-length convolutional codes) and is available in VLSI form. Viterbi decoding of convolutional codes is particularly attractive for satellite channels, and is finding increasing application in digital recording (Immink, 1989).

The Viterbi algorithm operates frame by frame over a finite number of frames and attempts to find the trellis path used by the coder. At any input frame the decoder does not know which node the coder reached and so it labels the possible nodes with metrics – the running Hamming distance between the trellis path and the input sequence. In the next frame the decoder uses these metrics to deduce the most likely path.

Example 3.29

Consider the trellis in Fig. 3.24 and assume the following information and code sequences:

$$\mathbf{i} = 1 \quad 0 \quad 1 \quad 1 \quad 0 \quad 0 \quad 0 \quad 1 \quad 0 \quad 0 \quad 1$$

$$\mathbf{v} = (11 \quad 01 \quad 00 \quad 10 \quad 10 \quad 11 \quad 00 \quad 11 \quad 01 \quad 11 \quad 11)$$

$$\mathbf{r} = (11 \quad \underline{11} \quad 00 \quad 10 \quad 10 \quad \underline{01} \quad 00 \quad 11 \quad 01 \quad 11 \quad 11)$$

Note that $\mathbf{r}$ is assumed to have two errors. Decoding starts by computing the metrics of possible coder paths up to the first *remerging node*, i.e. for the first K branches. This is shown in Fig. 3.31(a). For minimum distance (maximum likelihood) decoding we retain only those paths with minimum Hamming distance, and the 'surviving' paths are extended by one frame, as shown in Fig. 3.31(b). The new survivors are then extended for the following frame, Fig. 3.31(c), and so on. Where the two metrics at a node are identical, a random choice can be made, see Figs. 3.31(f) and (g). Note that in these two diagrams errors generate erroneous paths but, providing enough frames are examined, these paths are eventually eliminated, as in Fig. 3.31(i). At this stage in the decoding process a *unique path* covering both erroneous bits in **r** has been generated, and it defines the corrected decoder output, as indicated.

Example 3.29 has illustrated the error correction capability of a Viterbi decoder for the [5, 7] code. Clearly, the decoder observes a long code sequence (longer than one constraint length) and so d_∞ rather than d_{min} is of interest. For this particular code Table 3.9 gives $d_\infty = 5$ and so, according to (3.93), any pattern of two channel errors can be corrected. In fact, for this code, (3.93) implies that any two errors are correctable whilst three *closely spaced*, e.g. adjacent, errors are not. The implication here is

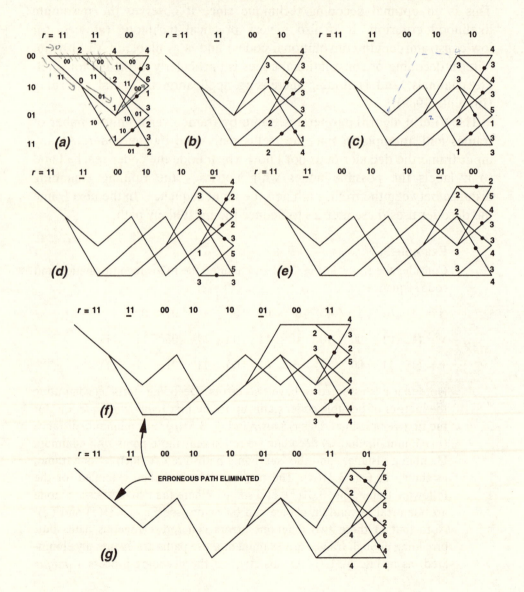

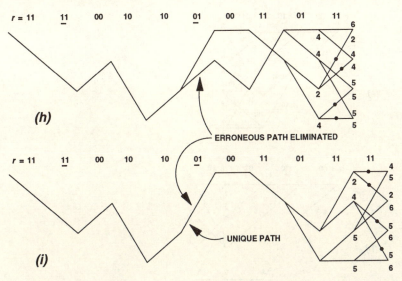

Fig. 3.31 Viterbi decoding: frame by frame analysis (the dots indicate the selected survivors).

that we cannot state the error correction capability of a convolutional code as succinctly as we can for a block code since it depends upon the error distribution. All we can say from (3.93) is that t errors within a few constraint lengths can be corrected, and that the error correction capability will increase as d_∞ increases. In the error correction sense the constraint length is therefore roughly analogous to the block length of a block code.

Example 3.30

The failure of the $[5, 7]$, $R = 1/2$, $K = 3$ code for three *closely spaced* errors is easily demonstrated. Assume the following sequences

$$\mathbf{v} = (00 \ 00 \quad 00 \quad 00 \quad 00 \quad 00 \quad \ldots)$$

$$\mathbf{r} = (00 \ \underline{10} \quad \underline{01} \quad \underline{10} \quad 00 \quad 00 \quad \ldots)$$

Decoding proceeds as in the previous example, by keeping only the minimum distance paths at each new frame, Fig. 3.32. Note, however, that the correct path (shown dotted) has been eliminated in Fig. 3.32(c), implying an inevitable decoding error. The solution here is to use a code with a larger d_∞, e.g. $d_\infty = 7$. It is also worth noting that the Viterbi decoder does not perform well against burst errors due to error propagation. In such cases it may be advantageous to use an outer code to combat the error bursts at the output of the Viterbi decoder, or to use interleaving.

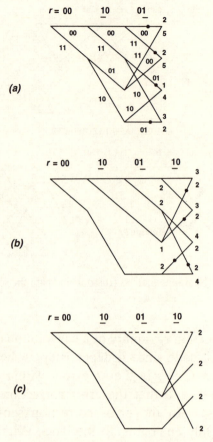

Fig. 3.32 Viterbi decoding: demonstration of decoding failure (the dots indicate the selected survivors).

Decoding window

A practical Viterbi decoder has a finite memory which can be visualized as a fixed window holding a finite portion of a moving trellis. For example, Figs. 3.31(*f*) and (*g*) could be redrawn in a fixed 8-frame window as shown in Fig. 3.33(*a*). Clearly, the depth W of this window or 'chainback' memory must be large enough to ensure that a well-defined decision can be made in the oldest frame. In other words, there should be a high probability that all paths merge to a single branch for the oldest frame, as illustrated in Fig. 3.33(*a*). The oldest frame is then decoded and effectively the trellis is shifted one frame to the left, giving a new frame on the right. A decoding failure occurs when paths fail to merge for the oldest frame.

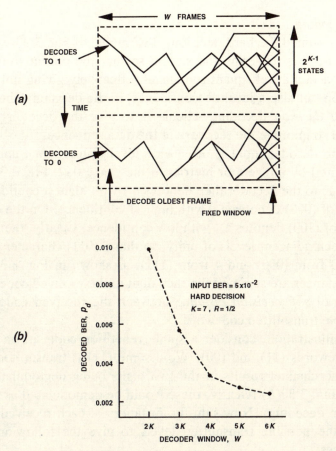

Fig. 3.33 Viterbi decoding window: (*a*) two consecutive windows from Fig. 3.31; (*b*) simulation to determine a practical window depth, W.

Unfortunately, the depth at which path merging takes place is a random variable and so cannot be guaranteed. However, computer simulation shows that a decoding depth W of 4 or 5 times the constraint length usually gives negligible degradation compared to an 'infinite' memory decoder. Figure 3.33(*b*) shows averaged results of a typical simulation for the [133, 171] code, assuming hard decision Viterbi decoding and a (random) BER of 5×10^{-2} at the decoder input, i.e. we are assuming a BSC. In this case there appears to be little advantage in increasing W beyond about $5K$. Looked at another way, it is generally found that increasing W beyond $4K$ or $5K$ will give only marginal improvement in E_b/N_o for a specified BER (Section 3.6.2), and this at the expense of increased decoding delay.

Soft decision decoding

So far we have assumed the hard decision, BSC model of Fig. 3.30(*a*). An alternative approach is to design the system so that the output from the demodulator is a multilevel, quantized signal, effectively giving not only the hard decision but also a level of 'confidence' in the decision. The idea behind this *soft decision* decoding scheme is to give the decoder more information and so improve the recovery of the data sequence.

In practice the decoder input is often an 8-level (3-bit) signal and this can be represented by integers or metrics in the range 0–7, Fig. 3.30(*b*). An input of (111) to the decoder then denotes a 1 with highest confidence, whilst an input of (000) denotes a 0 with highest confidence. On the other hand, an input of (100) denotes a 1 with low confidence. Clearly, there is a 'distance' between adjacent levels of unity, so that a (011) character has a distance d_s of 3 from (000) and 4 from (111), as shown in Fig. 3.30(*b*). Appropriate distances are summed for each digit in the received codeword to give an overall *soft decision distance* between the received codeword and each possible transmitted codeword.

As a simple illustration, consider a single repetition code, i.e. a code comprising codewords (11) and (00). We assume (11) is transmitted and that noise in the channel results in the two digits being demodulated at levels 6 and 3, Fig. 3.30(*b*) (clearly, these would be demodulated as (1, 0) in hard decision decoding). Now sum the distances of each received digit from each of the possible transmitted levels to give the following soft decision distances:

for a (11) transmission: $1 + 4 = 5$
for a (00) transmission: $6 + 3 = 9$

The decoder would then take the minimum soft decision distance and output the correct codeword (11) – despite the error in the last digit. The significant point is that, given a (1, 0) input, a hard decision decoder could only *detect* an error and has no way of correcting it.

It will be apparent that the above technique is easily extended to Viterbi decoding. All that is required is a change of metric from Hamming distance (for hard decoding) to a soft decision metric, all other decoding operations remaining the same. The performance advantage of soft decision decoding depends on a number of factors, i.e. the characteristics of the transmission channel, the number and spacing of the quantization levels and on the decoding algorithm. In practice there is little advantage in using more than 3-bit quantization, and performance advantage tends to be smallest on a (non-fading) Gaussian noise channel and better on a

Rayleigh fading channel. For a Gaussian channel and constant BER, 3-bit soft decision decoding enables E_b/N_o (as in Fig. 3.20) to be reduced by approximately 2 dB compared to that for hard decision decoding.

3.8 Transmission codes

In this section we are primarily concerned with those baseband signal processing techniques which are used to 'match' the binary data stream to a transmission channel. Essentially, this means spectrum shaping via linear filtering and via simple coding techniques, and a generalized model for baseband processing (excluding clock recovery) is shown in Fig. 3.34(*a*). We might regard this diagram as part of the data link and physical layer of the OSI layered architecture, or as a radio-data link with the modulation-demodulation processes omitted. Note that, in general, the processing involves channel coding (ECC and transmission coding) and transmit and receive filters. However, data channels vary significantly between applications and not all of these processing stages need to be present. For example, simple systems may have just a transmit filter and no coding.

Consider first the transmit and receive filters. Their basic role is to minimise the probability of error in the detector by minimizing the combined effect of ISI and channel noise. A common approach is to

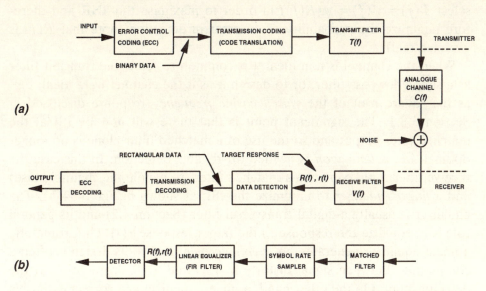

Fig. 3.34 Baseband data communications: (*a*) generalized model; (*b*) alternative receiver model.

assume the channel response $C(f)$ to be ideal or flat since it is usually unknown. In fact, this assumption is not unreasonable for some broadcasting and satellite channels. Now consider an isolated impulse at the input to the transmit filter $T(f)$ in Fig. 3.34(a). When the channel is ideal the spectrum or Fourier transform of the pulse $r(t)$ at the detector input is simply $R(f) = T(f)V(f)$. This means that the combined effect of the transmit and receive filters can be selected to give an optimum pulse shape $r(t)$ in order to minimize ISI at the detector input. As we shall see in Section 3.8.1, very often this *target response* would be a *Nyquist pulse*, $r(t)$, having a *raised cosine* type spectrum, $R(f)$, and in this case the resulting ISI will be zero.

The remaining question is how to divide the overall filtering task between $T(f)$ and $V(f)$. It is well known that the optimum detector for an isolated pulse is the *matched filter* since it maximises the SNR at the detector sampling instant. In fact, since we are assuming an ideal channel this concept can be extended to a sequence of pulses because there is no risk of ISI, i.e. the optimum receive filter will still be a matched filter. More specifically, under the assumptions of an ideal channel and a flat channel noise spectrum, maximum SNR at the detector sampling instant can be achieved by dividing the overall magnitude filtering *equally* between transmitter and receiver, half the filtering being in the transmit filter and half in the receive matched filter (Lucky, Salz and Weldon, 1968). In other words, given an impulse input to the transmit filter we select $T(f) = V(f) = \sqrt{(R(f))}$ in order to maximise the SNR and therefore minimize the probability of an error in detection (note that $R(f)$ is assumed real).

When the channel is non-ideal it is common to select the transmit filter as a simple lowpass filter, or to design it as if the channel *were* ideal, e.g. as the square root of the *raised cosine frequency* response discussed in Section 3.8.1. The significant point is that there will now be ISI at the matched filter output and so the use of a matched filter alone is no longer optimal, i.e. *equalization* is needed for reliable detection. In this case the receive filter could comprise a matched filter (to handle the channel noise) and a *linear equalizer* to minimize the ISI, as shown in Fig. 3.34(b). The equalizer is usually a digital transversal filter (Section 4.2) and its general role is to equalize the response to the target response $r(t)$. Unfortunately, a linear equalizer generates *noise enhancement* and so its design is often a compromise between signal distortion (ISI) and SNR degradation at the detector input. On the other hand, a linear equalizer can correct extensive amounts of ISI provided there is adequate SNR.

In practice the design of the matched filter may also be a compromise. Its design requires a knowledge of both $T(f)$ and the channel transfer function $C(f)$ and, as previously noted, in practice $C(f)$ may not be known precisely, or it may vary. In other words, very often the matched filter is not accurately defined. A realistic approach would then be to replace the matched filter in Fig. 3.34(*b*) with a lowpass filter, its role simply being that of a bandlimiting filter to reject out of band noise (simple systems might omit this filter altogether). When this is done the transversal equalizer is often made *adaptive* with the objective of minimizing the noise and ISI at the detector input (adaptive transversal filters are discussed in Section 4.4.6). Yet another form of receive filter consists of an equalizer (possibly adaptive) that shapes the channel characteristic to a specific *partial response (PR)* characteristic, followed by a filter matched to the PR characteristic.

Turning to signal detection, this is most easily performed *symbol by symbol* and the detector would then usually be some form of *threshold* circuit involving sampling and slicing (see also Section 3.6). In this case, the detector input is usually equalized to generate Nyquist pulses. It is well known, however, that detection based on symbol *sequences* (maximum-likelihood sequence detection, MLSD) yields optimum performance in the presence of ISI, albeit at the expense of substantially increased complexity. This type of detector would be implemented via the Viterbi algorithm discussed in Section 3.7.6; essentially, it decides on an entire sequence of inputs, or path through a 'trellis', rather than on a symbol by symbol basis.

Finally, it is important to realize that the use of transmit and receive filters is not the only way to achieve spectrum shaping. In fact, when transmission coding is used it is usually the dominant mechanism for controlling the transmitted spectrum and the transmit filter may then simply be a lowpass filter. Essentially, the output of the transmission coder is a signal (sometimes multilevel) having a spectrum matched to the transfer function of the analogue channel. In this section we will examine various coding techniques for achieving spectral shaping, such as the use of *dc-free codes*, *high-density bipolar codes*, and *partial response (PR)* signalling. PR schemes, for example, deliberately generate redundancy which can be exploited by maximum-likelihood detectors to give near optimal performance. Note that, in Fig. 3.34(*a*), the roles of ECC and transmission coding may 'blur' since many transmission codes have some error detecting capability, whilst the roles of detection and transmission decoding can also be blurred, as in the case of memory-less decoding of PR signals (Section 3.8.7).

3.8.1 *Nyquist pulses (binary transmission)*

Before discussing transmission coding techniques – which often result in multilevel transmission – it is first necessary to examine transmission using so-called Nyquist pulses. Here we are concerned with *binary* transmission and the need to minimise pulse distortion, specifically ISI.

Suppose the input to the transmit filter in Fig. 3.34(*a*) is a rectangular gate pulse of spectrum or Fourier transform $D(f)$. Assuming a linear channel, the pulse spectrum at the detector input is then

$$R(f) = D(f)T(f)C(f)V(f) \tag{3.106}$$

When $V(f)$ is essentially an equalizing filter the *target response* $R(f)$ will correspond to a pulse shape $r(t)$ which yields minimal ISI. The pulse shape $r(t)$ centred on $t = 0$ at the detector input is usually designed to follow *Nyquist's first criterion*, i.e.

$$r(t) = 0 \quad t = \pm kT \quad k = 1, 2, 3, \ldots \tag{3.107}$$

where T is the signalling period. Clearly, subsequent Nyquist pulses centred on $t = T, 2T, 3T, \ldots$ will not suffer any ISI from the pulse at $t = 0$ since the latter will have zero amplitude at the optimum sampling instants. Nyquist showed that the condition in (3.107) is possible provided the pulse spectrum $R(f)$ (its Fourier transform) has odd or skew symmetry about a frequency

$$f_1 = 1/2T \tag{3.108}$$

For a binary system, f_1 corresponds to half the bit rate. A class of functions having this symmetry are shown in Fig. 3.35, and are defined by

$$R(f) = \begin{cases} \dfrac{1}{2f_1} & f < f_1 - f_2 \\[2ex] \dfrac{1}{4f_1}\left\{1 - \sin\left[\dfrac{\pi}{2f_2}(f - f_1)\right]\right\} & f_1 - f_2 \leqslant f \leqslant f_1 + f_2 \\[2ex] 0 & f > f_1 + f_2 \end{cases} \tag{3.109}$$

where

$$r(t) = \frac{\sin(2\pi f_1 t)}{2\pi f_1 t} \frac{\cos(2\pi f_2 t)}{1 - (4 f_2 t)^2} \tag{3.110}$$

These pulses are usually called 'raised cosine' pulses due to the general

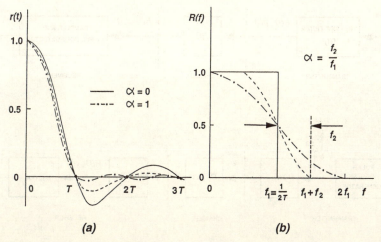

Fig. 3.35 Nyquist data pulses: (*a*) pulse shape; (*b*) skew symmetric spectrum.

shape of $R(f)$. The significant point is that these pulses are *bandlimited*, and so could be transmitted over an ideal but finite bandwidth channel without loss of pulse shape. This being the case, (3.107) means that zero ISI and 100% eye height could be achieved at the receiver.

Strictly speaking (3.110) is unrealizable (it spans all time) although it can be well approximated in practice. Note that, when $f_2 = 0$, the transmitted pulses are of $(\sin x)/x$ shape and the bandwidth requirement is minimal: only f_1 Hz is required to transmit at a rate of $2f_1$ symbols/s (see Nyquist's bandwidth–signalling speed relationship, Theorem 1.2). Unfortunately, the rectangular spectrum is also unrealizable and, in the absence of transmission coding techniques, practical systems have to settle for a finite roll-off factor $\alpha = f_2/f_1$, or finite *excess bandwidth* (typically 50%). For 100% roll-off pulses ($\alpha = 1$), Equation (3.109) reduces to a true raised cosine shape, i.e. of the form

$$R(f) = 1 + \cos(\pi fT) \tag{3.111}$$

Nyquist pulses are often generated in data broadcasting applications, and by the receive filter (raised cosine equalizer) for magnetic recording applications. In each case, filtering usually takes the form of quite simple direct form FIR structures. An alternative approach is to use pulses having a raised cosine *shape*.

Example 3.31

In Teletext it is common for all deliberate pulse shaping to be concentrated at the transmitter, as shown in Fig. 3.36(*a*). The idea is to shape

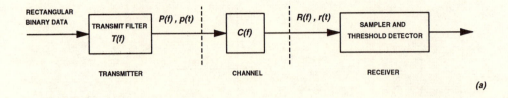

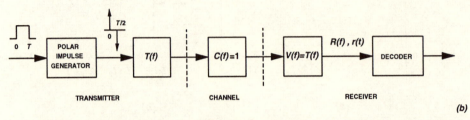

Fig. 3.36 (*a*) A typical pulse shaping scheme for Teletext ($P(f)$ often has a raised cosine spectrum); (*b*) effective pulse shaping scheme for the European RDS ($r(t)$ is a biphase symbol).

and bandlimit the rectangular binary data pulses so that minimal pulse distortion occurs during transmission. Using Nyquist pulses, the spectrum $P(f)$ of the transmitted pulse could be limited to the channel bandwidth f_c, corresponding to a roll-off factor of $(f_c - f_1)/f_1$. If the lowpass channel is ideal, the data at the detector input will then have zero ISI, i.e. $r(t) = p(t)$. As an example, the data rate in the UK Teletext system is 6.9375 Mbit/s, and this corresponds to a maximum roll-off of 44.1% if f_c is taken as 5 MHz (to be compatible with other video systems). In practice, it may be better to use a nominal 70% roll-off and to truncate the spectrum before transmission, despite a slight increase in ISI (Kallaway and Mahadeva, 1977).

In contrast, the CENELEC Standard for the European RDS specifies that spectrum shaping should be equally split between transmitter and receiver (in order to give optimum performance in the presence of random noise). The effective pulse shaping scheme for RDS is shown in Fig. 3.36(*b*). If the source has a signalling period T each filter has a transfer function of the form

$$T(f) = V(f) = \begin{cases} \cos\left(\dfrac{\pi f T}{4}\right) & f \leqslant 2/T \\[2ex] 0 & f > 2/T \end{cases}$$

and so the overall spectrum shaping is of the form $1 + \cos(\pi f T/2)$. For RDS, the basic data rate is 1187.5 bits/s and so the combined response of the shaping filters will fall to zero at 2375 Hz. Note that the effective signalling period is $T/2$ rather than T since a binary '1' pulse at the source generates an odd impulse pair separated by $T/2$ seconds (as indicated). This, in turn, generates a *biphase* or *Manchester* coded symbol $r(t)$ of nominal duration T (see Section 3.8.3). If we regard the signalling period as being $T/2$ then, according to (3.111), the overall data channel spectrum shaping corresponds to 100% cosine roll-off.

3.8.2 Fundamental requirements of transmission codes

Fig. 3.34 shows that, besides pulse shaping filters, baseband transmission also usually involves a *code translation* or *mapping* process. The broad objective is to match the digital signal to the baseband channel by using *coding* to shape the transmitted spectrum. As we shall see, this approach to spectrum shaping opens up the possibility of practical systems operating near the Nyquist limit of $2f_1$ symbols/s in a bandwidth of only f_1 Hz (Section 3.8.7). When we are dealing with an actual communications system the final translated code is usually called a *transmission* or *line* code. In contrast, the 'communications system' could be a magnetic recording channel, in which case the code used for transmission (writing) is usually called a *modulation* or *recording* code.

At this point it is instructive to examine the PSD or power spectrum for a random data stream applied to the transmission coding stage. Usually this is in non-return-to-zero (NRZ) unipolar format, in which case Fig. 3.37 shows that the PSD is essentially limited to $1/T$ Hz. This means, for example, that a 100 Mbit/s NRZ signal could be transmitted without serious degradation down a channel of 100 MHz bandwidth. On the other hand, this format is poorly matched to most transmission channels for several reasons.

First, the PSD is zero at the bit frequency and so it is impossible to phase lock a receiver clock *directly* to the signal. However, a timing component and therefore locking *can* be achieved by first performing some non-linear operation on the NRZ signal (a common approach is to generate a short pulse at each transition or zero-crossing). Unfortunately, there will still be no clock or timing information during long periods of 1 or 0. Secondly, provided the data are random, a *polar* NRZ code, e.g. $1 \equiv +1v$, $0 \equiv -1v$ will remove the dc component and the need for dc restoration or a dc coupled channel, but the PSD will still be significant near 0 Hz. This is inappropriate since most PCM and recording channels have a *bandpass* characteristic, with small low frequency response and a null at zero frequency.

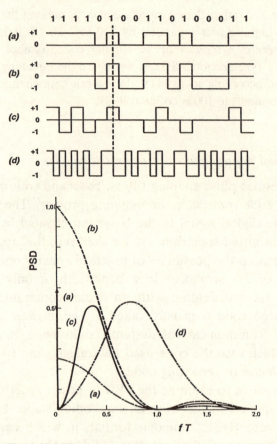

Fig. 3.37 PSD of some binary data formats: (*a*) NRZ unipolar, (*b*) NRZ polar, (*c*) AMI (bipolar), (*d*) Manchester (biphase).

Generally speaking, the polar NRZ code is also unsuitable for recording on *error propagation* grounds. Consider the basic longitudinal magnetic recording mechanism in Fig. 3.38. A polarity reversal in the NRZ write current occurs only if the present bit is not equal to the previous bit. A reversal generates a change in the direction of saturation and, upon readback, this transition generates a voltage pulse in the read head. Since a transition represents a switch from a string of 1s to a string of 0s (or vice-versa) then, clearly, failure to detect a readback pulse will give indefinite error propagation. For magnetic and optical recording, the usual solution is based upon the *modified* or *inverse* NRZ code, NRZI, in Fig. 3.38(*d*). Here a transition occurs only for a 1 and failure to detect a transition gives only a single bit error. In practice, *runlength constraints* are usually put on the basic NRZI code (Section 3.8.6).

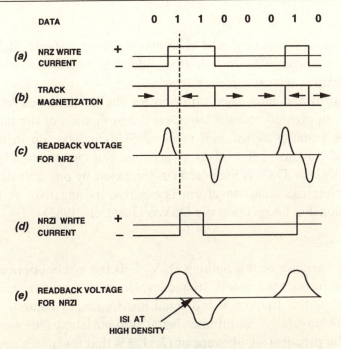

Fig. 3.38 Basic mechanism of saturated magnetic recording.

Design criteria

From the above discussion we can highlight some desirable properties of a transmission code:

A favourable PSD: it is often conjectured that the PSD of the coded data stream should 'match' the transfer function of the channel. For baseband (i.e. non-carrier, ac-coupled) channels this usually means that the PSD should be small at low frequencies with preferably a null at dc. Otherwise, a code with a significant low frequency content, or dc component, can give 'baseline (or dc) wander' – a form of ISI.

Adequate timing content: the transmission code should enable the data clock to be extracted from the signal itself, irrespective of the binary data pattern from the source.

Error propagation: the average number of errors in the decoded binary sequence per transmission error should be as low as possible.

Error detection: a good transmission code will have some error detection (and preferably, error correction) property.

DC-free codes

This type of code is frequently used in communications systems, including magnetic recording channels. In a dc-free code, the average power of the coded data stream is zero at zero frequency.

The best-known and most used criterion for the design of dc-free and low-frequency-suppressed codes is based on the evaluation of the range of values that the running *digital sum value (DSV)* can take up in the bit stream (the DSV is also referred to as *accumulated charge*, or *disparity count*). Basically, the DSV is increased or decreased by one unit depending upon whether the code waveform is positive or negative. A useful empirical relationship based upon the DSV is (Justesen, 1982)

$$\omega_c T \approx 1/2s^2 \tag{3.112}$$

where s^2 is the variance of the running DSV, T is the symbol period, and ω_c is the (low-frequency) cutoff frequency. In words, a low variance indicates a high cutoff frequency and small low-frequency content, i.e. a 'dc-free' code. Conversely, an infinite digital sum variation indicates a dc component. The practical significance of (3.112) is that it is much easier to calculate the variance than the spectrum of the coded signal, and so minimization of s^2 is a valid criterion in code design.

Example 3.32

Figure 3.39 illustrates a typical modulation code used in magnetic recording. Binary data have been mapped onto a *constrained* binary code sequence in which two consecutive 1s are separated by at least one and at most three 0s. This sequence is then converted into a NRZI waveform using a transition for a 1 and no transition for a 0. The net effect is to keep the running DSV within ±3 units, indicating that the code is dc-free (see also Example 3.35).

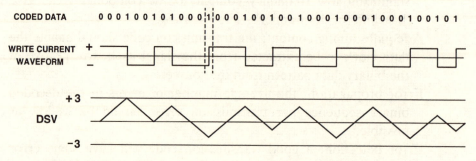

Fig. 3.39 Illustrating the running DSV for a dc-free modulation code.

Multilevel signalling

The transmitted symbol rate (baud rate) is also a fundamental considera-tion when selecting a transmission code. Because of hysteresis effects in the media, digital magnetic recording is invariably restricted to *saturation recording* whereby the media is magnetized into two stable states. In this case the coding process is usually a binary–binary translation where m data bits are translated to n code bits $(n > m)$ using a higher clock frequency. On the other hand, a binary to L level translation $(L \geq 3)$ is often employed in baseband communications systems since there is no such restriction. A binary data stream of R bits/s could be translated via an n-bit DAC to a stream of R/n symbols/s. The output of the DAC would then be a 2^n level analogue signalling waveform. This is attractive since, from Nyquist's bandwidth–signalling speed relationship a reduction in the symbol rate by a factor n reduces the required channel bandwidth by the same factor. The number of levels is restricted by the need to maintain acceptable error performance in the presence of channel noise.

3.8.3 Manchester (biphase) code

Here it is useful to consider the transmitted power spectrum in a little more detail. A convenient artifice is to assume that the transmitted waveform is the output of a filter whose unit impulse response is the required pulse shape $p(t)$ (using the notation in Fig. 3.36(a)). The input to this filter would be a random impulse train (corresponding to the coded sequence) where the strength of each impulse scales $p(t)$ to give the required transmission waveform. If $S_\delta(\omega)$ is the PSD of the impulse train, and if $P(\omega)$ is the Fourier transform of $p(t)$, it follows that the PSD of the transmitted waveform is

$$S(\omega) = S_\delta(\omega)|P(\omega)|^2 \tag{3.113}$$

where

$$P(\omega) = \int_{-\infty}^{\infty} p(t)e^{-j\omega t}\, dt \tag{3.114}$$

Equation (3.113) implies that, in practice, $S(\omega)$ can be controlled by changing either the transmission code or the pulse shape. Usually $S(\omega)$ is controlled by selecting a suitable transmission code, but the Manchester code is an exception in that control is exercised by adjusting $p(t)$. In particular, to achieve a null at zero frequency (no dc component) *irrespective of the incoming NRZ data pattern* we simply require

$$P(0) = \int_{-\infty}^{\infty} p(t)\,dt = 0 \qquad (3.115)$$

The most convenient way of doing this is to use a 'biphase' pulse shape $p(t)$ which has a positive to negative transition in the middle of a 1, as shown in Fig. 3.37(d). The shape $-p(t)$ is assigned to each binary 0. Note that the high timing content is paid for by the high bandwidth requirement (twice that of the NRZ code). The code is found in the European RDS, low-bit-rate weather satellite systems, local area networks and some digital recording systems.

3.8.4 *AMI and HDBn* codes

The *alternate mark inversion (AMI)* code in Fig. 3.37 (also called the *bipolar* or *pseudo-ternary* code) is generated by inverting alternate 1s or marks. This process removes the dc component, and so the resulting code goes some way to satisfying the requirements of a transmission code. In addition:

It is easy to generate and decode.

It is bandwidth efficient, requiring only half the bandwidth of biphase signalling.

It has an SED capability, since a single error will cause a violation of the alternating mark rule.

The timing content of the bipolar code is perhaps less obvious since its PSD does not possess a component at the bit frequency (see Fig. 3.37). Even so, the bit period can still be deduced by simply inspecting the waveform and so it must be possible to recover the required component.

In practice a clock component can be obtained by rectifying the bipolar waveform, but unfortunately this is not the end of the matter because a long sequence of binary 0s will fail to give any timing transitions.

To obtain higher timing content the AMI code can be modified by replacing strings of zeros with a *filling sequence*. The BnZS code for example (Croisier, 1970) follows the AMI rule unless n zeros occur together, in which case a filling sequence is used. Taking $n = 6$, this sequence is 0VB0VB where V denotes a violation of the AMI rule and B denotes a mark obeying the AMI rule.

Example 3.33

Binary 1 0 1 0 0 0 0 0 0 1 1 0 1 0 0 0 0 0 0 0 1 0 1 1 1

AMI   − 0 + 0 0 0 0 0 0 − + 0 − 0 0 0 0 0 0 0 + 0 − + −

B6ZS − 0 + 0 + − 0 − + − + 0 − 0 − + 0 + − 0 + 0 − + −

            0 V B 0 V B   0 V B 0 V B

(+ and − represent positive and negative marks)

The *high-density bipolar n* (HDB*n*) codes are somewhat more complex because there are four possible filling sequences compared with two in B*n*ZS. An HDB*n* code permits a string of *n* zeros but a string of *n* + 1 zeros is filled with one of two sequences, each *n* + 1 symbols long. The sequence will be B0 ... 0V or 0 ... 0V and the choice of sequence is determined by the rule that successive violations should be of alternate polarity. This rule ensures a null at dc in the PSD. Clearly, the timing content increases as *n* is reduced and usually *n* = 3.

Example 3.34

Binary   1 0 1 0 0 0 0 0 1 0 1 1 0 0 0 0 1 1 1 0 0 0 0 0 0 0 0 0 1 0 1

AMI   + 0 − 0 0 0 0 0 + 0 − + 0 0 0 0 − + − 0 0 0 0 0 0 0 0 0 + 0 −

HDB3   + 0 − 0 0 0 − 0 + 0 − + 0 **0** 0 + − + − 0 0 0 − + 0 0 + 0 − 0 +

         0 0 0 V   0 0 0 V   0 0 0 V B 0 0 V

           F1   F2   F3   F4

Calculation of the PSD for HDB3 is tedious but for a random binary source (independent 1 and 0 digits, with $P(0) = P(1) = 0.5$) the PSD turns out to be similar to that of the AMI code, i.e. nulls at $\omega = 0$ and $\omega = 2\pi/T$, where T is the bit period. As might be expected, this similarity increases as *n* increases. The HDB*n* code also retains the error detection property of the AMI code. For example, a negative mark in place of the bold symbol **(0)** in Example 3.34 effectively removes filling sequence F2, and sequence F3 will now have the same violation polarity as the immediately preceding sequence F1.

Table 3.13. *A 4B3T coding alphabet.*

Binary word	Ternary word Mode 1	Ternary word Mode 2	DSV
0000	0−+	0−+	0
0001	−+0	−+0	0
0010	−0+	−0+	0
1000	0+−	0+−	0
1001	+−0	+−0	0
1010	+0−	+0−	0
0011	+−+	−+−	1
1011	+00	−00	1
0101	0+0	0−0	1
0110	00+	00−	1
0111	−++	+−−	1
1110	++−	−−+	1
1100	+0+	−0−	2
1101	++0	−−0	2
0100	0++	0−−	2
1111	+++	−−−	3

3.8.5 Alphabetic ternary codes

This class of transmission code divides the binary source sequence into blocks of specified length and assigns a block of ternary symbols to each binary block using a look-up table. In general the transmission symbol rate will be less than the binary digit rate, thereby reducing the transmission bandwidth compared with that for the binary signal.

A good binary-to-ternary translation takes advantage of the relatively large number of possible ternary combinations in order to reduce the symbol rate, and therefore the bandwidth. For example, the 6B4T code (Catchpole, 1975) translates a block of six binary digits to four ternary symbols, whilst the MS43 and 4B3T codes translate blocks of four binary digits to three ternary symbols. Considering the 4B3T translation, three ternary symbols (27 combinations) are more than adequate to represent four binary digits (16 combinations) and ideally the transmission bandwidth will be scaled by three-quarters. This bandwidth efficiency makes the 4B3T code attractive for use as a line code on long distance repeatered networks. Similarly, a translation of, say, three binary digits to one of eight possible levels will scale the symbol rate and transmission bandwidth by one-third. A common example is the representation of three binary digits by one phase of an eight-phase phase shift keying (PSK) signal. In this case the digital carrier signal has a (zero-crossing) bandwidth of one-third of that for a binary PSK signal.

Table 3.13 shows a 4B3T coding alphabet (Jessop and Waters, 1970) and a PSD null at dc is achieved by selecting the mode in such a way as to minimize the running DSV (or disparity count). In other words, if the signal has recently contained an excess of positive marks, then ternary words with a negative bias are chosen to restore the balance (Example 3.35). Another point to note is the absence of the ternary character 000 in Table 3.13. This improves the timing content of the code, and can be utilized to check for loss of block synchronization. Should synchronization be lost, then the illegal character 000 will occasionally occur and synchronization can be corrected.

Example 3.35

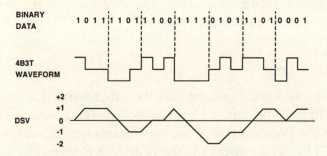

3.8.6 Runlength-limited (RLL) codes

This family of codes is used almost universally in optical and magnetic digital recording. Consider the basic NRZI recording code in Fig. 3.38(*d*). Here each logical 1 generates a flux transition on the media whilst a logical 0 corresponds to no transition, i.e. all the information content of the code is in the *position* of the transitions. Clearly, as the (on-track) recording density increases, the readback pulses from adjacent 1s will start to overlap (Fig. 3.38(*e*)) and, in general, this ISI can result in a significant change in both pulse position and pulse amplitude. In severe cases, this on-track *peak shift* will cause the peaks of the readback pulses to occur outside the clock window and the raw error rate will increase.

The usual solution for high-density recording is to use a modulation code which *preserves the minimum distance between transitions*, thereby minimizing ISI. Assuming two-state recording, this will be a binary-to-binary translation where m data bits are mapped to n code bits, corresponding to a code rate $R = m/n$ ($R \leq 1$). In an $R(d, k)$ RLL code sequence, two logical 1s are separated by a run of at least d consecutive zeros and any run of consecutive zeros has a length of at most k. Constraint d controls the highest transition density and resulting ISI, whilst constraint k ensures a *self-clocking* property or adequate frequency of transitions for synchronization of the read clock (this is a major failing of the *basic* NRZI code in Fig. 3.38(d)). Once a constrained or coded sequence has been generated it is recorded in the form of a NRZI waveform, as described in Section 3.8.2. In practice this requires a simple precoding step which converts the constrained coded sequence to a RLL *channel sequence* (a string of $\pm$ 1s) representing the positive and negative magnetization of the recording medium, or pits and lands in optical recording. The RLL channel sequence will then be constrained by minimum and maximum runlengths of $d + 1$ and $k + 1$ channel bits respectively.

An important parameter in digital recording is the ratio of data bits to recorded transitions, since this represents the efficiency of the code in utilizing recording space. This is called the *density ratio (DR)* and is given by

$$DR = R(d + 1) \tag{3.116}$$

The ratio increases as d increases but, on the other hand, the required timing accuracy for the detection of transitions increases exponentially with $d + 1$, while the clock content of the code decreases. In practice, d is typically 0, 1 or 2. Generally speaking, the code rate R should be as high as possible in order to maximise DR and other factors. The maximum value of R for specified values of d and k is called the *capacity $C(d, k)$* of the code, and is measured in information bits per code bit. Rate R is therefore usually selected as a rational number m/n close to $C(d, k)$.

Example 3.36

The $\frac{1}{2}(2, 7)$ RLL code (Franaszek, 1972) is a popular, high-performance, variable-length block code. Coding essentially amounts to partitioning the data sequence into blocks of 2, 3 or 4 bits, and applying the following translation:

Data	Codeword
10	0100
11	1000
000	000100
010	100100
011	001000
0010	00100100
0011	00001000

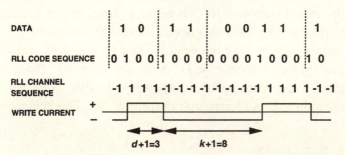

Fig. 3.40 Typical recording waveform for the $\frac{1}{2}$(2, 7) RLL code.

Fig. 3.40 shows the $\frac{1}{2}$(2, 7) sequence for arbitrary data, together with the precoding step to turn this sequence into a RLL NRZI channel sequence. Precoding ensures that 1s in the RLL code sequence indicate the positions of a $1 \rightarrow -1$ or $-1 \rightarrow 1$ transition in the RLL channel sequence. It is also worth noting that Example 3.32 is actually describing a RLL (1, 3) modulation code.

3.8.7 PR signalling

In practice a data communications system is often designed for a specific or *target* response at the detector, given a rectangular binary symbol (pulse) at the input, Fig. 3.41(a). We could regard $r(t)$ as the overall response per binary symbol. In a data broadcasting (e.g. Teletext) system,

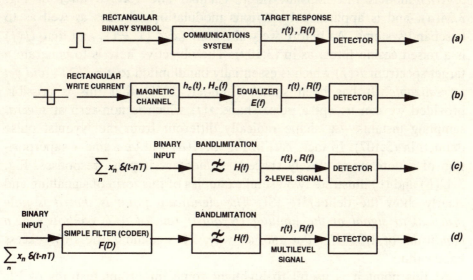

Fig. 3.41 Channel models: (a) defining target response for a pulse input; (b) basic magnetic recording channel; (c) a model for straight binary signalling; (d) a model for PR signalling.

for example, the target response $R(f)$ is often a raised cosine spectrum, as discussed in Example 3.31. In a magnetic recording system, Fig. 3.41(b), the input is a rectangular write current pulse and an equalizer delivers the target or desired (pulse) response $R(f)$. Also, the magnetic channel itself is often assigned a *pulse response* $h_c(t) \Leftrightarrow H_c(f)$, that is, $h_c(t)$ is the head/medium response to a rectangular current pulse. This means we can write

$$E(f) = R(f)/H_c(f) \tag{3.117}$$

and the equalizer can be designed if $H_c(f)$ can be determined e.g. experimentally.

In order to develop the theory of PR signalling, it is helpful to model the binary input as an impulse train, as in Fig. 3.41(c). Here we will assume the efficient polar signalling format and so take x_n as a sequence of independent binary symbols having values ±1 (logic $1 \equiv +1$ and logic $0 \equiv -1$). The signalling rate is $1/T$ symbols/s. Spectrum shaping is modelled by filter $H(f)$, and the target responses $R(f)$ and $r(t)$ could be defined by (3.109) and (3.110) respectively. The limiting case in Fig. 3.41(c) is when $R(f)$ has a rectangular spectrum cutting off at $f = 1/2T$, but, as previously discussed, this is impractical due to the discontinuity at $f = 1/2T$.

Now suppose we precede the bandlimiting filter in Fig. 3.41(c) with a simple linear filtering (coding) operation, as in Fig. 3.41(d). In effect, Fig. 3.41(d) models the transmit filter, channel and receive filter of Fig. 3.34(a), and is applicable to various modulation schemes as well as to baseband systems. As before we will assume $x_n \in \{1, -1\}$, and that $H(f)$ is a raised cosine filter, as in (3.109). The objective here is to generate a target spectrum $R(f)$ which is essentially bandlimited to $f = 1/2T$, *and yet is realizable*. It has been shown (Lender, 1966) that this is possible provided we can accept a pulse shape $r(t)$ which is non-zero at *several* sampling instants – a shape radically different from the Nyquist pulse defined in (3.107). In fact, $r(t)$ in Fig. 3.41(d) will be a linear superposition of a number of closely spaced, elementary impulse responses. Fig. 3.42(b) and (c) illustrate two useful examples of this form of signalling and clearly show the deliberate ISI. *The significant point is that it is now practical to signal at the limiting (Nyquist) rate of $1/T$ symbols/s in a bandwidth of $1/2T$ Hz!* Put another way, PR signalling offers 0% excess bandwidth.

At this point it is useful to highlight some important features of PR signalling:

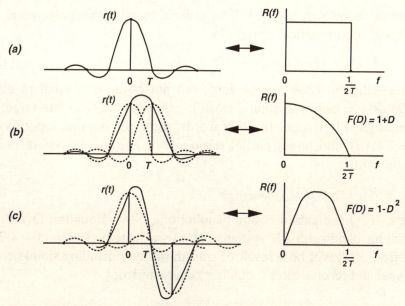

Fig. 3.42 Comparison of straight binary and PR signalling: (*a*) straight binary (ideal); (*b*) PR1; (*c*) PR4.

Since $r(t)$ is a linear superposition of impulse responses, a binary input will result in a *multilevel* output. For basic threshold/peak detection systems this generally requires an increase in SNR with respect to straight binary signalling for the same error rate, and PR systems exhibiting the smallest SNR degradation tend to be preferred (Kabal and Pasupathy, 1975). Typically three levels are used in practice.

PR signalling deliberately generates ISI and a degree of *correlation* between transmitted symbols (or signal levels), and so it is sometimes referred to as *correlative coding*. This correlation or constraint on level transitions can be exploited by detectors based on MLSD (or Viterbi detection) to give improved noise performance compared to symbol-by-symbol decoding of PR signals. For example, PR signalling with Viterbi decoding to perform MLSD is particularly attractive for high-density magnetic recording (Dolivo, Hermann and Olcer, 1989).

PR signalling is not just applicable to baseband systems, and it has been used in various carrier schemes, including FM and SSB.

Figure 3.41(*d*) shows that it is usual to define the main spectrum shaping filter or coder by a polynomial in D, where D is the delay operator (equivalent to z^{-1}). The general form of this polynomial which embraces most practical systems is

$$F(D) = (1 - D)^m (1 + D)^n \qquad (3.118)$$

where either m or n can be zero, but not both. It is usual to describe $F(D)$ as a *system* polynomial since it essentially defines the target spectrum of the PR system. In other words, using the impulse input model of Fig. 3.41(d), the target partial response, or end-to-end system function, can be written

$$R(f) = H(f)F(D)|_{D=e^{-j2\pi fT}} \qquad (3.119)$$

where $H(f)$ is a raised cosine bandlimiting filter (Equation (3.109)). For minimum bandwidth PR systems, $H(f)$ will cutoff close to $f = 1/2T$. In practice, $R(f)$ will be a result of transmit filter (including simple coding), channel and receive filter (equalizer) characteristics.

Example 3.37
Consider the PR system

$$F(D) = 1 + D \qquad (3.120)$$

This comprises a single delay and summer (Fig. 3.43) and its transfer function is obtained by evaluating $F(D)$ at $D = e^{-j2\pi fT}$, i.e.

$$H_F(f) = 1 + e^{-j2\pi fT}$$

$$|H_F(f)| = 2|\cos(\pi fT)| \qquad (3.121)$$

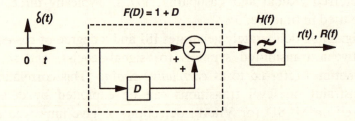

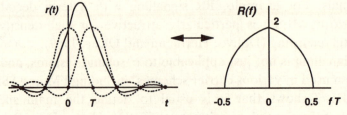

Fig. 3.43 Principle of duobinary coding (PR1).

We have already discussed this type of filter in Example 1.5 under sub-Nyquist sampling; the magnitude response is a comb filter with a first zero at $f = 1/2T$. When this is cascaded with the bandlimiting filter $H(f)$, the ideal minimum bandwidth target response becomes

$$R(f) = \begin{cases} 2\cos(\pi fT) & |f| \leq 1/2T \\ 0 & \text{otherwise} \end{cases} \tag{3.122}$$

In this ideal case the impulse response of $H(f)$ will be of the form $Sa(x) = (\sin x)/x$ and so the target impulse response will be

$$r(t) = Sa\left(\frac{\pi}{T}t\right) + Sa\left[\frac{\pi}{T}(t - T)\right] \tag{3.123}$$

Note that this has a baseline width equal to three clock periods (Fig. 3.43) and so generates considerable ISI. It is left to the reader to show that a random sequence of positive and negative impulses at the input to $F(D)$ in Fig. 3.43 will result in a three-level output signal, with transitions occurring only between adjacent levels. For ± 1 input symbols, the output levels will be $0, \pm 2$.

The significant point about this form of signalling is that, even for a practical, finite roll-off filter $H(f)$, the PSD for a random binary input will be negligible above $f = 1/2T$ (Lender, 1963). This means that, compared with 100% roll-off binary signalling, the correlative technique in Fig. 3.43 offers half the bandwidth or *double* the bit rate (which gave rise to the term 'duobinary').

Precoding
Fig. 3.44(*a*) represents the basic duobinary coding process in terms of signal samples. The coder output is the arithmetic sum

$$c_k = b_k + b_{k-1} \tag{3.124}$$

and this suggests that the binary data can be decoded as

$$\hat{b}_k = c_k - \hat{b}_{k-1} \tag{3.125}$$

where $\hat{b}_k$ is an estimate of b_k obtained by binary slicing the term $c_k - \hat{b}_{k-1}$. Unfortunately, if $\hat{b}_{k-1}$ is decoded incorrectly then the error will tend to propagate – a property we wish to avoid in transmission codes. An alternative decoder could use the MLSD technique (Viterbi decoding), but practical PR systems may employ a much simpler solution to the error propagation problem. Clearly, error propagation will be impossible if the decoder is made 'memoryless' and this can be achieved by employing a *precoder*, as in Fig. 3.44(*b*).

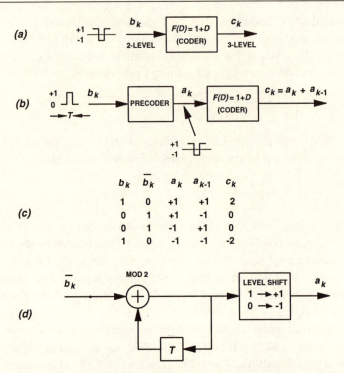

Fig. 3.44 Design of a precoder for duobinary signalling: (*a*) basic duobinary system; (*b*) use of precoder; (*c*) precoder design; (*d*) precoder circuit.

Consider the design of a precoder for duobinary coding. We know that, given symbols ± 1 at the coder input (i.e. a polar NRZ input) this coder generates symbols 0, ± 2 at its output. However, each output symbol c_k only conveys 1 bit of information and so we could use the memoryless decoding rule

$$
\begin{array}{cc}
c_k & b_k \\
\pm 2 \Rightarrow & 1 \\
0 \Rightarrow & 0
\end{array}
$$

This simple rule was used to construct the table in Fig. 3.44(*c*), where a_k is now the polar input to the coder. Ignoring the polar format (regarding -1 as 0) we can see from this table that $a_k = \bar{b}_k + a_{k-1} \bmod 2$, and this gives the precoder circuit in Fig. 3.44(*d*).

Example 3.38

The D-MAC television standard specifies that data and sound packets are to be transmitted at 20.25 Mbit/s in a baseband bandwidth of 8.5 MHz. The use of straight binary coding and raised cosine (Nyquist) pulse

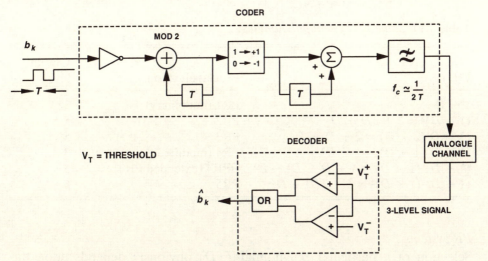

Fig. 3.45 Practical duobinary system (PR1).

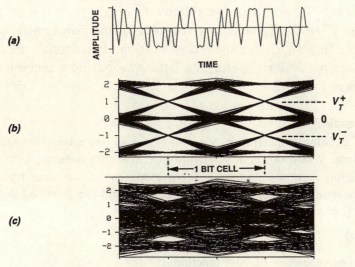

Fig. 3.46 Simulated results for the PR1 system in Fig. 3.45: (*a*) PR1 waveform, $f_c = 8.5$ MHz; (*b*) eye diagram for $f_c = 8.5$ MHz; (*c*) eye diagram for $f_c = 6$ MHz.

shaping, for example, would require typically 15 MHz, and so some form of multilevel transmission is required.

Fig. 3.45 shows a suitable duobinary coding scheme involving precoding and memoryless decoding. For a 20.25 Mbit/s input, the lowpass filter will ideally cut off at $f_c = 10.125$ MHz, but in practice reliable decoding (large eye height) can still be achieved for $f_c = 8.5$ MHz, as shown in Fig. 3.46(*b*). On the other hand, the eye height is virtually zero for a 6 MHz bandwidth. Note that this form of signalling is also known as *class 1 partial response* or PR1.

Table 3.14. *Some PR system polynomials*.

F(D)	PR classification	Output levels
$1 + D$	PR1 (duobinary)	3
$(1 + D)^2 = 1 + 2D + D^2$	PR2	5
$(1 + D)(2 - D) = 2 + D - D^2$	PR3	5
$(1 + D)(1 - D) = 1 - D^2$	PR4 (modified duobinary)	3
$(1 + D)^2(1 - D) = 1 + D - D^2 - D^3$	EPR4 (extended PR4)	5
$(1 + D)^2(1 - D)^2 = 1 - 2D^2 + D^4$	PR5	5

PR systems

Selection of a particular polynomial $F(D)$ obviously depends upon the required spectrum shaping, but it can also depend upon the permissible degradation in SNR and data eye relative to binary transmission (Kabal and Pasupathy, 1975). Table 3.14 gives some of the polynomials cited in the literature, together with their classification. Polynomials PR1, PR4 and EPR4 are particularly useful, the latter two having a spectral null at dc as well as at $f = 1/2T$.

Example 3.39

The duobinary spectral response in Fig. 3.43 has a large low-frequency content. Whilst this is of no significance in carrier systems, for baseband signalling it would require a dc coupled channel or some form of dc restoration. If a null at dc is required we could use *modified* duobinary (PR4) where

$$F(D) = (1 + D)(1 - D) = 1 - D^2 \tag{3.126}$$

The corresponding transfer function is

$$H_F(f) = 1 - e^{-j4\pi fT}$$

$$= j2 \sin (2\pi fT) e^{-j2\pi fT}$$

$$|H_F(f)| = 2|\sin (2\pi fT)| \tag{3.127}$$

and the bandlimited target response $R(f)$ will appear as in Fig. 3.42(c). Equation (3.126) suggests that $F(D)$ is realized by subtracting a two-sample delayed version of the binary signal from the current sample, and so the minimal bandwidth target impulse response for PR4 is

$$r(t) = Sa\left(\frac{\pi}{T}t\right) - Sa\left(\frac{\pi}{T}(t - 2T)\right) \tag{3.128}$$

The overall shape of $r(t)$ for both PR4 and PR1 is shown in Fig. 3.42. Since the number of output levels is still three, the precoder for PR4 can be designed using the same decoding rule as for PR1, see Fig. 3.47(b). Using the same argument as before (ignoring level shifting) we can write $a_k = b_k + a_{k-2}$ mod 2, and so the precoder is as shown in Fig. 3.47(c). In fact, to design a precoder for a general multilevel PR system, all we need to do is to divide the output levels into two sets, one corresponding to $b_k = 0$, and the other to $b_k = 1$. Fig. 3.48 shows typical power spectrum and eye diagram measurements for the output of the coder in Fig. 3.47(c), given random binary data at the input.

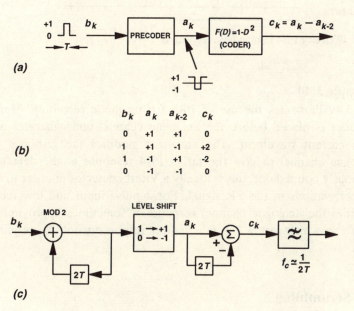

(a)

(b)

b_k	a_k	a_{k-2}	c_k
0	+1	+1	0
1	+1	-1	+2
1	-1	+1	-2
0	-1	-1	0

(c)

Fig. 3.47 PR4 signalling: (a) basic system; (b) precoder design; (c) practical coder.

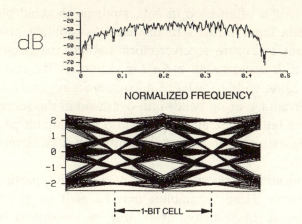

Fig. 3.48 Simulated PSD and eye diagram for PR4 data ($f_c = 0.42/T$).

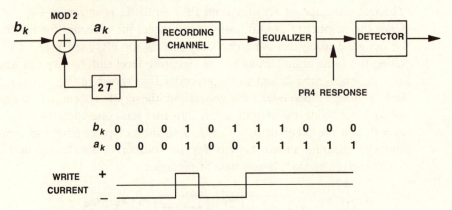

Fig. 3.49 Use of PR4 in magnetic recording.

Example 3.40

Fig. 3.49 illustrates the use of PR4 for magnetic recording. Here, the precoder is placed before the analogue channel and generates a NRZ write current waveform. The equalizer modifies the response of the analogue channel to give the target PR4 response at the detector. As previously pointed out, this is ideally a Viterbi detector in order to exploit the correlations in the PR signal. For a pulse input and low recording densities the analogue channel response is sometimes approximated by $1 - D$, in which case a $1 + D$ equalizer response would be required for PR4.

3.9 Scrambling

There are several reasons for using scrambling in digital communications systems. For example, it is often used in data modems to avoid placing restrictions on the data input format. In this case, scrambling provides enough data transitions to ensure receiver clock stability irrespective of the modem input (which could be all zeros for example). We could, of course, use a transmission code to achieve adequate timing content, but scrambling has the advantage of no bandwidth overhead in the sense that it does not increase the bit rate. Scrambling will also break up bit patterns which might otherwise cause accumulative clock jitter in a PCM repeater chain (Byrne *et al*, 1963).

Another major application of scrambling is in data security, particularly *conditional access systems*. Here, scrambling may be used on its own to provide a modest degree of security, or it may be controlled by an *encryption* algorithm in order to provide a secure system.

3.9.1 PRBS generation

This is the heart of any scrambling system. Consider the general linear feedback shift register device in Fig. 3.50. This consists of a binary n-stage shift register, coefficients c_i (which can be 1 or 0), and modulo 2 addition. The device is 'linear' in the sense that only addition is used in the feedback path. Suppose that we examine the binary sequence from any stage of the register. The sequence will start to repeat when the vector in the register repeats and, since there are only 2^n possible vectors, we would expect the sequence length to be less than or equal to 2^n. However, the simple logic in Fig. 3.50 will not cope with the all-0s state, since this prevents a change of state, and so the maximum possible sequence length before it repeats must be $2^n - 1$. Binary sequences of this maximum length are called M-sequences or PRBSs since they pass several statistical tests for randomness. For example, inspection of an M-sequence shows that there are always 2^{n-1} 1s and $2^{n-1} - 1$ 0s, implying equal probabilities for large n, i.e. $P(0) \approx P(1) \approx 0.5$. Also, when n is large, say $n > 10$, the auto-correlation function of an M-sequence closely approximates that of white noise. It is the pseudo-random nature of these sequences which makes them attractive for such applications as scrambling, noise generation, spread spectrum generation and simulation experiments such as Monte Carlo analysis.

The design problem is to select the shift register taps which generate an M-sequence. The underlying theory (Zierler, 1959; Elspas, 1959) is based upon finite fields and so involves algebraic polynomials and finite field (modulo 2) arithmetic. Considering Fig. 3.50, the binary value of digit a_r may be written as a linear recurrence relation:

$$a_r = c_1 a_{r-1} + c_2 a_{r-2} + \ldots + c_n a_{r-n} \tag{3.129}$$

and clocking the shift register generates the binary sequence

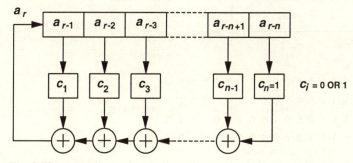

Fig. 3.50 n-stage linear feedback shift register.

$\{a_r\} = \{a_0 a_1 a_2 \ldots\}$. The *characteristic polynomial* of this sequence (the tap polynomial of the shift register) is

$$\phi(x) = 1 - \sum_{k=1}^{n} c_k x^k \qquad (3.130)$$

and over $GF(2)$ this can be written

$$\phi(x) = 1 + \sum_{k=1}^{n} c_k x^k \qquad (3.131)$$

It is this polynomial which holds the key to the design since it must be *primitive* over $GF(2)$ in order to generate a binary M-sequence.

Definition 3.6 *A polynomial $\phi(x)$ of degree n is primitive over $GF(2)$ if it is irreducible, i.e. has no factors except 1 and itself, and if it divides $x^k + 1$ for $k = 2^n - 1$ and does not divide $x^k + 1$ for $k < 2^n - 1$.*

From this definition it is apparent that a polynomial may be irreducible but not necessarily primitive.

Example 3.41
Consider the four-stage shift register device in Fig. 3.51. We have

$n = 4 \quad 2^n - 1 = 15$

$c_1 = c_2 = 0 \quad c_3 = c_4 = 1$

Therefore, from (3.131)

$$\phi(x) = x^4 + x^3 + 1 \qquad (3.132)$$

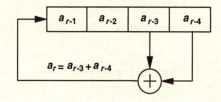

$$a_r = a_{r-3} + a_{r-4}$$

a_{r-1}	1	0	0	0	1	0	0	1	1	0	1	0	1	1	1	1
a_{r-2}	1	1	0	0	0	1	0	0	1	1	0	1	0	1	1	1
a_{r-3}	1	1	1	0	0	0	1	0	0	1	1	0	1	0	1	1
a_{r-4}	1	1	1	1	0	0	0	1	0	0	1	1	0	1	0	1

REPEATS

Fig. 3.51 M-sequence generator for $n = 4$.

This is irreducible over $GF(2)$. It also divides $x^{15} + 1$ over $GF(2)$ since

$$(x^4 + x^3 + 1)(x^{11} + x^{10} + x^9 + x^8 + x^6 + x^4 + x^3 + 1) = x^{15} + 1$$

In addition, $\phi(x)$ does not in fact divide $x^k + 1$ for $k < 15$, and so (3.132) satisfies the conditions for a maximum length sequence. In other words, Fig. 3.51 is an M-sequence generator of length 15 and the sequence shown assumes an initial 'all 1s' condition. Note that there are eight 1s and seven 0s and that the PRBS can be taken from any stage.

Example 3.42

The device in Fig. 3.52 has the characteristic polynomial

$$\phi(x) = x^6 + x^4 + x^2 + x + 1$$

This is irreducible but unfortunately $\phi(x)$ divides $x^{21} + 1$ over $GF(2)$, i.e.

$$(x^6 + x^4 + x^2 + x + 1)(x^{15} + x^{13} + x^{10} + x^7 + x^6 + x^3 + x + 1)$$

$$= x^{21} + 1$$

Therefore, $k = 21 < 2^n - 1$ and $\phi(x)$ is not primitive. The generator will in fact give three short cycles of length 21 in the ideal cycle length of 63. In general, the smallest integer k for which $\phi(x)$ divides $x^k + 1$ determines the period.

A list of primitive polynomials, i.e. irreducible polynomials which generate M-sequences, is given in Table 3.15. Most of these have been extracted from a more general table of irreducible polynomials not all of which generate M-sequences (Church, 1935). It will be seen that, in general, there are many alternative M-sequences for each value of n and since the number of alternative sequences increases rapidly with n only selected polynomials are given for high n. As we shall see, the polynomial resulting in the fewest feedback connections (fewest 1 entries) is often the most attractive. For $n = 4$ the table shows that the alternative M-sequence generator to that given in Example 3.41 is defined by the polynomial

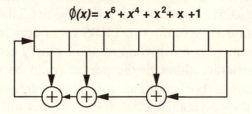

Fig. 3.52 Six-stage linear feedback shift register.

Table 3.15. *Primitive polynomials or M-sequence irreducible polynomials, sequence length $2^n - 1$ (polynomial $c_n x^n + c_{n-1} x^{n-1} + \ldots + c_1 x + 1$ is written $c_n c_{n-1} \ldots c_1 1$).*

n		n		n		n	
1	11	7	10011101	8	110000111	11	$1 + x^2 + x^{11}$
2	111		10100111		110001101		
3	1011		10101011		110101001	12	$1 + x + x^4 + x^6 + x^{12}$
	1101		10111001		111000011		
4	10011		10111111		111001111	13	$1 + x + x^3 + x^4 + x^{13}$
	11001		11000001		111100111		
5	100101		11001011		111110101	14	$1 + x + x^6 + x^{10} + x^{14}$
	101001		11010011	9	1000010001		
	101111		11010101		1000011011	15	$1 + x + x^{15}$
	110111		11100101		1000100001		
	111011		11101111		1000101101	16	$1 + x + x^3 + x^{12} + x^{16}$
	111101		11110001		1000110011	17	$1 + x^3 + x^{17}$
6	1000011		11110111		1001011001		
	1011011		11111101		1001011111	18	$1 + x^7 + x^{18}$
	1100001	8	100011101		1001101001		
	1100111		100101011		. . .	19	$1 + x + x^2 + x^5 + x^{19}$
	1101101		100101101				
	1110011		101001101	10	10000001001	20	$1 + x^3 + x^{20}$
7	10000011		101011111		10000011011	21	$1 + x^2 + x^{21}$
	10001001		101100011		10000100111		
	10001111		101100101		10000101101	22	$1 + x + x^{22}$
	10010001		101101001		10001100101		
			101110001		10001101111	23	$1 + x^5 + x^{23}$
					10010000001		
					10010001011	24	$1 + x + x^2 + x^7 + x^{24}$
					. . .		

$x^4 + x + 1$, i.e. feedback could be taken from the first and fourth stages instead of from the third and fourth stages. This generates a completely different PRBS to that in Fig. 3.51. Similarly, for $n = 9$ the two minimal-feedback entries in the table directly give the two PRBS generators in Fig. 3.53. Finally, note that the sequences are *pseudo-random* simply because they are deterministic and periodic, although the period could be made infinite for all practical purposes. For example, a 60-stage register will generate an *M*-sequence for the polynomial $x^{60} + x + 1$ and this will have a period of over 3600 years when clocked at 10 MHz!

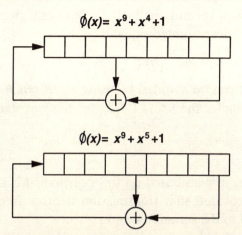

$\phi(x) = x^9 + x^4 + 1$

$\phi(x) = x^9 + x^5 + 1$

Fig. 3.53 Minimum-feedback *M*-sequence generators for $n = 9$.

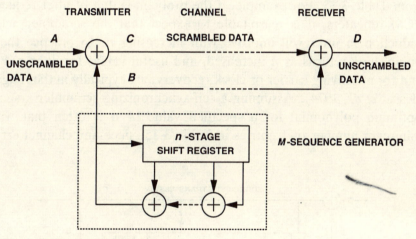

Fig. 3.54 Simple scrambler system.

3.9.2 Scrambler systems

Perhaps the most obvious scrambling system is simply to modulo 2 add an *M*-sequence to both data and scrambled-data streams, as shown in Fig. 3.54. The scrambling action can be demonstrated by considering the probability $P_C(1)$ of a 1 in the scrambler output:

$$P_C(1) = P_A(1)P_B(0) + P_A(0)P_B(1) \tag{3.133}$$

Since $P_B(0) \approx P_B(1) \approx 0.5$ for large n, then $P_C(1) \approx 0.5$ and this is independent of the input data. Also note that this process does not alter the bit rate. At the receiver it is necessary to subtract the same *M*-sequence, and in Fig. 3.54 this is sent on a second channel. In this case, provided there are no channel errors, the descrambled signal is

$$D = C + B = A + B + B = A \quad \text{over} \quad GF(2)$$

The need for an extra channel can be avoided by using a *self-synchronizing* scrambler system. For example, the CCITT V22 bis modem standard specifies the polynomial

$$\phi(x) = 1 + x^{14} + x^{17} \tag{3.134}$$

and the basic self-synchronizing system based on this polynomial is shown in Fig. 3.55(*a*). As before, provided that transmission is error free, the descrambler output is identical to the scrambler input.

In order to implement a particular scrambler system it is necessary to select the shift register length n, and then a suitable primitive polynomial from Table 3.15. For example, if the problem is that of jitter reduction in PCM repeaters, it is reasonable to expect that the scrambler action or reduction in jitter will improve with increasing n. In practice there are diminishing returns as n increased, and useful values for jitter reduction, and for improving carrier or clock recovery, are typically in the range 5–15 (Kasai *et al*, 1974). Assuming a self-synchronizing scrambler system, the optimum polynomial for a specified value of n is often that with the minimum number of 1 entries in Table 3.15, since any channel error will

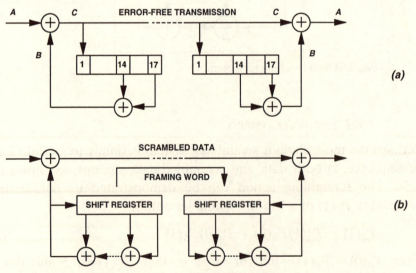

Fig. 3.55 Practical scrambler systems: (*a*) basic self-synchronizing system for CCITT V22 bis; (*b*) basic set–reset system.

then cause minimum *error extension* in the descrambled sequence. For example, considering the minimum tap registers in Fig. 3.55(a), a single channel error will generate an immediate error and two subsequent errors as the error emerges from the register. This corresponds to an error extension factor of 3 and in general the extension factor is the number of 1 entries in the polynomial (Table 3.15). In this respect, $n = 7$ or 9 is a better choice then $n = 8$ because there are no M-sequence polynomials of order 8 with fewer than five non-zero entries.

Error extension in self-synchronizing scramblers can upset error control (FEC) schemes if the error correction follows the descrambler. In this case it is necessary to perform error correction before descrambling, and error control coding after scrambling. Alternatively, we could avoid error extension altogether by using a *set–reset* scrambler, Fig. 3.55(b). Here the descrambler is reset to a defined state upon detection of some framing word in the data stream. It then regenerates the same PRBS as at the scrambler i.e. a synchronous sequence, and applies it to the modulo 2 adder.

Video scrambling and conditional access

Video scrambling is used when conditional access (pay as you view) is required on a commercial television system. Usually the active part of each video line is digitized and stored to enable some form of *time scrambling* to be applied. For example, lines could be reversed at random, or lines within a large block of lines could be transmitted in random sequence (line dispersal).

Another common technique might be referred to as *segment swapping*. Here, a number of digital video lines are split into variable-sized segments and segments are read out from the store in shuffled order. To achieve a *secure* system, control of the segment shuffling is often based on the US Data Encryption Standard (DES) or similar encryption algorithm. The DES algorithm has also been used to control line dispersal scrambling (Kupnicki & Moote, 1984). In its simplest form the segment swapping technique may be applied to single lines. Typically, the active part of a line is randomly cut at one of 256 (8-bit) equally spaced positions and the two segments are interchanged prior to transmission. The cut position also varies in a pseudo-random manner down the picture. Note that an *abrupt* cut will generate rapid, wideband transitions and so some 'cross fading' between cuts is required to avoid any bandwidth increase. On the other hand, no signal level shifts or inversions are involved, and so no artifacts will be generated by channel non-linearities.

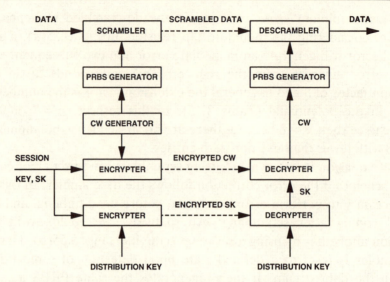

Fig. 3.56 Use of scrambling in a conditional access system.

A typical conditional access system achieves the required security by using encryption to 'lock' the scrambling process. In other words, descrambling is only possible if a correct 'unlocking key' is received. Fig. 3.56 shows a basic conditional access system for DBS television based upon this principle. Here, a long control word, CW, (typically 60-bits) seeds the PRBS generator (which has a long cycle time) which, in turn, could provide 8-bit numbers for segment swapping. An encrypted form of the CW is also sent to the receiver using an additional channel and it can only be decrypted by a secret 'session key' SK (which is transmitted on a third channel). In turn, decryption of SK is only possible by a properly authorized user.

4

Digital filters

We start by defining some important terms in digital filter theory.

A *sampled data filter* is an algorithm for converting a sequence of *continuous amplitude* analogue samples into another (analogue) sequence. The filter output sequence could then be lowpass filtered to recover a continuous time or analogue signal, and the overall filtering effect will be similar to that of conventional analogue filtering in the sense that we can associate a frequency response with the sampled data filter. In Section 4.1 we develop a formal discrete time linear filter theory based upon the concept of sampled data filtering.

Usually the signal is both sampled and quantized and the filter input is actually a sequence of *n*-bit words. The filter is then referred to as a *digital filter*, and it could be implemented in digital hardware or as a software routine on a DSP chip. For generality, the basic digital filter theory developed in Section 4.1 neglects quantization effects and treats the digital filter as a sampled data filter.

A practical example of a true sampled data filter is the CCD *transversal* filter used for ghost reduction in television receivers. Here, CCD rather than digital technology is sometimes preferred because it offers a long delay time in a small chip area with low power consumption. Examples of digital filters abound. *Adaptive* digital filtering can be used to reduce the noise level in aircraft passenger compartments (using adaptive acoustic noise cancellation), and adaptive transversal digital filters can be used for equalization of telephone lines. *Recursive* digital filters are used as tone detectors in data modems and as adaptive *temporal* digital filters for reducing random noise in broadcasting systems. *Linear phase* digital filters are used in high quality audio systems, for the digital generation of SSB and VSB signals in communications systems, and as *spatial* filters for improved decoding of PAL and NTSC television signals.

The importance of digital filters lies in their significant advantages compared to analogue filters:

> Performance is relatively insensitive to temperature, ageing, voltage drift and external interference.

> The filter response is completely reproducible and predictable, and software simulation can exactly reflect product performance.

> Systems requiring exact linear phase or exactly balanced processing operations are relatively easy to implement.

> Accuracy can be precisely controlled by adjusting wordlengths within the filter.

> Digital technology offers relatively high reliability, small dimensions, complex processing capability and low cost.

4.1 Discrete-time linear systems

The basic digital filter is a *time-invariant*, *discrete-time linear system*. Fortunately, there is a close correspondence between discrete-time linear system theory and continuous-time linear system theory, and so readers familiar with the latter should find the transition to digital systems quite straightforward. The correspondence is illustrated in Fig. 4.1. The continuous-time (or analogue) system utilizes the s-plane and the Laplace transform, and its transfer function $H(s)$ is the Laplace transform of the Dirac impulse response $h(t)$, i.e.

$$H(s) = \int_{-\infty}^{\infty} h(t)e^{-st}\,dt \qquad (4.1)$$

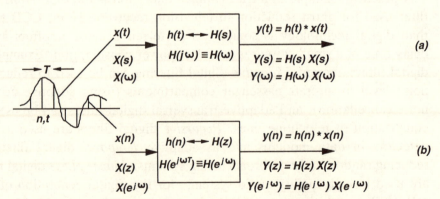

Fig. 4.1 Correspondence between continuous-time and discrete-time linear systems (or filters).

Similarly, the discrete-time system utilizes the *z-plane* and the *z-transform*, and its transfer function $H(z)$ is the *z*-transform of the *digital impulse response* $h(n)$, i.e.

$$H(z) = \sum_{n=-\infty}^{\infty} h(n)z^{-n} \tag{4.2}$$

where z is a complex variable. The system in Fig. 4.1(b) can be assumed to be a linear, time-invariant (i.e. non-adaptive) digital filter with sampled data sequence $x(n)$ as input. Remember that we are assuming each input sample to be unquantized in order to simplify the theory, but, in practice, $x(n)$ will be a stream of digital words each separated by the sample interval T seconds.

A linear digital filter is characterized by $H(z)$ or, equivalently, by $h(n)$. The impulse response is simply the filter output sequence when $x(n)$ is a digital impulse or single unit sample at $n = 0$ (see Fig. 4.2). Clearly, if the filter is to be physically realizable or *causal* then $h(n) = 0$ for $n < 0$. If $h(n)$ is of finite duration then the filter is an FIR filter, otherwise it is an IIR filter.

We will denote the *z*-transform of sequence $x(n)$ by $x(n) \Leftrightarrow X(z)$ where

$$X(z) = \sum_{n=-\infty}^{\infty} x(n)z^{-n} \tag{4.3}$$

The *z*-transform, like the Laplace transform, has a number of properties and the following three properties are particularly relevant to digital filter theory:

(1) Linearity property. If

$$x_1(n) \Leftrightarrow X_1(z)$$

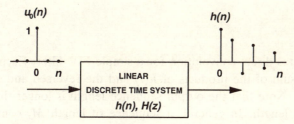

Fig. 4.2 Characterizing a linear discrete-time system by its digital impulse response.

$$x_2(n) \Leftrightarrow X_2(z)$$

then

$$a_1 x_1(n) + a_2 x_2(n) \Leftrightarrow a_1 X_1(z) + a_2 X_2(z) \tag{4.4}$$

where a_1 and a_2 are real constants.

(2) Time shift property. If

$$x(n) \Leftrightarrow X(n)$$

then

$$x(n - n_0) \Leftrightarrow z^{-n_0} X(z) \quad \forall n_0 \tag{4.5}$$

Note that a one sample delay ($n_0 = 1$) multiplies the z-transform by z^{-1}. We will use this concept when drawing filter structures.

(3) Convolution property. In Fig. 4.1(b), the output sequence $y(n)$ is the convolution of the input sequence $x(n)$ with the digital impulse response $h(n)$ of the filter. In the transform domain the filter output becomes a multiplication of transforms

$$Y(z) = H(z)X(z) \tag{4.6}$$

4.1.1 Discrete convolution

Equation (4.6) means that digital filtering could be performed *indirectly* in the transform domain, i.e. by multiplication of transforms, and $y(n)$ could then be obtained by inverse transforming $Y(z)$. This technique can be advantageous and is sometimes called *fast convolution* (Section 5.6). For now, let us stay with direct, aperiodic discrete convolution as defined by

$$y(n) = x(n)*h(n) = h(n)*x(n) = \sum_{i=0}^{n} h(i)x(n-i) \tag{4.7}$$

Example 4.1
Fig. 4.3 illustrates direct convolution for a simple FIR filter. Each value of $y(n)$ is the sum of the products of $h(i)$ and the reversed and shifted input sequence. Note that the output sequence length is longer than the input sequence length. In general, a sequence of length M_1 convolved with a sequence of length M_2 yields a sequence of length $M_1 + M_2 - 1$, and this fact is useful when performing fast (FFT based) convolution.

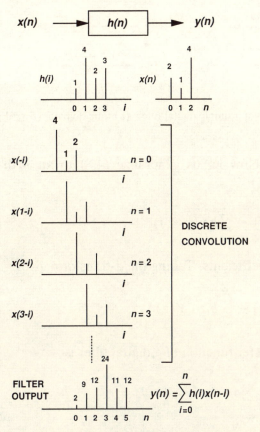

Fig. 4.3 Example of direct, discrete convolution (digital filtering).

4.1.2 *Filter transfer function*

A continuous time filter is described by a *linear differential equation* and its solution is usually via the Laplace transform. Similarly, a discrete time filter is described by a *linear difference equation* and its solution is usually via the z-transform. For example, the simple discrete time filter in Fig. 4.4(a) is described by the first-order equation

$$y(n) = x(n) - b_1 y(n-1) \tag{4.8}$$

Here, b_1 is a *filter coefficient* and 'T' is a one sample delay. Taking the z-transform and using (4.5) gives

$$Y(z) = X(z) - b_1 z^{-1} Y(z) \tag{4.9}$$

and so the digital filter could also be represented in the z-domain, as in Fig. 4.4(b) (in practice this is often more useful). Note that the delay is

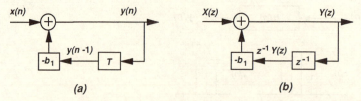

Fig. 4.4 Representations of a simple digital filter: (*a*) time domain; (*b*) z-domain.

now represented by z^{-1}. Now let us generalize (4.8) to an Mth-order equation (assuming $M \geq N$)

$$y(n) = \sum_{i=0}^{N} a_i x(n-i) - \sum_{i=1}^{M} b_i y(n-i) \tag{4.10}$$

where a_i and b_i are filter coefficients. Taking the z-transform gives

$$Y(z) = \sum_{i=0}^{N} a_i z^{-i} X(z) - \sum_{i=1}^{M} b_i z^{-i} Y(z) \tag{4.11}$$

and so the generalized transfer function of a digital filter is

$$H(z) = \frac{Y(z)}{X(z)} = \frac{\displaystyle\sum_{i=0}^{N} a_i z^{-i}}{1 + \displaystyle\sum_{i=1}^{M} b_i z^{-i}} \tag{4.12}$$

4.1.3 Filter stability

Discrete-time linear systems can be unstable and so, just as for continuous-time systems, stability criteria have to be obeyed. If a linear system is to be stable then it must have a defined frequency response or, put another way, the Fourier transform of its impulse response must exist or *converge*. Formally, we say that, for a stable continuous-time system, the *region of convergence* of $H(s)$ must include the $j\omega$-axis of the s-plane (which corresponds to the Fourier transform). Similarly, for a stable discrete-time system the region of convergence of $H(z)$ must include the *unit circle* of the z-plane, since letting $z = e^{j\omega T}$, i.e. $|z| = 1$, reduces $H(z)$ to the Fourier transform (see also Section 4.1.4). On the other hand, if the system is to be physically realizable or *causal*, its impulse response must be zero for t or $n < 0$, i.e. it must be 'right-sided', which in turn means that the region of convergence must not embrace any poles (Oppenheim and Willsky, 1983). Combining these points, we have the following criteria:

> For causal and stable continuous time systems all poles of $H(s)$ must lie in the *left half* of the s-plane. For causal and stable discrete time systems, all poles of $H(z)$ must lie *inside* the unit circle of the z-plane.

Suppose we write (4.12) as

$$H(z) = \frac{a_0 + a_1 z^{-1} + a_2 z^{-2} + \ldots + a_N z^{-N}}{1 + b_1 z^{-1} + b_2 z^{-2} + \ldots + b_M z^{-M}} \tag{4.13}$$

The numerator and denominator can always be written as a product of factors

$$H(z) = a_0 \frac{(z - z_1)(z - z_2) \ldots (z - z_N)}{(z - p_1)(z - p_2) \ldots (z - p_M)} z^{M-N} \tag{4.14}$$

Here, the z_i, $i = 1, 2, \ldots, N$ are the zeros of $H(z)$, e.g. $z = z_1$ makes $H(z)$ zero, and p_i, $i = 1, 2, \ldots, M$ are the poles of $H(z)$, e.g. $z = p_1$ makes $H(z)$ infinite. The factor z^{M-N} corresponds to an $(M - N)$-fold zero at $z = 0$ $(M > N)$, or to an $(N - M)$-fold pole at $z = 0$ $(N > M)$. We will see in Section 4.1.5 that this factor simply represents a shift of $h(n)$ in time, and so generally it can be neglected. As an example, the filter in Fig. 4.4(b) has the transfer function

$$H(z) = \frac{1}{1 + b_1 z^{-1}} = \frac{z}{z + b_1} \tag{4.15}$$

This corresponds to a zero at $z = 0$ and a pole at $z = -b_1$, and the filter is

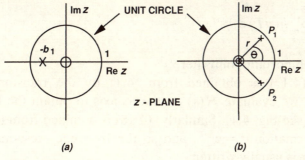

Fig. 4.5 Pole-zero plots for causal and stable filters: (a) first-order; (b) second-order.

stable provided $|b_1| < 1$ (Fig. 4.5(a)). Similarly, from (4.14), we could write an *all-pole* (i.e. $N = 0$) second-order transfer function as

$$H(z) = \frac{a_0 z^2}{(z - p_1)(z - p_2)} \tag{4.16}$$

which has poles at $z = p_1$ and $z = p_2$. For the second-order filter in Fig. 4.5(b) the poles are, in fact, *complex conjugate*, i.e.

$$p_1 = r\,\mathrm{e}^{\mathrm{j}\theta} \qquad p_2 = r\,\mathrm{e}^{-\mathrm{j}\theta}$$

and the filter is stable if $|r| < 1$. In general, as far as stability is concerned, the zeros may lie inside the circle, on the circle, or outside the circle. The most significant practical implication of this section is that a digital filter becomes unstable if *quantization* of filter coefficients takes poles outside the unit circle.

4.1.4 *Filter frequency response*

In the introduction to this chapter we stated that a sampled data filter (and here we include the digital filter) can have a steady state sinusoidal frequency response just like any analogue filter. The fact that we are using signal samples simply means that the response becomes *periodic*. We know from continuous-time filter theory that the filter frequency response is given by the Fourier transform of its impulse response

$$H(\omega) = \int_{-\infty}^{\infty} h(t)\,\mathrm{e}^{-\mathrm{j}\omega t}\,\mathrm{d}t \tag{4.17}$$

This gives the frequency response in magnitude and phase. Similarly, the frequency response of the digital filter is given by the *discrete-time* Fourier transform of $h(n)$ and could be written

$$H(\omega) = \sum_{n=-\infty}^{\infty} h(nT)\,\mathrm{e}^{-\mathrm{j}\omega nT} \tag{4.18}$$

There are several important points here:
(1) Equation (4.17) is obtained from (4.1) by the transformation $s = \mathrm{j}\omega$, i.e. we evaluate $H(s)$ on the $\mathrm{j}\omega$-axis to obtain the Fourier transform (see Fig. 4.1). Similarly, (4.18) is obtained from (4.2) by the transformation $z = \mathrm{e}^{\mathrm{j}\omega T}$, and so the frequency response of a digital filter is usually written

$$H(\mathrm{e}^{\mathrm{j}\omega T}) = H(z)\big|_{z=\mathrm{e}^{\mathrm{j}\omega T}} \tag{4.19}$$

In other words, we evaluate $H(z)$ on the unit circle to obtain the Fourier transform or frequency response. Note that the $j\omega$-axis and the unit circle play the same role in that they both yield the Fourier transform, and both are fundamental to stability criteria.

(2) The exponential term in (4.18) is periodic (period $2\pi/T$) and so the frequency response of a digital filter will *repeat*. In fact, (4.18) is simply a *Fourier series* expansion in the sense that it describes a periodic function in the frequency domain (rather than in the more usual time domain). Taking this view, the $h(nT)$ terms can then be described as Fourier series coefficients. Referring to Fig. 4.1(*b*) and to (4.19) we can therefore write the steady state frequency response of a digital filter in terms of a Fourier series pair

$$H(e^{j\omega T}) = \sum_{n=-\infty}^{\infty} h(nT) e^{-j\omega nT} \tag{4.20}$$

$$h(nT) = \frac{1}{\omega_s} \int_{-\omega_s/2}^{\omega_s/2} H(e^{j\omega T}) e^{j\omega nT} \, d\omega \tag{4.21}$$

where $\omega_s = 2\pi/T$. Note that whenever T is omitted in (4.20) or (4.21) we are assuming a normalized sampling frequency, i.e.

$$f_s = \frac{1}{T} = \frac{\omega_s}{2\pi} = 1 \text{ Hz}$$

(3) We can generalize (4.20) to the Fourier transform of an aperiodic discrete time sequence $x(n)$, giving

$$X(e^{j\omega}) = \sum_{n=-\infty}^{\infty} x(n) e^{-j\omega n} \tag{4.22}$$

Again, this is periodic and follows from ideal sampling theory (Section 1.3).

Fig. 4.6 illustrates idealized magnitude responses for lowpass, highpass, and bandpass filters on normalized frequency scales. Since the responses repeat, it is usual to display only the response in the range $0 \leqslant \omega \leqslant \pi$. In practice, the repeated response is removed by the sharp cutoff anti-imaging lowpass filter following the DAC. It is worth noting that sometimes it is better to realize this filter by a combination of oversampling, lowpass digital filtering, and a simple lowpass analogue filter following the DAC, as discussed in Section 1.4.2.

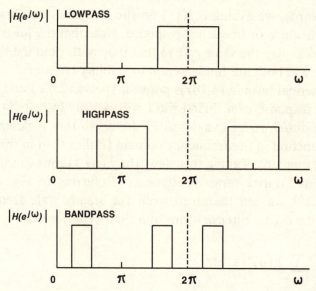

Fig. 4.6 Illustrating the repetitive frequency response of a digital filter (2π corresponds to the normalized sampling rate).

4.1.5 Linear phase filters

Consider a linear continuous-time system $H(\omega)$ which simply scales an input signal $x(t) \Leftrightarrow X(\omega)$ by a factor k and delays it by α seconds. Its output is

$$y(t) = kx(t - \alpha) \Leftrightarrow kX(\omega)\,\mathrm{e}^{-j\alpha\omega} = H(\omega)X(\omega) \qquad (4.23)$$

Clearly then, apart from shaping the amplitude spectrum, a more general linear filter with a transfer function

$$H(\omega) = |H(\omega)|\mathrm{e}^{j\phi(\omega)}$$

where

$$\phi(\omega) = -\alpha\omega \qquad (4.24)$$

simply delays the signal by α s (the *phase delay*) and introduces no phase distortion whatsoever. If the signal occupies a small band of frequencies around a carrier ω_c it is usual to consider the *envelope* or *group* delay t_g. The delay of an envelope comprising a narrow group of frequencies centred on ω_c is

$$t_g = -\frac{d\phi}{d\omega}\bigg|_{\omega=\omega_c} \qquad (4.25)$$

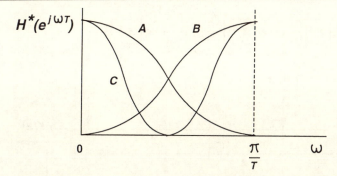

Fig. 4.9 Amplitude responses of three elementary linear phase filters.

4.2 Filter realization

A digital filter is essentially an algorithm for shaping a discrete-time sequence $x(n)$ to an output sequence $y(n)$. It could operate either in real time or non-real time, and be implemented using either hardware or software. Digital filters are frequently implemented using a dedicated DSP chip (i.e. hardware), or as a software routine on a general purpose signal processor. Let us first examine some basic hardware architectures or filter *structures*.

4.2.1 Recursive structures

The basic filter algorithm is defined by (4.12), and direct implementation of this leads to the *direct form* realization in Fig. 4.10. The feedback paths make the filter *recursive*, and in the absence of feedback ($b_i = 0$, $i = 1, 2, \ldots, M$), the filter is said to be *non-recursive*. A non-recursive structure will always yield an FIR filter, whilst a recursive structure *usually* yields an IIR filter. Occasionally, a recursive filter comprises a non-recursive structure cascaded with a recursive structure, the poles of the latter being located so as to cancel some of the zeros of the non-recursive structure. This particular combination yields a linear phase FIR filter which can be much more efficient than a non-recursive realization. Examples of this recursive type of FIR filter are *frequency sampling* filters, and a variant of frequency sampling filters called *difference routing* filters (Van Gerwen *et al*, 1975).

   We can use the idea of cascading recursive and non-recursive structures in order to simplify Fig. 4.10. Suppose we expand (4.12) as a cascade of two filters

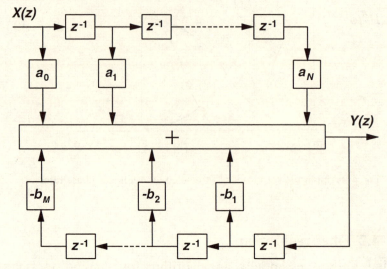

Fig. 4.10 Direct form realization of an Mth-order filter $(M \geqslant N)$.

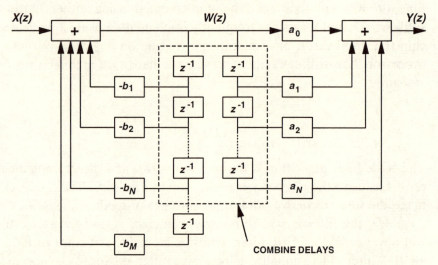

Fig. 4.11 Canonical direct form Mth-order filter.

$$H(z) = \frac{W(z)}{X(z)} \cdot \frac{Y(z)}{W(z)}$$

$$= \frac{1}{1 + \sum\limits_{i=1}^{M} b_i z^{-i}} \cdot \sum\limits_{i=0}^{N} a_i z^{-i} \qquad (4.37)$$

Direct implementation of (4.37) yields the structure in Fig. 4.11, and, clearly, we can combine delays to give a minimum delay or *canonical* type of filter. Neither of these two direct forms is recommended for high-order recursive filters since they have high coefficient sensitivity compared to other structures. Each coefficient a_i affects *all* of the zeros and each coefficient b_i affects *all* of the poles, which means that small errors arising from *coefficient quantization* can have a significant effect upon the frequency response. In general, a filter structure becomes less coefficient sensitive as each pole (zero) becomes dependent on only a few coefficients, and so a common structure for realizing high-order filters comprises a cascade of low-order sections. Consider a second-order section with the transfer function

$$H_i'(z) = \frac{a_{0i} + a_{1i}z^{-1} + a_{2i}z^{-2}}{1 + b_{1i}z^{-1} + b_{2i}z^{-2}}$$

$$= a_{0i}\frac{1 + A_iz^{-1} + B_iz^{-2}}{1 + C_iz^{-1} + D_iz^{-2}}$$

$$= a_{0i}H_i(z) \tag{4.38}$$

A high-order filter (Fig. 4.12(a)) can now be realized as a cascade of these second-order sections (together with possibly first-order sections) to give an overall transfer function

$$H(z) = a_0\prod_i H_i(z) \tag{4.39}$$

where

$$a_0 = \prod_i a_{0i} \tag{4.40}$$

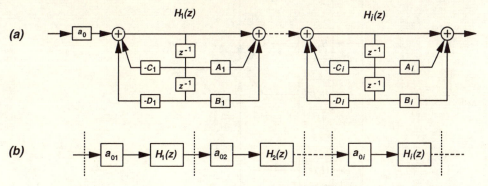

Fig. 4.12 Canonical cascade form for high-order recursive filters: (a) cascade of second-order sections; (b) use of scaling in fixed-point applications.

Clearly, error in any one coefficient can now only affect the pole(s) or zero(s) associated with that particular section. IIR filter design program-mes often assume the structure in Fig. 4.12(a); for example, they will generate a seventh-order filter as a cascade of a single gain constant, a_0, three second-order sections, and a first-order section. On the other hand, if the cascade structure is realized using *fixed point arithmetic* then it is important to pay attention to scaling between sections. In this case the cascade will have a scaling multiplier (< 1) for each section in order to prevent overflow in the following circuit (Fig. 4.12(b)). Sometimes scaling by a constant > 1 will be used if it appears that the most significant bits in the following circuit would otherwise remain unused. We will see in Example 4.11 that interstage scaling can also give unity passband gain for each section. Apart from scaling, it may also be necessary to control the ordering of sections and the pairing of the poles and zeros for best results (Rabiner and Gold, 1975).

Rather than express (4.12) as a product of low-order sections, we could express it as the partial fraction expansion

$$H(z) = A + \sum_i H_i(z) \tag{4.41}$$

where A is a constant (Fig. 4.13). For example, it is readily shown that the second-order transfer function

$$H(z) = \frac{3 + 3.6z^{-1} + 0.6z^{-2}}{1 + 0.1z^{-1} - 0.2z^{-2}} = \frac{3(z + 1)(z + 0.2)}{(z + 0.5)(z - 0.4)}$$

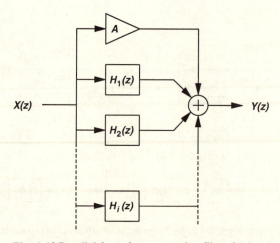

Fig. 4.13 Parallel form for a recursive filter (A is a constant).

can be decomposed to

$$H(z) = -3 - \frac{1}{1 + 0.5z^{-1}} + \frac{7}{1 - 0.4z^{-1}}$$

which is a parallel connection of first-order sections, with $A = -3$. Similarly, if we parallel

$$H_1(z) = (1 + b_1z^{-1} + b_2z^{-2})^{-1}$$

$$H_2(z) = (1 + b_3z^{-1} + b_4z^{-2})^{-1}$$

then

$$H(z) = H_1(z) + H_2(z) \tag{4.42}$$

which is of the form

$$H(z) = \frac{2 + c_1z^{-1} + c_5z^{-2}}{1 + c_1z^{-1} + c_2z^{-2} + c_3z^{-3} + c_4z^{-4}} \tag{4.43}$$

A fourth-order recursive filter can therefore be realized either directly, or as a cascade of two second-order sections, or as a parallel connection of two second-order sections. The parallel form has similar coefficient sensitivity to the cascade form, but it generates a lower level of *roundoff noise* (Section 4.7) than the cascade structure. Apart from cascade and parallel forms, low coefficient sensitivity can also be obtained from *ladder, lattice* and *wave* structures.

4.2.2 Non-recursive structures

A non-recursive structure has the transfer function

$$H(z) = a_0 + a_1z^{-1} + a_2z^{-2} + \ldots + a_Nz^{-N} \tag{4.44}$$

and, according to (4.14), we can write this as

$$H(z) = a_0(z - z_1)(z - z_2) \ldots (z - z_N)z^{-N} \tag{4.45}$$

This structure has N zeros and an N-fold pole at $z = 0$ (which means that it is always stable). As previously noted, poles at the origin simply represent a shift of $h(n)$ in time, and we usually refer to the structure as an *all-zeros* filter. For example, if $N = 2$, and assuming real coefficients, the filter will have two real zeros, or a pair of complex conjugate zeros. Implementing (4.44) directly gives the structure in Fig. 4.14.

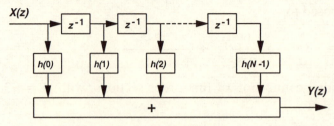

Fig. 4.14 Direct form transversal FIR filter.

Note that the coefficients a_i have been relabelled since they now correspond directly to the finite impulse response $h(n)$, which has length N. Thus, we have the convolution

$$y(n) = \sum_{i=0}^{N-1} h(i)x(n-i) \tag{4.46}$$

and the transfer function is

$$H(z) = \sum_{i=0}^{N-1} h(i)z^{-i} \tag{4.47}$$

This type of structure, where only the input signal is stored in the delay elements, is commonly referred to as a *transversal* filter. The *transposed* form of Fig. 4.14 is shown in Fig. 4.15 (transposed configurations are a useful family of filters that are easily generated from parent structures using a set of signal flow graph rules, see Appendix A). This alternative structure has an identical transfer function but is favoured for DSP applications since each delay z^{-1} acts as a *pipeline register* for the associated adder. In other words, only a single add is required after multiplication before the result can be stored (or output). In contrast, the transversal structure is more demanding since it requires $N-1$ additions after multiplication in order to generate a single output sample, $y(n)$. Also, the transposed filter has a simple repetitive structure or cell which can be cascaded without limit, and this makes it attractive for VLSI implementation. Dedicated FIR filter chips often use this configuration.

Non-recursive filters are usually designed for a true linear phase response $(h(n)_s = h(-n)_s, \ n \neq 0)$ and so (4.46) can be written

$$y(n) = h(0)[x(n) + x(n-(N-1))]$$
$$+ h(1)[x(n-1) + x(n-(N-2))] + \ldots$$

This leads to the efficient linear phase structure in Fig. 4.16. Note that, if

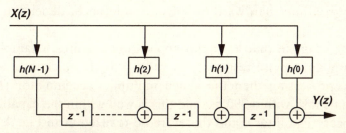

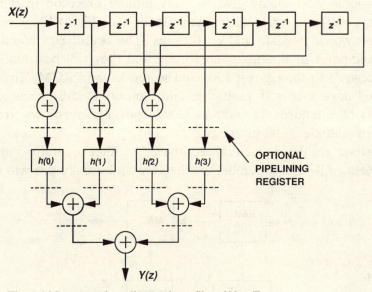

Fig. 4.15 Direct form transposed FIR filter.

Fig. 4.16 Structure for a linear phase filter ($N = 7$).

N is odd, only $(N + 1)/2$ multipliers are now required, compared to N multipliers in Fig. 4.14. Also, if filter delay is not critical, pipelining latches can be inserted to increase the word rate (since the multiplication time can now approach the sample period).

4.2.3 *Computation considerations*

We will stay with non-recursive filters in order to illustrate some basic computation requirements for a digital filter. The filter must perform the convolution in (4.46) for each output sample $y(n)$, and, assuming real-time operation, this means that N multiplications and additions (accumulations) must occur within one sample period. We could achieve this using any of the *parallel* architectures discussed in the previous section, or we may adopt a *sequential* hardware or software approach, depending upon the problem.

Suppose for example that an audio application requires a sample rate $f_s = 50$ kHz and an impulse response length $N = 51$. If we choose sequential hardware processing then the single multiplier-accumulator (MAC) must perform its task within 392 ns, and this is well within the capability of practical devices. The filter could therefore be realized as in Fig. 4.17(*a*), where the RAM stores N data words and the PROM stores the $(N + 1)/2$ coefficients (assuming N odd and $h(n)$ is symmetrical). Note that filter design programmes usually give virtually infinite precision in the filter coefficients, whilst a practical implementation like Fig. 4.17(*a*) typically uses 16-bit words for data and coefficients. The coefficients could be in either fixed-point or floating-point format and they will be rounded off (not truncated) to the nearest LSB and stored in the PROM. The MAC device will have a 16×16 multiplier and typically a 35-bit accumulator, i.e. 3 bits of extended precision to handle possible overflows from the addition of multiple 32-bit products.

In contrast to the sequential approach, a filter operating at, say, $f_s = 20$ MHz, will need a parallel architecture and usually this will be the

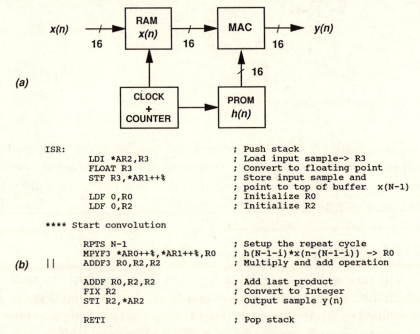

```
ISR:                              ; Push stack
        LDI  *AR2,R3              ; Load input sample-> R3
        FLOAT R3                  ; Convert to floating point
        STF  R3,*AR1++%           ; Store input sample and
                                  ; point to top of buffer  x(N-1)
        LDF  0,R0                 ; Initialize R0
        LDF  0,R2                 ; Initialize R2

**** Start convolution

        RPTS N-1                  ; Setup the repeat cycle
        MPYF3 *AR0++%,*AR1++%,R0  ; h(N-1-i)*x(n-(N-1-i)) -> R0
||      ADDF3 R0,R2,R2            ; Multiply and add operation

        ADDF R0,R2,R2             ; Add last product
        FIX  R2                   ; Convert to Integer
        STI  R2,*AR2              ; Output sample y(n)

        RETI                      ; Pop stack
```

Fig. 4.17 Two approaches to real-time audio rate FIR filtering: (*a*) sequential hardware processing; (*b*) use of a DSP chip: a TMS320C30 interrupt service routine.

transposed structure in Fig. 4.15. In this case only a single multiply–add operation would be required in the available 50 ns sample period.

Many workers have tackled the multiplication problem in digital filters. If at all possible, it is advantageous to reduce the coefficients to small integers or to integer powers of 2, since then the convolution in (4.46) reduces to simple *bit-shift and add* operations. Clearly, this can also result in significant deviation from the desired frequency response, but, on the other hand, filters based upon such techniques can be very useful (Lynn, 1980; Annegarn, Nillesen and Raven, 1986; Noonan and Marquis, 1989; Wade, Van-Eetvelt and Darwen, 1990). Even when the coefficients can-not be approximated to integer powers of 2, it is still possible to reduce (4.46) to a sequence of bit-shift and add operations by using *distributed arithmetic* (Burrus, 1977; Tan and Hawkins, 1981). Suppose that the input samples $x(n)$ are scaled to fractional 2's complement form with a resolu-tion of m bits including the sign bit (see Appendix B). Sample $x(n - i)$ is now represented by

$$x(n - i) = \sum_{k=1}^{m-1} x_{n-i,k} 2^{-k} - X_{n-i} \tag{4.48}$$

where $x_{n-i,k}$ is either 0 or 1 and X_{n-i} is the sign bit. Therefore, (4.46) becomes

$$y(n) = \sum_{i=0}^{N-1} h(i) \left(\sum_{k=1}^{m-1} x_{n-i,k} 2^{-k} - X_{n-i} \right)$$

$$= \sum_{k=1}^{m-1} \sum_{i=0}^{N-1} h(i) x_{n-i,k} 2^{-k} - \sum_{i=0}^{N-1} h(i) X_{n-i} \tag{4.49}$$

Letting

$$C_{n,k} = \sum_{i=0}^{N-1} h(i) x_{n-i,k}$$

and

$$C_n = \sum_{i=0}^{N-1} h(i) X_{n-i}$$

then

$$y(n) = \sum_{k=1}^{m-1} C_{n,k} 2^{-k} - C_n \tag{4.50}$$

Both $C_{n,k}$ and C_n can be obtained from a look-up table containing all the combinations of a sum of N coefficients (the address for the table could be generated from shift registers which store the N delayed inputs). Clearly, (4.50) is a sequence of shift and add operations and true multiplication has been avoided.

Use of DSP chips

Very often a digital filter will be implemented using a general purpose DSP chip. These have a dedicated high-speed hardware multiplier and are quite capable of realizing high-order filters at audio rates. The convolution in (4.46) would be evaluated sequentially (one coefficient at a time) and the ideal DSP solution will run as fast as one instruction cycle per coefficient. An N-tap FIR filter could then be performed by the following instructions:

    (1) repeat the next instruction N times

    (2) multiply, accumulate and data move

where instruction (2) is accomplished in just one instruction cycle. Assuming the architecture of a DSP chip is optimized for single cycle multiply–accumulate–move operations, and taking the instruction cycle time as 100 ns, the foregoing argument implies that the chip could support FIR filter tap lengths approaching 200 for an audio rate of 50 kHz (neglecting ancillary cycles). For an *adaptive* filter (Section 4.6) the N coefficients are usually updated for each new input sample (as in (4.157)). In this case the best DSP solution will run the filter as fast as two instruction cycles per coefficient: one for performing multiplication and accumulation, and the other for updating the coefficient.

Example 4.5

The TMS320C30 processor is a general purpose floating-point device with a 60 ns instruction cycle time. It has eight extended precision registers R0–R7 and eight auxiliary registers AR0–AR7, and the use of these for implementing an N-tap FIR filter is illustrated in the assembly code in Fig. 4.17(b). This shows an interrupt service routine for real time audio rate filtering.

Assuming all necessary processor set-up instructions have been performed prior to this routine, the code will service an interrupt (load sample $x(n)$), perform the convolution in (4.46), and output a sample $y(n)$. The first five instructions are concerned with input buffering (circular addressing) and the following three instructions essentially evaluate (4.46). Clearly, the total number of cycles required by this routine must be less than the number of cycles between interrupts, and so this limits the maximum sampling frequency f_s for a fixed N. For example, using one wait-state RAM, the interrupt service routine shown requires some 426 cycles for $N = 99$. The maximum sampling frequency is then approximately

$$f_s(\text{max}) = \frac{1}{426 \times 60 \times 10^{-9}} \approx 39 \text{ kHz}$$

4.3 IIR filter design

IIR filters are attractive for several reasons:

For the same filter specification, an IIR filter generally requires a significantly lower-order transfer function than an FIR filter. Put another way, the usual non-recursive realization of an FIR filter (as in (4.45)) has no poles near the unit circle and so requires a large value of N and significantly more computation in order to achieve sharp cutoff.

IIR filters can be designed by transforming a classical continuous-time filter, such as a Butterworth or Chebyshev filter, into a digital equivalent. This means that we can utilize well-established analogue filter theory in the design procedure.

4.3.1 Pole-zero placement method

It is possible to design *simple* IIR filters without resort to sophisticated software. All we need to do is place poles and zeros in the z-plane so as to achieve the desired frequency response. The significant point is that the frequency response (Fourier transform of $h(n)$) is represented by the unit circle, and that an anti-clockwise rotation round the circle from $z = 1$ to $z = -1$ spans the normalized frequency range 0–0.5 (or 0–π), as shown in Fig. 4.18. For example, the pair of complex conjugate poles at

$$z = r\,e^{\pm j\theta_1}, \quad \theta_1 = \omega_1 T$$

will yield a simple resonant peak at $\omega_0 \approx \omega_1$. The effect of the pole (or zero) on the response increases the closer it is to the circle, and poles are often located very close to the circle (e.g. $r = 0.999$) in order to achieve high Q. Similarly, a pair of complex conjugate zeros lying *on* the circle at

$$z = e^{\pm j\theta_2}, \quad \theta_2 = \omega_2 T$$

will give a zero in the magnitude response at $\omega = \omega_2$. Conversely, a zero well away from the circle will give only a small attenuation at this

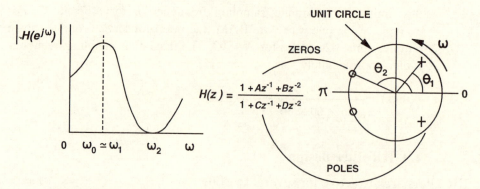

Fig. 4.18 Typical pole-zero plot and frequency response for a general second-order filter.

frequency, and can be useful for 'flattening' a bandpass response, for example. Taking specific examples, if $\theta_1 = 45°$ and $\theta_2 = 135°$, then the response will have a resonant peak close to the normalized frequency $f = 0.125$ and a zero at $f = 0.375$.

When designing filters in this way, it is helpful to bear in mind some general rules on pole-zero locations:

Assuming the filter coefficients to be real, then the poles and zeros can only be real (lie on the real axis), or occur in complex conjugate pairs. The pole-zero plot is therefore always symmetrical with respect to the real axis. Also, we say that a zero (pole) at $r\, e^{j\theta}$ is *reflected* in the unit circle as $r^{-1} e^{j\theta}$.

If all zeros are outside the unit circle and each pole inside the circle is a reflection of a zero outside the circle, then the filter is an *all-pass* (i.e. constant amplitude) or *phase shift* network. A simple example is illustrated in Fig. 4.19(*a*). All-pass filters are useful for phase correction.

If all zeros lie *inside* the unit circle, then the network has *minimum phase* or no *excess phase*. This means that the phase variation with frequency is small compared to the variation when all the zeros are reflected to occur outside the circle (the magnitude response being the same in both cases).

If a filter has only zeros (except for the origin) and if these are reflected in pairs in the unit circle (possibly also with zeros *on* the circle) then the filter has exactly linear phase over the band $0-\pi$. Fig. 4.19(*b*) shows the z-plane plot of a simple, linear phase filter.

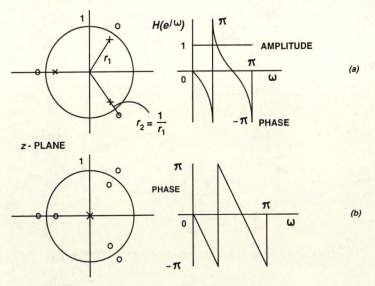

Fig. 4.19 Special pole-zero plots: (*a*) all-pass or phase-shift network; (*b*) linear phase network.

Example 4.6

An electronic reverberation system could be based upon the Nth-order recursive filter in Fig. 4.20(*a*). Since $|g| < 1$ a pulse at the input appears at the output as a decaying pulse train, the pulses being separated by an interval of NT seconds. The amplitude response of the filter is a series of peaks spaced at intervals of $2\pi/NT$, which gives rise to the term *comb filter*. Unfortunately, the regularity of the pulses generates a rather 'metallic' sound, and one way to improve the reverberation is to generate extra 'reflections' (pulses) by adding a phase shift network in series with the comb filter. A basic phase shifter is shown in Fig. 4.20(*b*), and for this filter we have

$$Y(z) = -gW(z) + z^{-N}W(z) \tag{4.51}$$

where

$$W(z) = X(z) + gz^{-N}W(z)$$

giving

$$H(z) = \frac{Y(z)}{X(z)} = \frac{-g + z^{-N}}{1 - gz^{-N}} \tag{4.52}$$

The poles of (4.52) are given by the roots of $z^N = g$. Letting $g = r^N$ and noting that $e^{j2\pi m} = 1$, $m = 0, 1, 2, \ldots$ we have

$$z = r\,e^{j2\pi m/N}, \quad m = 0, 1, 2, \ldots, N-1$$

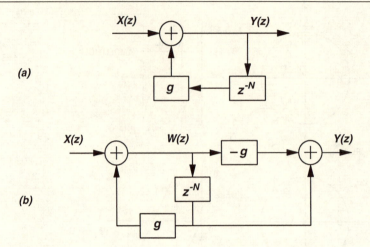

Fig. 4.20 (*a*) Comb filter; (*b*) *N*th-order phase shift network.

This corresponds to N poles on a circle of radius r, $(r < 1)$. It is similarly shown that there are N zeros at

$$z = r^{-1} e^{j2\pi m/N}, \quad m = 0, 1, 2, \ldots, N - 1$$

Note that each zero is a reflection of a corresponding pole, as expected for a phase shift network. The frequency response of the network is given by (4.19)

$$H(e^{j\omega T}) = H(z)|_{z=e^{j\omega T}}$$

$$= \frac{-g + e^{-j\omega NT}}{1 - ge^{-j\omega NT}}$$

$$= e^{-j\omega NT}\left(\frac{1 - ge^{j\omega NT}}{1 - ge^{-j\omega NT}}\right) \tag{4.53}$$

It is readily shown from (4.53) that the filter has a constant amplitude response and a phase shift

$$\phi(\omega) = -\omega NT - 2\tan^{-1}\left[\frac{g\sin(\omega NT)}{1 - g\cos(\omega NT)}\right] \tag{4.54}$$

All-pole second-order section

The simple recursive filter in Fig. 4.21 is quite common. Typical applications include symbol clock recovery in digital communications systems, and narrow bandpass filtering for tone detection in FSK modems (which requires two similar cascaded sections per tone). From (4.13), the transfer function is simply

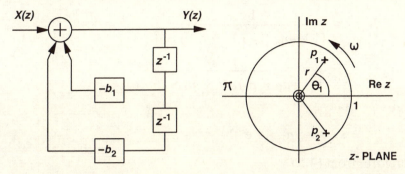

Fig. 4.21 All-pole second-order filter.

$$H(z) = (1 + b_1 z^{-1} + b_2 z^{-2})^{-1} \tag{4.55}$$

and this has complex poles at

$$z = -\frac{b_1}{2} \pm \frac{j}{2}\sqrt{(4b_2 - b_1^2)}, \quad 4b_2 > b_1^2$$

From (4.19) we can write the frequency response as

$$H(e^{j\theta}) = (1 + b_1 e^{-j\theta} + b_2 e^{-j2\theta})^{-1}, \quad \theta = \omega T \tag{4.56}$$

The filter is capable of simple high-Q resonance and so we are primarily interested in its magnitude response. Expanding (4.56), we have

$$H(e^{j\theta})^{-1} = 1 + b_1 \cos \theta + b_2 \cos (2\theta) - j[b_1 \sin \theta + b_2 \sin (2\theta)]$$

$$|H(e^{j\theta})^{-1}|^2 = D(\theta) = [1 + b_1 \cos \theta + b_2 \cos (2\theta)]^2$$

$$+ [b_1 \sin \theta + b_2 \sin (2\theta)]^2 \tag{4.57}$$

and simplifying gives

$$D(\theta) = 1 + b_1^2 + b_2^2 + 2b_1(1 + b_2) \cos \theta + 2b_2 \cos (2\theta) \tag{4.58}$$

The magnitude response is a maximum when $D(\theta)$ is a minimum, and so differentiating (4.58) gives the angle θ_0 corresponding to resonance

$$\cos \theta_0 = \frac{-b_1(1 + b_2)}{4b_2}, \quad \theta_0 = \omega_0 T \tag{4.59}$$

Provided that the poles are close to the unit circle, the angular position of the poles (θ_1 in Fig. 4.21) is virtually the same as the resonance angle, θ_0. The exact relationship between θ_0 and θ_1 can be obtained as follows. Using (4.16), the transfer function for Fig. 4.21 can be written

$$H(z) = \frac{1}{(1 - p_1 z^{-1})(1 - p_2 z^{-1})}$$

$$= \frac{1}{1 - (p_1 + p_2)z^{-1} + p_1 p_2 z^{-2}}$$

where

$$p_1 = r\,e^{j\theta_1}, \quad p_2 = r\,e^{-j\theta_1}$$

Therefore, from (4.55)

$$b_1 = -(p_1 + p_2) = -2r\cos\theta_1 \qquad (4.60)$$

$$b_2 = p_1 p_2 = r^2, \quad r < 1 \qquad (4.61)$$

and substituting in (4.59) gives

$$\cos\theta_0 = \frac{(1 + r^2)}{2r}\cos\theta_1 \qquad (4.62)$$

Clearly, $\theta_0 \approx \theta_1$ when $r \approx 1$, and in this case it can be shown that the filter Q is

$$Q \approx \frac{\pi f_0}{1 - r} \qquad (4.63)$$

where f_0 is the normalized resonant frequency. The frequency response at resonance is therefore inversely proportional to the distance from the pole to the unit circle.

Example 4.7

The filter in Fig. 4.21 is useful for symbol clock recovery in a communications system. A particular application required the filter to resonate at 100 kHz and hence to 'ring' when excited with a random binary bit-stream which had a strong component at 100 kHz. The filter was to be implemented using a gate array device, and so it was advantageous to reduce the multiplications to 'bit-shift and add' operations. For example, in order to achieve a reasonable Q, we see from (4.61) and (4.63) that $b_2 \approx 1$, and so a reasonable bit-shift and add approximation is

$$b_2 = \frac{127}{128}, \quad r = 0.99608609$$

A further criterion when using a gate array device is that the routing delays between logic blocks should be well below the filter clock period. This delay also depends upon the word length used in the filter. In practice, a nominal sampling frequency of 2.5 MHz proved adequate for a

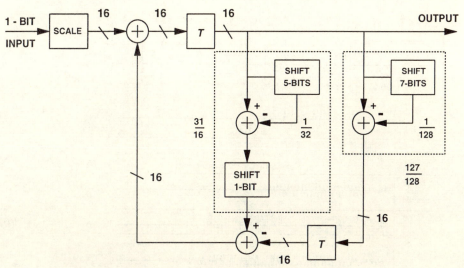

Fig. 4.22 A practical form of second-order recursive filter.

16-bit word length, in which case a little manipulation of (4.59) yields $b_1 \approx -2$. A reasonable bit-shift and add approximation to b_1 is therefore

$$b_1 = -\frac{31}{16}$$

and the practical circuit is shown in Fig. 4.22 (Smithson, 1992).

The above approximations will, of course, give errors in the frequency response, and in order to achieve resonance at *exactly* 100 kHz it is necessary to define the sampling frequency accurately. Substituting for b_1 and b_2 in (4.59) gives

$$f_s = 2.6761245 \text{ MHz}$$

corresponding to a normalized resonant frequency $f_0 = 0.0373675$. Finally, note that the excitation to the practical circuit is only one bit wide, and so this must be scaled to a 16-bit word for the first adder. Ideally, scaling gives a filter output virtually filling the 16-bit dynamic range when the filter excitation is a maximum, i.e. a 101010 . . . bit-stream.

Example 4.8
Fig. 4.23(*a*) shows the symbol clock recovery system for a practical data modem. Input *A* is a random NRZ data stream at 1200 baud, and because the PSD of this signal is zero at the symbol rate (see Fig. 3.37) a little processing is required before applying it to the filter. A strong component at the symbol rate can be generated by using a half-symbol delay and modulo-2 addition, thereby generating a pulse at each transi-

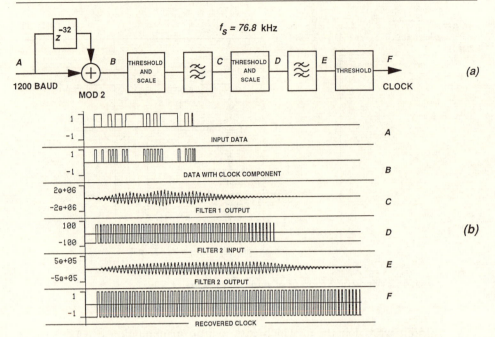

Fig. 4.23 Symbol clock recovery in a data modem: (*a*) circuit, (*b*) computer simulation.

tion (see Fig. 4.23(*b*)). The modem filters sample at 76.8 kHz, i.e. 64 samples/symbol and so a half-symbol delay corresponds to z^{-32}.

Assuming second-order filters as in Fig. 4.21, the filter coefficients can be determined from (4.59). For example, setting

$$b_2 = 255/256 \quad \text{and} \quad \theta_0 = 2\pi 1.2/76.8$$

gives $b_1 = -1.9864744$ and $Q \approx 25$. The modem specification requires that a clock is generated for at least 40 symbols after a signal break, and the simulation in Fig. 4.23(*b*) indicates that this can be achieved by cascading two identical filters. Scaling at the filter input ensures that the signal fills the dynamic range of the fixed-point processor used to realize the filters.

4.3.2 Bilinear transform method

The bilinear transform method is one of the most popular techniques for designing high-order IIR filters since there are a number of factors in its favour:

Classical analogue filter design techniques can be utilized to generate a *prototype* analogue filter, thereby solving the problem of finding the pole and zero locations. The prototype transfer function $H_p(s)$ of order N is then *transformed* to $H(z)$ of the same order to give the digital filter.

There is no danger of *aliasing* errors in the frequency response of the digital filter. This means, for example, that the method can design a lowpass digital filter with a cutoff frequency approaching half the sampling frequency. Other transformation techniques, such as the *impulse invariance method*, suffer from significant aliasing unless the cutoff frequency (or any significant amplitude response) is well below half the sampling rate. In other words, impulse invariance would be unsuitable for high-pass and bandstop filters because of aliasing.

The bilinear transform method preserves any equiripple characteristics of the analogue prototype. It also maintains (and for some filters, even enhances) the sharp roll-off characteristics of the prototype.

Stable prototype filters are transformed to stable digital filters.

A possible disadvantage of the technique is that the impulse response, phase response and detailed magnitude response of the prototype filter are *not* preserved in the transformation.

Classical analogue filters

The general transfer function for a lowpass analogue filter of order k is

$$H(s) = \frac{a_0}{b_0 + b_1s + b_2s^2 + \ldots + b_ks^k} \tag{4.64}$$

and if the filter is to have unity gain at zero frequency we select $a_0 = b_0$. The denominator is usually factorized into first-order and second-order polynomials since this facilitates a digital filter with a cascade architecture, as in Fig. 4.12. Factored forms of the denominator for some Butterworth (maximally flat) and Chebyshev (equiripple) lowpass filters are given in compact form in Table 4.2. The first column gives the first-order polynomial, where appropriate, whilst the second column gives second-order polynomials, less redundant information. Thus, a fifth-order Butterworth lowpass filter has the transfer function

$$H(s) = \frac{1}{(1 + s)(1 + 0.6180340s + s^2)(1 + 1.6180340s + s^2)} \tag{4.65}$$

Note that the normalized radian cutoff frequency (i.e. when $\omega = 1$ rad/s) corresponds to the 3 dB point for Butterworth filters, whilst the Chebyshev filters are 0.5 dB and 1 dB respectively below the maximum response at this frequency. Given lowpass filters in this normalized form it is easy to apply standard transformations in order to generate lowpass, highpass, bandpass and bandstop filters having arbitrary cutoff frequencies. These transformations are as follows:

Lowpass: To scale the unity cutoff frequency to give a lowpass prototype with cutoff frequency ω_{pc} use the transformation

$$s \Rightarrow s/\omega_{pc} \qquad (4.66)$$

Table 4.2. *Factored forms of lowpass filter denominator polynomials. The filters are normalized to a radian cutoff frequency of 1 rad/s (based upon tables from L.A. Weinberg, 'Network Analysis and Synthesis', McGraw-Hill, 1962, with permission).*

Butterworth polynomials

k		Second-order polynomials: $1 + b_1 s + s^2 \to b_1 s$			
1	$1 + s$				
2		$1.4142136s$			
3	$1 + s$	$1.0000000s$			
4		$0.7653668s$	$1.8477590s$		
5	$1 + s$	$0.6180340s$	$1.6180340s$		
6		$0.5176380s$	$1.4142136s$	$1.9318516s$	
7	$1 + s$	$0.4450418s$	$1.2469796s$	$1.8019378s$	
8		$0.3901806s$	$1.1111404s$	$1.6629392s$	$1.9615706s$
9	$1 + s$	$0.3472964s$	$1.0000000s$	$1.5320888s$	$1.8793852s$
10		$0.3128690s$	$0.9079810s$	$1.4142136s$	$1.7820130s$	$1.9753766s$

Chebyshev polynomials (0.5 dB ripple)

k		Second-order polynomials: $b_0 + b_1 s + s^2 \to b_0 + b_1 s$			
1	$2.8627752 + s$				
2		$1.5162026 + 1.4256244s$			
3	$0.6264565 + s$	$1.1424477 + 0.6264564s$			
4		$1.0635187 + 0.3507062s$	$0.3564119 + 0.8466796s$		
5	$0.3623196 + s$	$1.0357841 + 0.2239258s$	$0.4767669 + 0.5862454s$		
6		$1.0230227 + 0.1553002s$	$0.5900102 + 0.4242880s$	$0.1569974 + 0.5795880s$	
7	$0.2561700 + s$	$1.0161074 + 0.1140064s$	$0.6768836 + 0.3194388s$	$0.2538781 + 0.4616024s$	
8		$1.0119319 + 0.0872402s$	$0.7413338 + 0.2484390s$	$0.3586503 + 0.3718152s$	
				$0.0880523 + 0.4385858s$	
9	$0.1984053 + s$	$1.0092110 + 0.0689054s$	$0.7893647 + 0.1984052s$	$0.4525405 + 0.3039746s$	
				$0.1563424 + 0.3728800s$	
10		$1.0073355 + 0.0557988s$	$0.8256997 + 0.1619344s$	$0.5318072 + 0.2522188s$	
			$0.2379146 + 0.3178144s$	$0.0562789 + 0.3522998s$	

Table 4.2. (*cont.*)

Chebyshev polynomials (1 dB ripple)

k	Second-order polynomials: $b_0 + b_1 s + s^2 \rightarrow b_0 + b_1 s$		
1	$1.9652267 + s$		
2	$1.1025104 + 1.0977344s$		
3	$0.4941706 + s$ $0.9942046 + 0.4941706s$		
4	$0.9865049 + 0.2790720s$ $0.2793981 + 0.6737394s$		
5	$0.2894933 + s$ $0.9883149 + 0.1789168s$ $0.4292978 + 0.4684100s$		
6	$0.9907329 + 0.1243620s$ $0.5577196 + 0.3397634s$ $0.1247069 + 0.4641254s$		
7	$0.2054141 + s$ $0.9926793 + 0.0914178s$ $0.6534554 + 0.2561472s$ $0.2304500 + 0.3701434s$		
8	$0.9941408 + 0.0700164s$ $0.7235427 + 0.1993900s$ $0.3408592 + 0.2987408s$		
			$0.0702612 + 0.3519966s$
9	$0.1593305 + s$ $0.9952326 + 0.0553348s$ $0.7753861 + 0.1593304s$ $0.4385620 + 0.2441084s$		
			$0.1423640 + 0.2994434s$
10	$0.9960584 + 0.0448288s$ $0.5205301 + 0.2026332s$ $0.8144228 + 0.1300986s$		
			$0.2266375 + 0.2553328s$ $0.0450019 + 0.2830386s$

Highpass: To transform a normalized lowpass response to a high-pass prototype with cutoff frequency ω_{pc} use the transformation

$$s \Rightarrow \omega_{pc}/s \qquad\qquad (4.67)$$

Bandpass: To transform a normalized lowpass response to a band-pass prototype having lower and upper cutoff frequencies ω_{pl} and ω_{pu} respectively,

$$s \Rightarrow \frac{s^2 + \omega_{pl}\omega_{pu}}{s(\omega_{pu} - \omega_{pl})} \qquad\qquad (4.68)$$

Bandstop: To transform a normalized lowpass response to a band-stop prototype with cutoff frequencies ω_{pl} and ω_{pu},

$$s \Rightarrow \frac{s(\omega_{pu} - \omega_{pl})}{s^2 + \omega_{pl}\omega_{pu}} \qquad\qquad (4.69)$$

Bandlimitation and frequency warping
In order to avoid aliasing, the bilinear transform approach first compresses the entire frequency response of the prototype filter (the frequency range $-\infty < \omega_p < \infty$) into the band

$$-\frac{\pi}{T} < \omega < \frac{\pi}{T} \qquad\qquad (4.70)$$

of the digital filter. This *bandlimitation* or *frequency compression* is achieved using the transformation

$$s = \tanh\left(\frac{s_b T}{2}\right) = \frac{1 - e^{-s_b T}}{1 + e^{-s_b T}} \tag{4.71}$$

where s_b denotes an intermediate s-plane corresponding to the band-limited frequency response. For example, (4.71) shows that the maximum frequency limit on the imaginary axis, corresponding to $s_b = j\pi/T$, maps to ∞ in the s-plane, and vice-versa. The second stage in the bilinear transform is to move from the s_b-plane to the z-plane by letting

$$z = e^{s_b T} \tag{4.72}$$

and so the overall transformation can be written

$$s = \frac{1 - z^{-1}}{1 + z^{-1}} = \frac{z - 1}{z + 1} \tag{4.73}$$

We can regard (4.73) as mapping the infinite imaginary axis in the s-plane onto one revolution of the unit circle in the z-plane, and the practical outcome is to *warp* the frequency response. In particular, high-frequency parts of the prototype response are compressed into smaller and smaller bands in the digital filter response as $\omega \Rightarrow \pi/T$. This frequency compression can be expressed mathematically by letting $s = j\omega_p$ (a point on the imaginary axis of the s-plane) and $z = e^{j\omega T}$ (the corresponding point on the unit circle) in (4.73), giving

$$\omega_p = \tan(\omega T/2) \tag{4.74}$$

Clearly, very high frequencies in the prototype response are compressed to occur near $\omega = \pi/T$ in the digital filter response. We are now in a position to state the bilinear transform design procedure for a general IIR filter:

(1) Note the critical frequencies ω_i in the required digital filter response, e.g. the passband edge frequencies in a bandpass filter, and 'prewarp' them according to (4.74), i.e. setting $T = 1$

$$\omega_{pi} = \tan(\omega_i/2) \tag{4.75}$$

(2) Design a *prototype analogue filter* $H_p(s)$ using critical frequencies ω_{pi} (and using Table 4.2, if appropriate). A high-order denominator polynomial should be factorized into second- and first-order polynomials.

(3) Generate the digital filter using the transformation

$$H(z) = H_p(s)\big|_{s=\frac{z-1}{z+1}} \tag{4.76}$$

As previously noted, it is useful to express $H_p(s)$ in terms of low-order polynomials since this will facilitate the generation of a canonical cascade structure. Usually $H_p(s)$ will then be in terms of first- and second-order sections and the general second-order prototype filter is given by

$$H_p(s) = \frac{a_0 + a_1 s + a_2 s^2}{b_0 + b_1 s + b_2 s^2} \tag{4.77}$$

Applying (4.76) then gives the corresponding digital filter section

$$H(z) = \frac{(a_0 + a_1 + a_2) + 2(a_0 - a_2)z^{-1} + (a_0 - a_1 + a_2)z^{-2}}{(b_0 + b_1 + b_2) + 2(b_0 - b_2)z^{-1} + (b_0 - b_1 + b_2)z^{-2}} \tag{4.78}$$

Example 4.9

Suppose we require a second-order lowpass digital filter with a maximally flat amplitude response and a normalized 3 dB cutoff frequency $f_c = 0.1$. The required analogue filter is the second-order Butterworth section in Table 4.2:

$$H(s) = \frac{1}{1 + 1.4142136s + s^2} \tag{4.79}$$

Note that $H(s)$ gives unity gain at zero-frequency and has a normalized cutoff frequency of 1 rad/s. The critical frequency in the digital filter response is simply the normalized cutoff frequency $\omega_c = 0.2\pi$, and so the prototype cutoff frequency is, from (4.75),

$$\omega_{pc} = \tan\left(\frac{\omega_c}{2}\right) = 0.3249197$$

Using (4.66), the required prototype filter is therefore

$$H_p(s) = \frac{\omega_{pc}^2}{\omega_{pc}^2 + 1.4142136\omega_{pc}s + s^2}$$

$$= \frac{0.1055728}{0.1055728 + 0.4595058s + s^2} \tag{4.80}$$

Using (4.77) and (4.78) then gives the transfer function of the digital filter

$$H(z) = \frac{0.1055728(1 + 2z^{-1} + z^{-2})}{1.5650786 - 1.7888544z^{-1} + 0.6460670z^{-2}}$$

$$H(z) = \frac{0.0674553(1 + 2z^{-1} + z^{-2})}{1 - 1.1429805z^{-1} + 0.4128016z^{-2}} \tag{4.81}$$

In practice, non-integer filter coefficients are often scaled to enable them to be expressed in fractional 2's complement form. Scaling (4.81) by 2 (a simple bit-shift) for example, gives all coefficients less than one:

$$H(z) = \frac{0.0337276(1 + 2z^{-1} + z^{-2})}{0.5 - 0.5714903z^{-1} + 0.2064008z^{-2}} \tag{4.82}$$

and this is of the general form

$$H(z) = \frac{a_0 + a_1 z^{-1} + a_2 z^{-2}}{b_0^{-1} + b_1 z^{-1} + b_2 z^{-2}} \tag{4.83}$$

This generalized transfer function contains a coefficient scaling factor $b_0 = 2^m$, m being a positive integer, and corresponds to Fig. 4.24(a). Using this diagram, and comparing (4.82) with (4.83) gives the required filter structure in Fig. 4.24(b). The complex poles are at $z = 0.6425$ $\angle \pm 27.2°$.

A comparison of the frequency response of the digital filter in Fig. 4.24(b) with that of the prototype analogue filter will show that the digital filter has a significantly sharper roll-off, due to the frequency compression. Finally, note that the filter *dc gain* is 0 dB and is determined by setting $z = 1$ in $H(z)$.

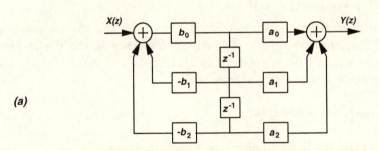

(a)

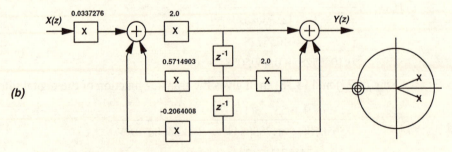

(b)

Fig. 4.24 (a) General second-order section, with coefficient scaling; (b) Butterworth filter and pole-zero plot for $f_c = 0.1$.

Example 4.10

Consider the design of a second-order highpass Chebyshev filter with a peak–peak passband ripple of 1 dB and a cutoff frequency $f_c = 0.3$. From Table 4.2 the normalized analogue lowpass filter with unity gain at zero frequency is

$$H(s) = \frac{1.1025104}{1.1025104 + 1.0977344s + s^2} \tag{4.84}$$

The prewarped cutoff frequency is

$$\omega_{pc} = \tan{(0.3\pi)} = 1.3763819$$

and applying (4.67) in order to transform to a highpass filter gives the prototype response

$$H_p(s) = \frac{1.1025104s^2}{1.8944272 + 1.5109178s + 1.1025104s^2} \tag{4.85}$$

This can be directly transformed to a digital highpass filter using (4.77) and (4.78), giving

$$H(z) = \frac{1.1025104(1 - 2z^{-1} + z^{-2})}{4.5078394 + 1.5838336z^{-1} + 1.4860358z^{-2}}$$

$$H(z) = \frac{0.2445762(1 - 2z^{-1} + z^{-2})}{1 + 0.3513509z^{-1} + 0.3296559z^{-2}} \tag{4.86}$$

The frequency response will be 1 dB below the maximum response at $f_c = 0.3$. The filter has a double zero at $z = 1$ and complex poles at 0.5742 $\angle \pm 107.8°$. The structure can be deduced from (4.83) and Fig. 4.24(*a*).

Example 4.11

Usually a filter specification involves a minimum stopband attenuation as well as a cutoff frequency. Suppose we require a lowpass Butterworth filter with $f_c = 0.125$ and an attenuation > 40 dB at $f = 0.28$. The prewarped critical frequencies are, from (4.75),

$$\omega_{pc} = \tan{(0.125\pi)} = 0.4142136$$

$$\omega_{p1} = \tan{(0.28\pi)} = 1.2087923$$

and so the prototype filter must have at least 40 dB attenuation at a normalized frequency of $\omega_{p1}/\omega_{pc} = 2.92$. Reference to normalized (unity cutoff frequency) curves for analogue Butterworth filters shows that a fifth-order filter is required, and so the normalized lowpass response is given by (4.65). Realizing the filter as a cascade of low-order sections,

$$H(s) = \prod_{i=0}^{2} H_i(s) \tag{4.87}$$

$$H(z) = \prod_{i=0}^{2} H_i(z) \tag{4.88}$$

$$H_i(z) = H_{pi}(s)\big|_{s=\frac{z-1}{z+1}} \tag{4.89}$$

According to (4.89), the bilinear transform is applied individually to each prototype section. The prewarped prototype response for the first-order section in (4.65) is, from (4.66),

$$H_{p0}(s) = \frac{1}{1 + (s/\omega_{pc})} = \frac{0.4142136}{0.4142136 + s} \tag{4.90}$$

$$\therefore H_0(z) = \frac{0.4142136(z + 1)}{0.4142136(z + 1) + (z - 1)}$$

$$= \frac{0.2928932(1 + z^{-1})}{1 - 0.4142136z^{-1}} \tag{4.91}$$

Similarly,

$$H_{p1}(s) = \frac{1}{1 + 0.6180340(s/\omega_{pc}) + (s/\omega_{pc})^2} \tag{4.92}$$

$$= \frac{0.1715729}{0.1715729 + 0.2559981s + s^2} \tag{4.93}$$

Using (4.77) and (4.78) gives

$$H_1(z) = \frac{0.1715729(1 + 2z^{-1} + z^{-2})}{1.4275709 - 1.6568543z^{-1} + 0.9155748z^{-2}}$$

$$= \frac{0.1201852(1 + 2z^{-1} + z^{-2})}{1 - 1.1606108z^{-1} + 0.6413515z^{-2}} \tag{4.94}$$

The same approach for the last second-order section gives

$$H_2(z) = \frac{0.0931558(1 + 2z^{-1} + z^{-2})}{1 - 0.8995918z^{-1} + 0.2722149z^{-2}} \tag{4.95}$$

Note that each section has unity gain at dc ($z = 1$) due to its scaling factor. The overall scaling or gain factor is

$$a_0 = 3.2792160 \times 10^{-3}$$

and the complete filter is shown in Fig. 4.25.

The design of a bandpass or bandstop IIR filter from a lowpass filter $H(s)$ is slightly more complex due to the nature of the analogue transformations

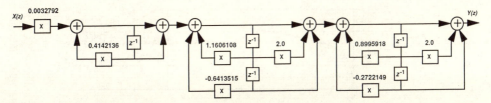

Fig. 4.25 Fifth-order lowpass Butterworth filter ($f_c = 0.125$).

(4.68) and (4.69). These require *both* cutoff frequencies ω_{pl} and ω_{pu} to be prewarped when designing the prototype filter $H_p(s)$. Also, note that these transformations will generate a *fourth-order* prototype $H_p(s)$ from a second-order lowpass filter $H(s)$, and so factorization of the denominator of $H_p(s)$ will normally be required for a canonical cascade architecture.

Example 4.12

We will use the second-order Chebyshev lowpass filter in (4.84) in order to generate a fourth-order bandpass digital filter. Suppose the passband edges are at $f_i = 14$ kHz and $f_u = 16$ kHz, and that $f_s = 44.1$ kHz. Therefore

$$\omega_{pl} = \tan(\omega_l/2) = 1.5483028$$

$$\omega_{pu} = \tan(\omega_u/2) = 2.1747680$$

Applying (4.68) to (4.84) gives a prototype filter $H_p(s)$ with a fourth-order denominator and this can be factorized as

$$H_p(s) = \left(\frac{0.4326896}{4.571334 + 0.3960015s + s^2}\right)\left(\frac{s^2}{2.480246 + 0.2916908s + s^2}\right)$$

(4.96)

Using (4.77) and (4.78) for each second-order section, and writing $H(z) = H_0(z)H_1(z)$, gives

$$H_0(z) = \frac{0.0725097\,(1 + 2z^{-1} + z^{-2})}{1 + 1.1969610z^{-1} + 0.8672769z^{-2}}$$

(4.97)

$$H_1(z) = \frac{0.2651158(1 - 2z^{-1} + z^{-2})}{1 + 0.7848732z^{-1} + 0.8453363z^{-2}}$$

(4.98)

Fig. 4.26 gives the magnitude and phase response for the fourth-order IIR filter.

In practice, the above design problems can be solved very quickly (including a display of the frequency response and pole-zero diagram) using standard DSP software. This software often assumes infinite digital

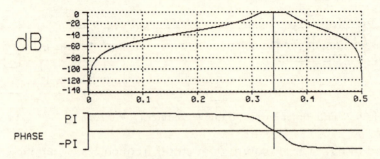

Fig. 4.26 Frequency response for a fourth-order Chebyshev bandpass filter
designed using the bilinear transform method (Example 4.12).

precision in the filter realization, and in this case the order of the cascaded
sections is not critical. Note, however, that the above cascaded structures
will often be modified for fixed-point arithmetic, as mentioned in Section
4.2.1.

4.4 FIR filter design

The attractive features of finite impulse response filters are:

> They can be (and usually are) designed for *exactly* linear phase
> over the normalized band $0-\pi$.
>
> The usual, non-recursive realization (Figs. 4.14 and 4.15) has no
> poles (except at $z = 0$) and so is always stable.
>
> Non-recursive FIR filters can have lower *roundoff* noise and lower
> coefficient sensitivity than IIR filters. The coefficient resolution
> required for FIR filters is typically 12–16 bits compared to 16–24
> bits for IIR filters.

4.4.1 The ideal lowpass filter

A study of the ideal lowpass filter serves as a good introduction to FIR
filter design. We require a filter which approaches the ideal constant
amplitude and linear phase response in Fig. 4.27. The structure will
probably be that of Fig. 4.14 or Fig. 4.15, and the main design problem is
to compute the coefficients $h(n)$. From (4.21)

$$h(n) = \frac{1}{\omega_s} \int_{-\omega_s/2}^{\omega_s/2} H(e^{j\omega T}) e^{j\omega n T} \, d\omega \qquad (4.99)$$

and in this case the integral is easily evaluated since the response is simple.
Over the passband

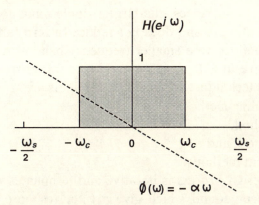

Fig. 4.27 Ideal lowpass filter.

$$H(e^{j\omega T}) = 1 \times e^{-j\alpha\omega} \qquad (4.100)$$

and substituting (4.100) into (4.99), and integrating, gives

$$h(n) = 2Tf_c \frac{\sin[(nT - \alpha)\omega_c]}{(nT - \alpha)\omega_c} \qquad (4.101)$$

Neglecting the filter phase delay gives the shifted or non-causal impulse response

$$h(n)_s = 2Tf_c \frac{\sin(n\omega_c T)}{n\omega_c T} \qquad (4.102)$$

The response is symmetrical, as expected, but the filter is unrealizable since n extends to infinity in either direction (see (4.27)). In order to create an FIR filter of impulse response length N, we could simply truncate (4.102) to

$$h(n)_s = 2Tf_c \frac{\sin(n\omega_c T)}{n\omega_c T}, \quad -\left(\frac{N-1}{2}\right) \leq n \leq \frac{N-1}{2}, \quad N \text{ odd}$$

$$(4.103)$$

and in this case the amplitude response would be given by the Fourier series expansion in (4.31). Truncation means, of course, that there are fewer coefficients in the expansion, and so we would expect the amplitude response to be only an *approximation* to the ideal 'brickwall' response. In fact, for the lowpass filter, the response would have significant passband ripple and poor stopband attenuation. The effect is called the *Gibbs phenomenon* and arises whenever a truncated Fourier series is used to approximate a discontinuous function (i.e. rectangular response). The significant point is that, as N increases, the largest ripple tends to about 9% of the size of the discontinuity and does *not* reduce to zero (although the ripple does become confined to a smaller frequency band around the discontinuity). One objective in FIR design is to reduce this error significantly and the main design techniques for linear phase filters are:

    the frequency sampling method;
    the windowing method;
    the equiripple method; and
    computation efficient methods.

Generally speaking, the design *process* is iterative and terminates when a set of coefficients have been found which give a filter response falling within specified tolerances.

4.4.2 Frequency sampling method

The *basic* idea behind frequency sampling design is quite straightforward and an understanding of this approach is useful for the windowing method. Frequency sampling is based upon the *discrete Fourier transform* or *DFT* (Section 5.2), and is illustrated in Fig. 4.28.

The desired frequency response $D(e^{j\omega})$ is sampled at equispaced intervals to give N complex valued samples $H(k)$, $k = 0, 1, \ldots, N - 1$. These values are then used in the inverse DFT (IDFT) to give the N-point impulse response $h(n)$

$$h(n) = \frac{1}{N} \sum_{k=0}^{N-1} H(k)\, e^{j2\pi kn/N}, \quad n = 0, 1, \ldots, N - 1 \qquad (4.104)$$

Since $h(n)$ is real, it follows from the properties of the Fourier transform that the frequency samples must have complex conjugate symmetry about $\omega = 0$, or, equivalently, about $\omega = \pi$. This means that the magnitude samples must be symmetric about π and one way of achieving this for N

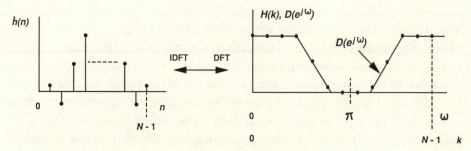

Fig. 4.28 Concept behind frequency sampling design.

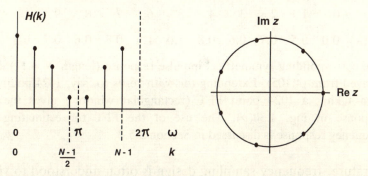

Fig. 4.29 A suitable sampling grid for frequency sampling (N odd).

odd is shown in Fig. 4.29. In this case, the magnitude symmetry can be expressed as

$$|H(k)| = |H(N - k)|, \quad k = 0, 1, \ldots, (N - 1)/2$$

Using this symmetry, and the fact that

$$H(k) = |H(k)|e^{j\phi(k)}$$

we can reduce (4.104) to give the shifted impulse response

$$h(n)_s = \frac{1}{N}\left[|H(0)| + 2\sum_{k=1}^{(N-1)/2} |H(k)| \cos\left(\frac{2\pi kn}{N}\right)\right], \quad N \text{ odd}$$

(4.105)

Since $h(n)_s$ is symmetrical and is centred on $n = 0$, then (4.105) need only be evaluated for $0 \leqslant n \leqslant (N - 1)/2$.

Equation (4.105) gives a *non-recursive* realization for a frequency sampled filter and can be useful where the desired amplitude response is too complex to perform the integration in (4.99). In fact, (4.105) can be used to derive an approximation to *any* desired continuous frequency response. It could be used, for example, to generate complex amplitude responses for speech processing (Cain, Kale and Yardin, 1988), where $|H(k)|$ is non-zero for most values of k. The *actual* amplitude response can be computed using (4.31) (or, equivalently, via an FFT) and it will be an interpolation through the points $H(k)$ defined in Fig. 4.28, with finite deviation from the desired response between points.

Example 4.13

Suppose that a smooth, non-brickwall type response is specified by the following points:

k	0	1	2	3	4	5	6	7	8	9	10
$\lvert H(k)\rvert$	0.0	0.5	0.7	0.6	0.8	1.0	1.0	0.8	0.6	0.7	0.5

The corresponding symmetrical impulse response (length $N = 11$) can be derived from (4.105). Extending this with zeros to, say, 1024 points, and then taking a 1024 point FFT (rectangular window) gives the filter response in Fig. 4.30(a). The use of the FFT for estimating filter frequency response is discussed in Section 5.5.2.

In the literature, frequency sampling design is often understood to yield a *recursive* type of FIR filter structure. Consider the general FIR transfer function

$$H(z) = \sum_{n=0}^{N-1} h(n)z^{-n}$$

Using (4.104), and letting $W = e^{-j2\pi/N}$,

$$H(z) = \sum_{n=0}^{N-1} \left[\frac{1}{N} \sum_{k=0}^{N-1} H(k)W^{-nk} \right] z^{-n}$$

$$= \frac{1}{N} \sum_{k=0}^{N-1} H(k) \sum_{n=0}^{N-1} W^{-nk} z^{-n} \tag{4.106}$$

Expressing the last summation in closed form

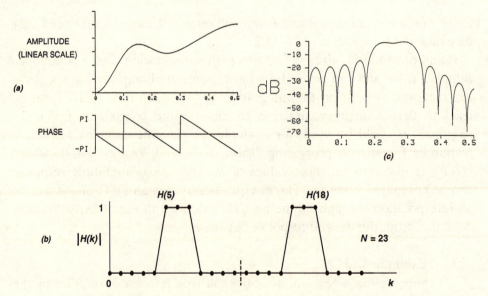

Fig. 4.30 Frequency sampling design: (a) for a non-brickwall response; (b) and (c) for a simple bandpass response.

$$H(z) = \frac{1}{N} \sum_{k=0}^{N-1} H(k) \left(\frac{1 - z^{-N}}{1 - W^{-k} z^{-1}} \right)$$

$$= \frac{1 - z^{-N}}{N} \sum_{k=0}^{N-1} \frac{H(k)}{1 - z^{-1} e^{j2\pi k/N}} \tag{4.107}$$

which is of the form

$$H(z) = H_1(z) H_2(z) \tag{4.108}$$

where

$$H_1(z) = \frac{1 - z^{-N}}{N} \tag{4.109}$$

The FIR filter structure is now a cascade of a non-recursive section $H_1(z)$ and a recursive section $H_2(z)$. Section $H_1(z)$ is a comb filter with zeros given by

$$z^N = 1 = e^{j2\pi k}, \quad k = 0, 1, 2, \ldots, N - 1$$

i.e. the kth zero lies on the unit circle at

$$z_k = e^{j2\pi k/N}, \quad k = 0, 1, 2, \ldots, N - 1 \tag{4.110}$$

Section $H_2(z)$ is a parallel structure (as in Fig. 4.13) and its individual terms can be paired-off to create second-order resonators. The kth term $(k \neq 0)$ and its complex conjugate at $N - k$ can be paired to give

$$H_2(z)|_k = \frac{H(k)}{1 - z^{-1} e^{j2\pi k/N}} + \frac{H(N - k)}{1 - z^{-1} e^{-j2\pi k/N}} \tag{4.111}$$

Assuming these samples are in the passband, we have, for linear phase and N odd,

$$H(k) = e^{j\phi_k}; \quad H(N - k) = e^{-j\phi_k}; \quad \phi_k = -\left(\frac{N-1}{2} \right) 2\pi k/N;$$

$$k = 0, 1, \ldots, (N - 1)/2$$

and (4.111) can be reduced to

$$H_2(z)|_k = \frac{2 \cos \phi_k - 2 \cos(\phi_k - 2\pi k/N) z^{-1}}{1 - 2 \cos(2\pi k/N) z^{-1} + z^{-2}} \tag{4.112}$$

Clearly, this is of the form

$$H_2(z)|_k = \frac{a_0 + a_1 z^{-1}}{1 + b_1 z^{-1} + b_2 z^{-2}}$$

where b_1 and b_2 determine two complex conjugate poles, corresponding to simple resonance.

For the elementary bandpass example in Fig. 4.30(b), this means that samples $H(5)$ and $H(18)$ correspond to a resonator (pole-pair) of the form given in (4.112), and the other two passband samples will require two additional resonators. The complete filter has three resonators in parallel and all cascaded with a comb filter having a delay z^{-23} (corresponding to $N = 23$). The significant point here is that, according to (4.107), all poles of $H_2(z)$ lie on the unit circle and occur at exactly the same positions as the zeros of $H_1(z)$. Where $H(k)$ is non-zero, the poles will therefore cancel the effect of the corresponding zeros and give a passband. Conversely, uncancelled zeros provide the stopband.

In practice, a typical resonator could have a pair of complex conjugate poles slightly inside the unit circle in order to ensure stability (the corresponding zeros also being inside the circle). This procedure was adopted for the filter specification in Fig. 4.30(b) (assuming linear phase). The coefficients of the resonators and comb filter were slightly modified to locate cancelling poles and zeros at a radius $r = 0.999$, and the resulting amplitude response is shown in Fig. 4.30(c). As previously pointed out, the response is an interpolation through the specified points and it will deviate from the ideal response between points. For sharp transition filters, the usual procedure is to introduce transition band samples and optimize them so as to achieve maximum ripple cancellation in the passband and/or stopband (Rabiner and Gold, 1975).

In conclusion, it should now be clear that the recursive type of FIR filter can be advantageous when most of the amplitude samples $|H(k)|$ are zero, as for example, in narrow passband filters. In this case, only a relatively few resonators are required, and the realization can be much more efficient than a non-recursive form. Noonan and Marquis (1989) describe a similar transversal–recursive structure for $H(z)$.

4.4.3 *Windowing method*

This design technique relies upon the ability to solve (4.99) for $h(n)$, either as a closed form expression (as in (4.103)), or as a summation approximation via the IDFT (in a similar way to frequency sampling). When the IDFT is used, windowing design can be applied to complicated frequency responses, just like the frequency sampling method. The desired frequency response is sampled at a large number of points, M, and the IDFT is computed. For large values of M, the IDFT approximates to the ideal impulse response as defined in (4.99), or, equivalently, to the infinite sequence $h(n)_s$ in Fig. 4.31(a).

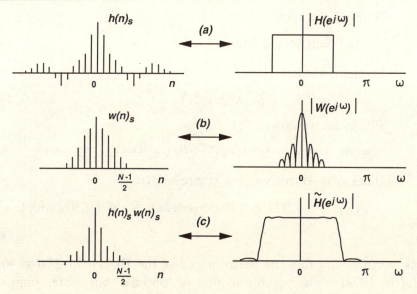

Fig. 4.31 Window design method (only one period of the frequency response is shown and N is assumed odd).

Assume that we have computed $h(n)_s$, either by direct integration or, approximately, via the IDFT. To obtain an FIR filter of impulse response length N $(N \ll M)$, the window method multiplies $h(n)_s$ by a *window function* $w(n)_s$ to give the windowed coefficients

$$\tilde{h}(n)_s = h(n)_s w(n)_s, \quad -\left(\frac{N-1}{2}\right) \leqslant n \leqslant \left(\frac{N-1}{2}\right) \quad (4.113)$$

The objective here is to moderate the Gibbs phenomenon through a less abrupt truncation of the Fourier series (note that simply truncating $h(n)_s$ corresponds to multiplication by a *rectangular* window function). Clearly, the final frequency response $\tilde{H}(e^{j\omega})$ will now be the (circular) convolution of $H(e^{j\omega})$ with the transform $W(e^{j\omega})$ of the windowing function, and so the choice of windowing function is important.

A good windowing function $W(e^{j\omega})$ has a narrow main lobe and small, rapidly decreasing sidelobes, giving rapid passband–stopband transitions and high stopband attenuation, respectively. Unfortunately, these are conflicting requirements and, in practice, the designer can select from a range of 'optimized' windows. Four popular and computationally simple windows spanning the range

$$-\left(\frac{N-1}{2}\right) \leqslant n \leqslant \left(\frac{N-1}{2}\right), \; N \text{ odd}$$

are defined as follows:

Hamming window:

$$w(n)_s = 0.54 + 0.46 \cos (2\pi n/N) \tag{4.114}$$

Hanning window:

$$w(n)_s = 0.5 + 0.5 \cos (2\pi n/N) \tag{4.115}$$

Blackman window:

$$w(n)_s = 0.42 + 0.50 \cos (2\pi n/N) + 0.08 \cos (4\pi n/N) \tag{4.116}$$

Blackman–Harris window (three-term):

$$w(n)_s = 0.42323 + 0.49755 \cos (2\pi n/N) + 0.07922 \cos (4\pi n/N)$$

$$\tag{4.117}$$

It is worth noting that the performance of the Blackman–Harris window (Harris, 1978) is comparable to that of 'optimal' but more computation intensive windows, such as the *Kaiser–Bessel* window. Figure 4.32 illus-

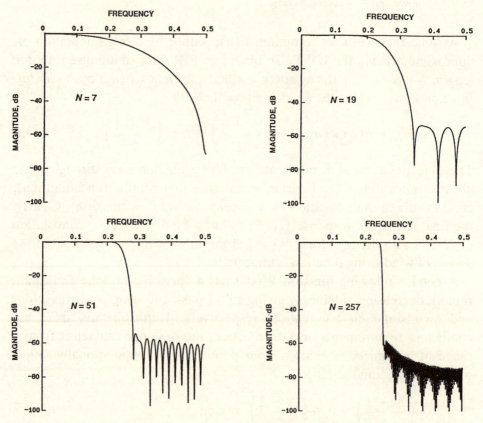

Fig. 4.32 Windowing design: effect of N: Hamming window, $f_c = 0.25$.

trates the use of the Hamming window for a lowpass filter, and shows the improvement in response as N is increased. The responses were computed using windowed coefficients in (4.31).

Generalized brickwall filter design

Consider the design of a bandpass filter with band edges at f_1 and f_2. Since this is a 'brickwall' response we can easily perform the integration and obtain a closed form for $h(n)_s$. We have

$$h(n)_s = \frac{1}{\omega_s} \int_{-\omega_s/2}^{\omega_s/2} H(e^{j\omega T})_s e^{j\omega n T}\, d\omega \qquad (4.118)$$

and assuming $h(n)$ symmetric, this can be written (see 4.28)

$$h(n)_s = \frac{1}{\omega_s} \int_{-\omega_s/2}^{\omega_s/2} H^*(e^{j\omega T}) e^{j\omega n T}\, d\omega \qquad (4.119)$$

As shown in Example 4.4, $H^*(e^{j\omega T})$ can be regarded as the amplitude response, and the idealized bandpass response is shown in Fig. 4.33. Therefore

$$
\begin{aligned}
h(n)_s &= \frac{1}{\omega_s}\left[\int_{-\omega_2}^{-\omega_1} e^{j\omega n T}\, d\omega + \int_{\omega_1}^{\omega_2} e^{j\omega n T}\, d\omega\right]\\[2mm]
&= \frac{1}{j2n\pi}[e^{-jn\omega_1 T} - e^{-jn\omega_2 T} + e^{jn\omega_2 T} - e^{jn\omega_1 T}]\\[2mm]
&= 2T\left[f_2\frac{\sin(n\omega_2 T)}{n\omega_2 T} - f_1\frac{\sin(n\omega_1 T)}{n\omega_1 T}\right]
\end{aligned}
\qquad (4.120)
$$

where

$$h(0)_s = 2T(f_2 - f_1) \qquad (4.121)$$

Note that (4.120) is general and can be used for the design of lowpass, bandpass and highpass filters. Similarly, for a *bandstop* filter, it is readily shown that the coefficients are given by

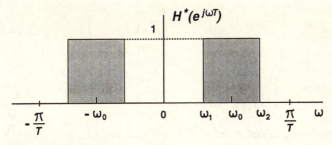

Fig. 4.33 Ideal bandpass response.

$$h'(n)_s = \begin{cases} - h(n)_s & n = 1, 2, 3 \dots \\ 1 - h(n)_s & n = 0 \end{cases} \qquad (4.122)$$

where $h(n)_s$ is given by (4.120) and (4.121), and f_1, f_2 define the stopband edges.

Fig. 4.34 shows the bandpass response for $f_1 = 0.2$ and $f_2 = 0.3$ using various windows. Note that the convolutional process inherent to windowing reduces stopband ripple at the expense of smearing the band edge. This tends to make the *design process* for 'brickwall' responses highly iterative since, generally speaking, f_1 and f_2 cannot be specified exactly. Also, windowing does not use any optimization criteria and invariably generates less computation efficient filters than other techniques. On the other hand, the design equations are simple and easily incorporated into a computer programme (Appendix C) for use in non-critical applications. Also, windowing is often used for designing special filters such as differentiators and Hilbert transformers.

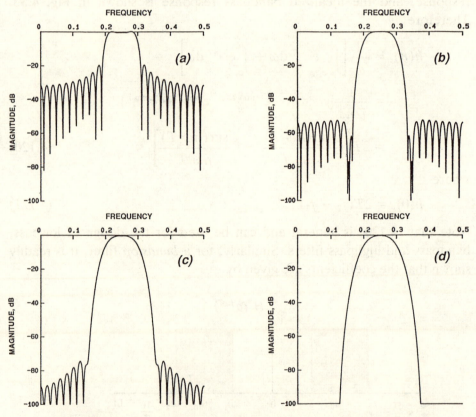

Fig. 4.34 Windowing design: effect of different windows: (*a*) rectangular, (*b*) Hamming, (*c*) Blackman, (*d*) Blackman–Harris. $N = 51$, $f_1 = 0.2$, $f_2 = 0.3$.

Example 4.14

The ideal digital Hilbert transformer has a flat amplitude characteristic and gives $-\pi/2$ phase shift to all positive frequency components up to π (and $\pi/2$ phase shift for all components from zero to $-\pi$). Assuming unity sampling period ($T = 1$), this can be expressed as

$$H(e^{j\omega})_s = \begin{cases} e^{j\pi/2} & -\pi < \omega < 0 \\ e^{-j\pi/2} & 0 < \omega < \pi \end{cases} \qquad (4.123)$$

Note that $H(e^{j\omega})_s$ is a non-causal response. Also, according to (4.33), the corresponding impulse response $h(n)_s$ will be anti-symmetric. Using (4.118)

$$h(n)_s = \frac{1}{2\pi}\left[\int_{-\pi}^{0} e^{j(\omega n + \pi/2)}\, d\omega + \int_{0}^{\pi} e^{j(\omega n - \pi/2)}\, d\omega\right] \qquad (4.124)$$

which reduces to

$$h(n)_s = \begin{cases} \dfrac{2}{n\pi} & n \text{ odd} \\ 0 & n \text{ even} \end{cases} \qquad (4.125)$$

This impulse response can be made causal by shifting it by $\alpha = (N - 1)/2$ sample periods, where N is odd and corresponds to the tap length of a transversal filter. Note, however, that this will generate a filter with an overall phase shift

$$\phi(\omega) = -(\pi/2 + \alpha\omega), \qquad \omega > 0$$

and this clearly deviates from the ideal Hilbert transformer. On the other hand, the linear phase term is easily compensated for, as illustrated in

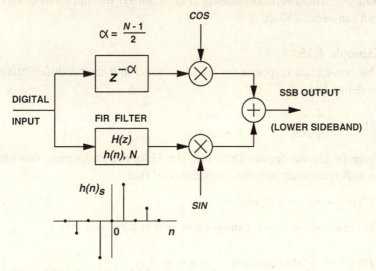

Fig. 4.35 Digital single sideband generation.

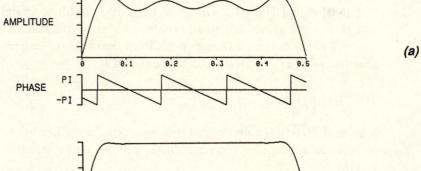

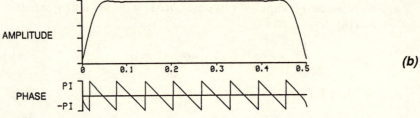

Fig. 4.36 Response of practical Hilbert transformers (linear scale): (*a*) $N = 15$ rectangular window, (*b*) $N = 33$ Hanning window.

Fig. 4.35. This shows how a practical FIR filter obeying (4.125) (or a windowed version) might be used in place of an ideal Hilbert transformer in order to generate a single-sideband (SSB) signal. The corresponding frequency response of this filter is shown in Fig. 4.36. If $h(n)_s$ is simply truncated the response suffers the usual Gibbs oscillation, as shown in Fig. 4.36(*a*). On the other hand, simulation shows that using $N = 33$ and Hanning windowed coefficients (Fig. 4.36(*b*)) the upper sideband rejection can exceed 50 dB.

Example 4.15

The non-causal response for an ideal wideband digital differentiator can be defined as

$$H(e^{j\omega})_s = \begin{cases} \omega e^{-j\pi/2} & -\pi < \omega < 0 \\ \omega e^{j\pi/2} & 0 < \omega < \pi \end{cases} \tag{4.126}$$

From (4.33) we deduce that, like the Hilbert transformer, this filter has an anti-symmetric impulse response, and that

$$H^*(e^{j\omega}) = \omega \quad -\pi < \omega < \pi \tag{4.127}$$

This real, odd function can be expressed as a Fourier series

$$H^*(e^{j\omega}) = \sum_{n=1}^{\infty} a(n) \sin(n\omega) \quad -\pi < \omega < \pi \tag{4.128}$$

where

$$a(n) = \frac{1}{\pi} \int_{-\pi}^{\pi} \omega \sin(n\omega) \, d\omega \qquad (4.129)$$

Integrating and simplifying gives

$$a(n) = -\frac{2}{n} \cos(n\omega) \qquad n = 1, 2, 3, \ldots \qquad (4.130)$$

$$\therefore \; H^*(e^{j\omega}) = -2 \sum_{n=1}^{\infty} \left(\frac{1}{n} \right) \cos(n\pi) \sin(n\omega) \qquad (4.131)$$

Comparing (4.131) with (4.35) gives, for N odd,

$$h(n)_s = \frac{1}{n} \cos(n\pi)$$

$$= \frac{1}{n}(-1)^n \qquad n = \pm 1, \pm 2, \pm 3, \ldots \qquad (4.132)$$

where $h(0)_s = 0$. Fig. 4.37 shows two practical (causal) filter responses for $N = 15$. They correspond to reasonably wideband differentiators and have a phase response

$$\phi(\omega) = \pi/2 - \alpha\omega$$

Note that similar results could have been obtained using the frequency sampling design method.

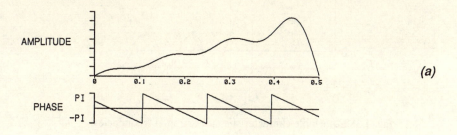

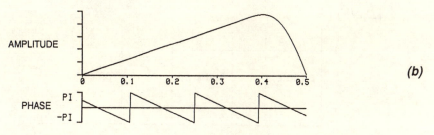

Fig. 4.37 Response of practical differentiators (linear scale): (*a*) N = 15, rectangular window; (*b*) N = 15, Hamming window.

4.4.4 Equiripple method

This is also sometimes referred to as the 'optimal', 'minimax', or 'Remez' method. We have seen that approximations made during the design process introduce ripples in the ideal characteristic. The significant point here is that some 'optimum' behaviour is often achieved when the ripples in the passband (stopband) all have the same magnitude, as, for example, in Fig. 4.38. McClellan, Parks and Rabiner (1973) have described a comprehensive FORTRAN programme for the design of FIR linear phase equiripple filters, and their technique minimizes an error function in order to generate an optimum design.

Suppose that we wish to obtain the 'best' approximation (in some optimization sense) to the bandpass filter in Fig. 4.39. In this case the desired *real* response $D(e^{j\omega})$ is specified only in frequency bands 1, 2 and 3 (a subset A of the total normalized frequency range) and it is left

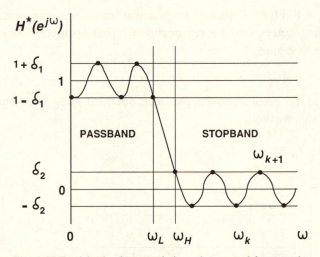

Fig. 4.38 Equiripple characteristic and extremal frequencies.

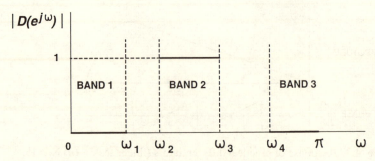

Fig. 4.39 Bandpass specification for equiripple design.

undefined in the transition bands. The optimization algorithm will generate an equiripple characteristic over these defined bands, whilst leaving the band edges well defined (in contrast to the window design method). Over the subset A we can define the weighted error function

$$E(e^{j\omega}) = W(e^{j\omega})|D(e^{j\omega}) - H^*(e^{j\omega})| \qquad (4.133)$$
$$\omega \in A$$

where

$W(e^{j\omega})$ = a positive weighting function

$H^*(e^{j\omega})$ = the approximation to $D(e^{j\omega})$

The original FORTRAN programme permits multiband design and each band has its own user-defined weighting function (usually a constant) to permit different approximation errors for different bands. In other words, δ_1 and δ_2 in Fig. 4.38 can be user-defined.

The design problem is to approximate $D(e^{j\omega})$ by minimizing the maximum approximation error over the defined frequency bands (hence the term 'minimax'). From (4.31) we can write the approximation to $D(e^{j\omega})$ as a sum of cosines

$$H^*(e^{j\omega}) = \sum_{n=0}^{r-1} a(n) \cos(\omega n) \qquad (4.134)$$

where $r - 1 = (N - 1)/2$. When $H^*(e^{j\omega})$ is expressed in this way, a *Chebyshev* or minimax approximation to $D(e^{j\omega})$ can be found via the *alternation theorem*. In this context, the theorem states that, if $H^*(e^{j\omega})$ is a linear combination of r cosine functions, then a necessary and sufficient condition that $H^*(e^{j\omega})$ be the unique best-weighted Chebyshev approximation to $D(e^{j\omega})$ on subset A is that $E(e^{j\omega})$ exhibits at least $r + 1$ *extremal* frequences in A. The extremal frequencies ω_k, $k = 0, 1, \ldots, r$ (see Fig. 4.38) are the points at which the peak error occurs, and the error at these points alternates in sign, i.e.

$$E[e^{j\omega_k}] = -E[e^{j\omega_{k+1}}] \qquad (4.135)$$

Assume an initial guess of $r + 1$ extremal frequencies for which error $E(e^{j\omega})$ is forced to have magnitude δ. In this case, using (4.133), the alternating peak errors can be expressed

$$W(e^{j\omega_k})[D(e^{j\omega_k}) - H^*(e^{j\omega_k})] = (-1)^k \delta; \quad k = 0, 1, \ldots, r \quad (4.136)$$

This is a system of $r + 1$ equations in $r + 1$ variables (the r unknown coefficients $a(n)$ and the unknown error δ), and so it has a definite solution. McClellan *et al* applied the *Remez multiple exchange algorithm* to solve (4.136). In essence, this algorithm is iterative and it evaluates $E(e^{j\omega})$ on a dense frequency grid in order to find the extremal frequencies. If, at some points

$$|E(e^{j\omega})| > \delta$$

then the extremal frequencies of $E(e^{j\omega})$ are used as the new extremal set for the next iteration. Each iteration increases δ and usually δ converges to an upper bound. After convergence, $H^*(e^{j\omega})$ is evaluated at equally spaced frequencies and the IDFT used to find the filter coefficients. The original FORTRAN programme provides fast solutions for lowpass, highpass, bandpass and bandstop filters, multiple passband–stopband filters, differentiators and Hilbert transformers. The use of this programme is illustrated by the following examples.

Example 4.16

Fig. 4.40(*a*) shows an amplitude specification for a low-order, linear phase, lowpass filter intended for the *Q*-channel of a digital NTSC colour television decoder. The actual sampling frequency is 3.579 MHz and so the band edges must be normalized as shown.

An acceptable response is found for the following input data:

filter impulse response length N: 13
number of bands: 2
grid density: 50
band edge cutoff frequencies: 0.0, 0.1676, 0.2235, 0.5
weight for each band: 1, 1

The resulting computer listing is given in Fig. 4.40(*c*), and the corresponding amplitude response is shown in Fig. 4.40(*b*). The passband ripple is ± 1 dB and the band edges are still well defined (in contrast to the window method).

Example 4.17

Suppose that an equiripple linear phase bandpass filter is required with the following specification.

sampling frequency: 3 MHz
centre frequency: 1 MHz
lower cutoff frequency: 0.9 MHz
upper cutoff frequency: 1.1 MHz
attenuation: > 45 dB at 0.75 MHz and 1.25 MHz

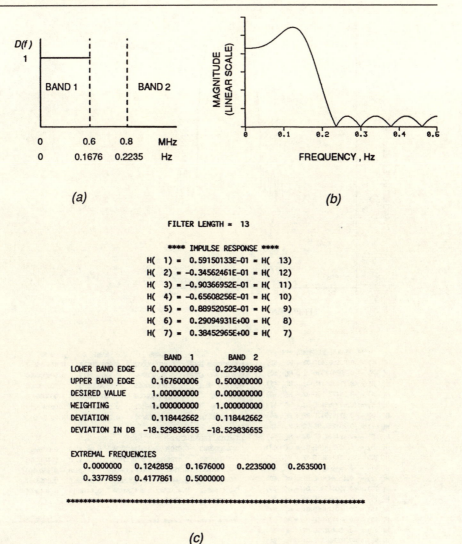

Fig. 4.40 Lowpass filter design using the equiripple method.

The normalized band edges are therefore

  0.00, 0.25 band 1, stopband
  0.30, 0.37 band 2, passband
  0.42, 0.50 band 3, stopband

After a few runs of the programme we find that $N = 51$ gives a satisfactory response (Fig. 4.41). Note from the computer listing that the weighting ratio of 5:1 between stopbands and passband forces the stopband ripple to be a factor five down on the passband ripple. Even so,

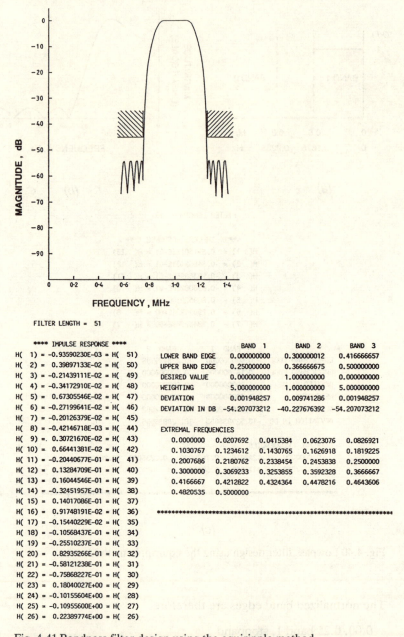

FILTER LENGTH = 51

```
**** IMPULSE RESPONSE ****
H(  1) = -0.93590230E-03 = H(  51)
H(  2) =  0.39897133E-02 = H(  50)
H(  3) = -0.21439111E-02 = H(  49)
H(  4) = -0.34172910E-02 = H(  48)
H(  5) =  0.67305546E-02 = H(  47)
H(  6) = -0.27199641E-02 = H(  46)
H(  7) = -0.20126379E-02 = H(  45)
H(  8) = -0.42146718E-03 = H(  44)
H(  9) =. 0.30721670E-02 = H(  43)
H( 10) =  0.66441381E-02 = H(  42)
H( 11) = -0.20440677E-01 = H(  41)
H( 12) =  0.13284709E-01 = H(  40)
H( 13) =  0.16044546E-01 = H(  39)
H( 14) = -0.32451957E-01 = H(  38)
H( 15) =  0.14017086E-01 = H(  37)
H( 16) =  0.91748191E-02 = H(  36)
H( 17) = -0.15440229E-02 = H(  35)
H( 18) = -0.10568437E-01 = H(  34)
H( 19) = -0.25510237E-01 = H(  33)
H( 20) =  0.82935266E-01 = H(  32)
H( 21) = -0.58121238E-01 = H(  31)
H( 22) = -0.75868227E-01 = H(  30)
H( 23) =  0.18040027E+00 = H(  29)
H( 24) = -0.10155604E+00 = H(  28)
H( 25) = -0.10955600E+00 = H(  27)
H( 26) =  0.22389774E+00 = H(  26)
```

	BAND 1	BAND 2	BAND 3
LOWER BAND EDGE	0.000000000	0.300000012	0.416666657
UPPER BAND EDGE	0.250000000	0.366666675	0.500000000
DESIRED VALUE	0.000000000	1.000000000	0.000000000
WEIGHTING	5.000000000	1.000000000	5.000000000
DEVIATION	0.001948257	0.009741286	0.001948257
DEVIATION IN DB	-54.207073212	-40.227676392	-54.207073212

```
EXTREMAL FREQUENCIES
   0.0000000   0.0207692   0.0415384   0.0623076   0.0826921
   0.1030767   0.1234612   0.1430765   0.1626918   0.1819225
   0.2007686   0.2180762   0.2338454   0.2453838   0.2500000
   0.3000000   0.3069233   0.3253855   0.3592328   0.3666667
   0.4166667   0.4212822   0.4324364   0.4478216   0.4643606
   0.4820535   0.5000000
```

****

Fig. 4.41 Bandpass filter design using the equiripple method.

the passband ripple is still less than ± 0.1 dB. Also note that the extremal frequencies in the stopbands correspond to the equiripple peaks in the response. A similar response could, of course, be achieved using the window method, by iteratively adjusting the band edges, although the overall response will be inferior for the same filter order.

4.5 Computation efficient FIR filters

Despite their popularity, classical FIR design techniques (such as the equiripple and windowing methods) are not necessarily the best approach for all applications. These techniques lead naturally to a direct form realization of $h(n)$, i.e. one multiplier for each value of $h(n)$, and if the filter has linear phase and an impulse response length N, this requires $N - 1$ adders and $(N + 1)/2$ multipliers. Clearly, the direct form structure allows for the $h(n)$ to be independent. On the other hand, in practice, the impulse response tends to have a smooth envelope and there is often a strong correlation between adjacent values of $h(n)$ – which suggests that such a versatile structure is unnecessary.

The need to minimize filter arithmetic (particularly multiplication) becomes important when working at high sample rates, or with VLSI or gate array technology. A number of filter optimization techniques are described in the literature and a common approach is to simplify the filter structure by using very computation efficient sections to help generate $h(n)$. A useful exception to this is the simple *half-band* filter, and this also enables an approximate binary relationship between the coefficients to be exploited in order to simplify the filter further.

4.5.1 Half-band FIR filters

A half-band filter has a lowpass response and a normalized cutoff frequency of 0.25. In this case (4.102) reduces to

$$h(n)_s = \frac{1}{2} \frac{\sin(n\pi/2)}{n\pi/2} \qquad (4.137)$$

and so $h(n)_s = 0$, $n = \pm2$, ±4, ±6, ... The lowpass response coupled with the fact that nearly half the coefficients are zero makes this type of filter particularly efficient for 2:1 decimation.

In addition, if we now apply a Hamming window to these coefficients it is found that many of the windowed coefficients $\tilde{h}(n)_s$ are simply related to each other, leading, in turn to simplified structures (Anastassopoulos *et al*, 1985). In particular, for a filter having impulse response length $N = 2M + 1$, where $M = 14 + 12k$, $k = 0, 1, 2, \ldots$ many of the coefficients are *binary* related, to a good approximation. For example, for $M = 14$

$$\tilde{h}(n)_s \approx -\frac{1}{2}\tilde{h}(n-2)_s, \quad n \geqslant 5$$

and for $M = 26$

$$\tilde{h}(n)_s \approx \frac{1}{2}\tilde{h}(n - 4)_s, \quad n \geqslant 11$$

The error in these approximations is sufficiently low (of the order of 10^{-4}) to retain good stopband attenuation, whilst their binary relationship enables relatively simple structures to be used.

4.5.2 RRS structures

As previously mentioned, a number of techniques for generating computation efficient filters utilize very efficient structures as part of the overall filter. Usually these techniques incorporate a *recursive running sum (RRS)* section, or one of its variants, since this type of structure is particularly efficient. The basic RRS section is shown in Fig. 4.42(*a*), and mathematically it is equivalent to a transversal structure having all coefficients equal to unity and an impulse response length $N = k$. In other words, it generates a running sum or average, and has the output

$$Y(z) = X(z) + z^{-1}X(z) + z^{-2}X(z) + \ldots + z^{-(k-1)}X(z)$$

$$\therefore H(z) = 1 + z^{-1} + z^{-2} + \ldots + z^{-(k-1)}$$

$$= \frac{1 - z^{-k}}{1 - z^{-1}} \tag{4.138}$$

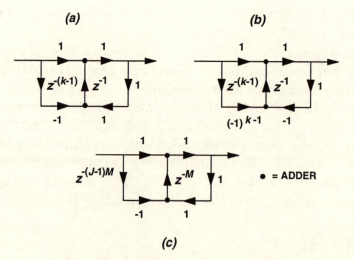

Fig. 4.42 RRS sections: (*a*) lowpass, (*b*) highpass, (*c*) sparse response generator.

Expanding (4.138) as

$$Y(z) = X(z) - z^{-k}X(z) + z^{-1}Y(z)$$

and minimizing delays gives the computation efficient structure in Fig. 4.42(a). Also, letting $z = e^{j\omega}$ in (4.138), the frequency response can be reduced to

$$H(e^{j\omega}) = \frac{\sin(\omega k/2)}{\sin(\omega/2)} e^{-j\omega(k-1)/2} \qquad (4.139)$$

This is a narrowband lowpass response (as expected for an averager) with first zeros at $\omega = 2\pi m/k$, $m = 1, 2, 3, \ldots$. A design technique utilizing RRS sections is discussed in Section 4.5.4.

Example 4.18
The lowpass RRS section in Fig. 4.42(a) is particularly computation efficient, and so a highpass RRS section would also be useful. The lowpass section (Equation (4.138)) has a pole at $z = 1$ whilst the highpass filter requires a pole at $z = -1$, corresponding to a transfer function of the form

$$H(z)|_{hp} = \frac{f(z^{-1})}{1 + z^{-1}}$$

Clearly, the highpass filter can be generated from the lowpass filter by using the transformation

$$z^{-1} \Rightarrow -z^{-1}$$

$$\therefore\ H(z)|_{hp} = \frac{1 - (-z^{-1})^k}{1 + z^{-1}}$$

$$= \frac{1 - (-1)^k z^{-k}}{1 + z^{-1}} \qquad (4.140)$$

Expanding (4.140) in terms of input and output signals gives

$$Y(z) = X(z) + (-1)^{k-1}z^{-k}X(z) - z^{-1}Y(z) \qquad (4.141)$$

and this readily gives the highpass section in Fig. 4.42(b). A simple transformation ($z^{-1} \Rightarrow -z^{-2}$) can also yield a bandpass RRS section centred at $\pi/2$.

Finally, note that a sparse (but finite) impulse response can be generated by adding more delays, as in Fig. 4.42(c). An application of this structure is discussed in the following section.

4.5.3 Prefilter–equalizer design

Suppose we require a computation efficient FIR bandpass filter centred on $\omega = \omega_0$ (see Fig. 4.33). It is easily shown that (4.120) can be expressed in the alternative form

$$h(n)_s = \frac{2\Delta}{\pi} \frac{\sin(n\Delta)}{n\Delta} \cos(n\omega_0) \qquad (4.142)$$

where $\Delta = (\omega_2 - \omega_1)/2$ and a normalized sampling frequency has been assumed. For very narrowband filters (and neglecting amplitude scaling) this becomes

$$h(n)_s \approx \cos(2\pi n f_0) \qquad (4.143)$$

If we assume f_0 is a rational number k/M where k and M are integers and $k \leqslant M/2$, then $h(n)_s$ becomes an infinite length, periodic sequence $h_p(n)$ with period M. The corresponding frequency response is a discrete periodic sequence $H_p(e^{j\omega})$, also of period M (a graphical proof is given in Fig. 5.3) and in this case the bandpass response centred on normalized frequency $f_0 = k/M$ becomes infinitely narrow. Conversely, if we select, i.e. window, just J periods of $h_p(n)$ (J integer) the filter will have an impulse response length $N = JM$ and the bandpass response will be broadened by convolution with the window transform. The effective window function is an impulse train of length N and so the convolved frequency response is

$$H(e^{j\omega}) = H_p(e^{j\omega}) * \frac{\sin(\omega N/2)}{\sin(\omega/2)} e^{-j\omega(N-1)/2} \qquad (4.144)$$

The window transform has a first zero at $\omega = 2\pi/N$ and so convolution gives a bandpass response with main lobe nulls at $\omega_0 \pm 2\pi/N$.

A computation efficient way of realizing J periods of $h_p(n)$ is to generate J impulses from an RRS-type section (Fig. 4.42(c)) and then

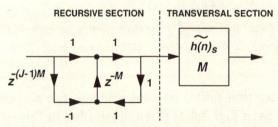

Fig. 4.43 Elementary bandpass structure of impulse response length $N = JM$.

apply them to a simple transversal section, as shown in Fig. 4.43. Here, $\tilde{h}(n)_s$ is one period of $h_p(n)$ (as defined by (4.143)), where the coefficients have been approximated to the nearest power-of-two.

Example 4.19

Assume a bandpass response is required at frequency

$$f_0 = \frac{k}{M} = \frac{2}{7} = 0.2857$$

Using (4.143) gives the following coefficients

n	$h(n)_s$	$\tilde{h}(n)_s$
0	1	1
1	−0.2225	−0.25
2	−0.9010	−1
3	0.6235	0.5

Setting $J = 3$ ($N = 21$) gives an elementary bandpass response centred on f_0 with a mainlobe width between nulls of 0.0952. The sidelobe attenuation can be increased by cascading two bandpass structures as shown in Fig. 4.44(a). The overall impulse response (now length 41) is still symmetrical and so the phase response is exactly linear. In all, the structure requires $2(M + 1) = 16$ adders and $2(4M − 1) = 54$ delays. Variations on this basic design are described by Cabezas and Diniz (1990).

Equalization

The response in Fig. 4.44(a) can be significantly improved by adding an equalizer $Q(z)$, as shown in Fig. 4.44(b). Here, $P_i(z)$, $i = 1, \ldots, r$ is a computation efficient structure, such as that in Fig. 4.43, and $Q(z)$ is designed to minimize the error between an ideal bandpass characteristic and that generated by $P(z)$. In this context $P(z)$ acts as a *prefilter* (Adams and Wilson 1983, 1984) and its task is to achieve the *basic* response in the most computation efficient way. The equalizer then has the task of 'tidying-up' this response, in particular, flattening the passband response.

In principle the design of $Q(z)$ is straightforward and can be based on the weighted error function defined in (4.133). For this case the function can be expressed as

$$E(e^{j\omega}) = \underset{\omega \in A}{W(e^{j\omega})} |D(e^{j\omega}) − P(e^{j\omega})Q(e^{j\omega})| \tag{4.145}$$

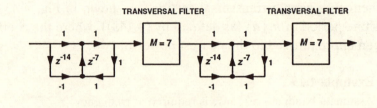

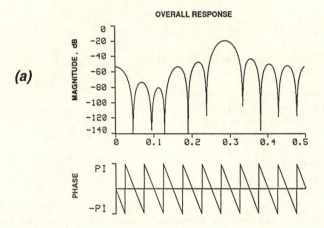

(a)

(b)

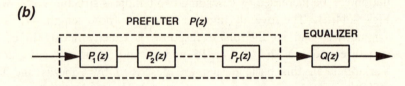

Fig. 4.44 Prefilter–equalizer design: (*a*) prefilter formed from two bandpass sections each having $J = 3$, $k = 2$, $M = 7$; (*b*) general prefilter–equalizer structure.

and $Q(z)$ could be found by minimizing the maximum value of $E(e^{j\omega})$ using some optimization technique, e.g. a modified Remez algorithm.

Finally, it is worth noting that *interpolated* FIR filters (Neuvo, Cheng-Yu and Mitra, 1984; Saramaki, Neuvo and Mitra, 1988) offer an alternative to the prefilter–equalizer approach for efficient filter design. As the name implies, this technique exploits the redundancy in $h(n)$ by first generating a sparse set of impulse response samples, and then using a second section to interpolate between samples. This approach has typically 1/2–1/8 the number of multipliers and adders of a direct form design. On the other hand, the design process is iterative and, like the prefilter–equalizer method, it usually involves modification and use of the Remez algorithm.

4.5.4 Multiplier-less design

Sometimes multiplication can be as fast as addition when using a general purpose processor to realize a digital filter. On the other hand, as previously mentioned, the need to reduce the number of multipliers becomes important when working at high sample rates, or with VLSI or gate array technology. A number of approaches to true multiplier-less design are described in the literature. Here, coefficients are usually reduced to simple integers, e.g. 0, ±1, ±2, ±4, ... (Van Gerwen *et al*, 1975; Lynn, 1980), or to simple combinations of powers of two (Agarwal and Sudhakar, 1983; Wade *et al*, 1990), or to simply 0, +1, −1 (Noonan and Marquis, 1989).

Cascading primitive sections

A rather obvious approach to multiplier-less design is to cascade primitive FIR sections where each section uses bit-shifting and only a few adders and delays. Clearly, some of these sections could be of the RRS-type for computation efficiency, and other possible sections are illustrated in Fig. 4.45. For example, the 2-adder section in Fig. 4.45(*b*) has the transfer function

$$H(z) = 1 + bz^{-k} + z^{-2k}$$

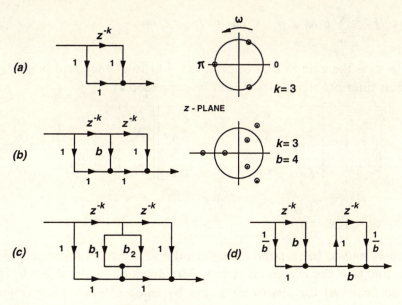

Fig. 4.45 Some primitive linear phase FIR sections.

and coefficient b will typically be a small integer or a simple combination of powers of two. The section can generate either on-circle or off-circle zeros, depending upon b. The section in Fig. 4.45(d) is useful for generating off-circle zeros in the passband.

Ideally any design algorithm based on a cascade of primitives should be non-interactive and one such approach is described by Wade *et al* (1990). Essentially, this technique searches for the optimum factorization of $H(z)$ for a given filter specification and a given set of n primitives. The optimization criterion is simply the total number of adders in the cascade since this represents the required computation. In other words, we seek to minimize the linear objective function

$$A = \sum_{j=1}^{n} a_j N_j, \quad N_j \geq 0 \tag{4.146}$$

where a_j is the number of adders associated with primitive j and N_j is the number of times the primitive is used. Since the N_j are integers, they correspond to the integer variables in a formal *integer programming* (IP) problem. Function A is to be minimized subject to m constraint equations derived from m equally spaced grid points in the frequency response specification (including transition bands). The ith constraint stipulates that the combined gain of an arbitrary cascade of primitives must be within an upper and lower bound, i.e.

$$b_{il} \leq \sum_{j=1}^{n} g_{ij} N_j \leq b_{iu}, \quad i = 1, 2, \ldots, m \tag{4.147}$$

where g_{ij} is the gain in dBs of primitive j at frequency grid point i. For a lowpass filter problem, (4.147) can be expressed as

$$\begin{bmatrix} b_{pl} \\ \vdots \\ b_{sl} \end{bmatrix} \leq \begin{bmatrix} g_{11} & g_{12} & \cdots & g_{1n} \\ g_{21} & g_{22} & \cdots & g_{2n} \\ & & \vdots & \\ g_{m1} & g_{m2} & \cdots & g_{mn} \end{bmatrix} \begin{bmatrix} N_1 \\ N_2 \\ \vdots \\ N_n \end{bmatrix} \leq \begin{bmatrix} b_{pu} \\ \vdots \\ b_{su} \end{bmatrix}$$

or

$$\mathbf{b}_l \leq \mathbf{GN} \leq \mathbf{b}_u \tag{4.148}$$

where b_p and b_s correspond to passband and stopband tolerances respectively, and $\mathbf{G}$ is the *constraint matrix*. Minimization of A subject to (4.148) can be achieved for low-order filters by using a 'branch-and-bound' IP method (which makes extensive use of *linear programming*).

Example 4.20

Applying the IP approach to the lowpass filter specification in Fig. 4.46(a) gives the following transfer function

$$H(z) = (1 + z^{-1})^2(1 + z^{-2})(1 + z^{-1} + z^{-2})(1 + 1.75z^{-2} + z^{-4})$$

$$\times\,(1 - 4z^{-2} + z^{-4})(1 - 4z^{-3} + z^{-6})(1 + z^{-1} + \ldots + z^{-4})$$

$$(4.149)$$

This corresponds to the structure in Fig. 4.46(c), and the actual filter response is shown in Fig. 4.46(b). Note that the last term in (4.149) corresponds to a lowpass RRS section with $k = 5$. The filter requires 14 adders and 25 delays, compared to 16 adders, 9 multipliers and 16 delays for an equivalent filter designed using the equiripple method (Section 4.4.4). Generally speaking, this 'primitive cascade' approach reduces the overall computation at the expense of increased storage (delay) and sometimes can lead to useful saving in VLSI chip area or gate array size.

An alternative optimization technique for primitive cascade synthesis is based on the *genetic algorithm (GA)* (Holland, 1975; Davis, 1991). This is a powerful parallel-search technique suitable for large search spaces and so is probably more appropriate than linear programming for higher-order filters. In the present context, the name refers to the fact that each trial cascade is represented as a *chromosome* and each primitive in the cascade is defined (coded) by a hierarchy of *genes* (Wade, Roberts and Williams, 1994). By applying the concepts of cross-over and mutation to chromo-

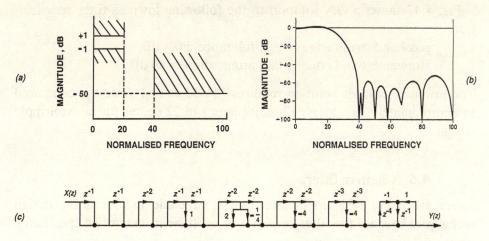

Fig. 4.46 Illustrating multiplier-less lowpass filter design: (a) specification, (b) actual response, (c) primitive cascade structure. (100 corresponds to half the sampling frequency.)

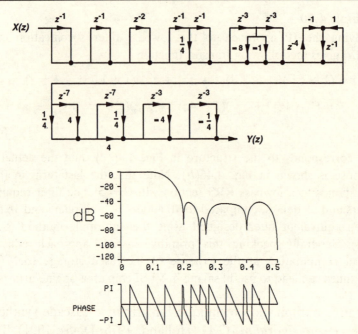

Fig. 4.47 Multiplier-less lowpass filter designed using a GA (see text).

somes, coupled with the concept of 'the survival of the fittest', each generation of chromosomes becomes progressively 'fitter' in the sense that more and more chromosomes are good solutions. After a number of generations, the fittest chromosome is selected as a sub-optimal solution to the problem.

Fig. 4.47 shows a GA solution to the following lowpass filter specification:

passband: band edge = 0.1, tolerance ± 0.1 dB
stopband: band edge = 0.2, attenuation > 40 dB

The primitive cascade solution requires 14 adders and 37 delays compared to approximately 22 adders, 12 multipliers and 22 delays for an equiripple design.

4.6 Adaptive filters

There are many DSP applications which preclude the foregoing design techniques because the filter response or coefficients cannot be specified *a priori*.

Consider the problem of reflection (or ghost) cancellation in television receivers, and in particular the simple 'postghost' model in Fig. 4.48(*a*). If

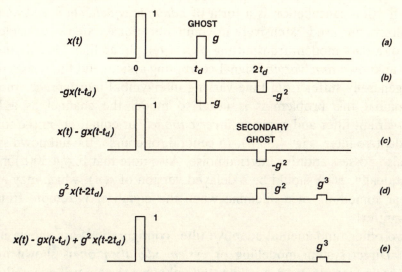

Fig. 4.48 Principle of ghost cancellation in television systems.

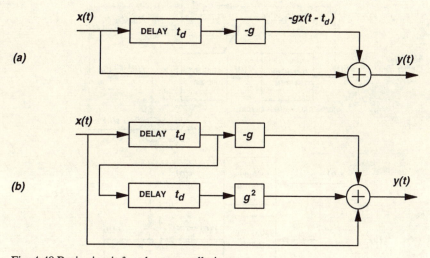

Fig. 4.49 Basic circuit for ghost cancellation.

we use the filter in Fig. 4.49(a), the primary ghost g will be cancelled, but a secondary ghost $-g^2$ will be generated (Fig. 4.48(c)). This secondary ghost could be cancelled by extending the filter to that of Fig. 4.49(b), and this will generate a yet smaller ghost (g^3), as shown in Fig. 4.48(e). The structure in Fig. 4.49(b) is simply a transversal FIR filter, and, clearly, the coefficients should be adaptable to cope with varying reflections. Moreover, many adaptive filters use this type of structure in order to minimize the possibility of instability.

Reflection cancellation is a form of *adaptive equalization* and adaptive equalizers are used extensively in communications systems. Consider the case of a data modem transmitting over a telephone line. Usually the line is an unknown *time-varying* signal-corrupting system and the received data will generally suffer from time-varying intersymbol interference. In order to combat this problem it is usual to model the channel as a linear time-variant filter and apply an *inverse model*, or equalizer, in the form of an adaptive filter, Fig. 4.50(a). In general, of course, the unknown system will also possess additive internal noise. Also note that Fig. 4.50(a) implies that, ideally, $r(n)$ should be a delayed version of $s(n)$, which may at first appear something of a dilemma when the equalizer is remote from the transmitter!

Two other fundamental adaptive filter configurations are shown in Fig. 4.50. Direct system modelling or *system identification* is shown in Fig. 4.50(b). As for inverse modelling, the unknown system will generally have

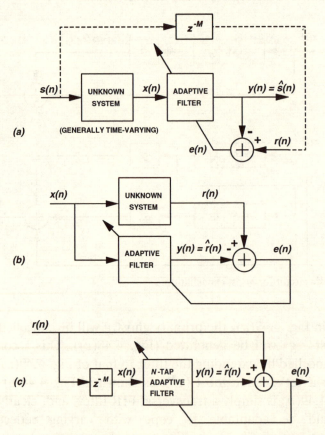

Fig. 4.50 Adaptive filter configurations: (a) inverse system modelling; (b) direct system modelling or system identification; (c) linear prediction.

internal noise but when the noise component of $r(n)$ is relatively small the transfer function of the adaptive filter will become similar to that of the unknown system. Direct modelling is the basis of *echo cancellation* in telephone systems. Fig. 4.50(*c*) illustrates *linear prediction*, whereby the filter generates a good estimate of input sample $r(n)$ based on previous samples of $r(n)$. This configuration finds application in source coding and in signal enhancement/interference suppression systems.

4.6.1 Wiener filtering

Fig. 4.51 shows the common transversal realization of the adaptive filters in Fig. 4.50. In each case the filter output $y(n)$ is subtracted from a reference signal $r(n)$ to generate an error signal $e(n)$, and an adaptive algorithm adjusts the filter coefficients in order to minimize some error criterion. For Fig. 4.51 we have

$$e(n) = r(n) - \mathbf{h}^{\mathrm{T}}(n)\mathbf{X}(n) \tag{4.150}$$

where

$$\mathbf{h}^{\mathrm{T}}(n) = [h_0(n) \quad h_1(n) \quad h_2(n) \quad \ldots \quad h_{N-1}(n)]$$

$$\mathbf{X}^{\mathrm{T}}(n) = [x(n) \quad x(n-1) \quad x(n-2) \quad \ldots \quad x(n-N+1)]$$

and 'T' denotes the transpose of the vector. The subsequent mathematical analysis depends upon whether we take a *statistical* or *deterministic* view of the time series involved, and this depends upon the application. A statistical (or *Wiener*) formulation of the problem assumes $x(n)$ to be some stationary random or stochastic signal and uses the *mmse* criterion, i.e.

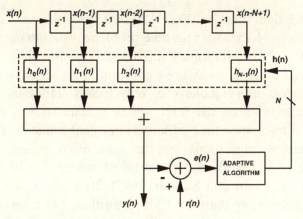

Fig. 4.51 General adaptive transversal filter.

$$\text{minimize} \quad E[e^2(n)] \qquad (4.151)$$

For example, the unknown system in Fig. 4.50(*a*) might simply add a random noise component to wanted (generally random) signal $s(n)$ and the goal of the filter is then to obtain a good estimate of $s(n)$ from a noisy input $x(n)$. Using (4.150) in (4.151), i.e. assuming a transversal filter structure, it turns out that $E[e^2(n)]$ may be viewed as an N-dimensional paraboloid with a *single* minimum, and so by seeking this minimum we naturally end up with a unique and optimum vector $\mathbf{h}_{opt}$. In fact, for a stationary signal the adaptive part of Fig. 4.51 is not essential (we can use fixed coefficients) and, using an analysis similar to that in Section 2.2.3, we find

$$\mathbf{h}_{opt} = \mathbf{R}_{xx}^{-1}\mathbf{R}_{rx} \qquad (4.152)$$

where

$$\mathbf{R}_{xx} = E[\mathbf{X}(n)\mathbf{X}^{\mathrm{T}}(n)] \qquad (4.153)$$

$$\mathbf{R}_{rx} = E[r(n)\mathbf{X}(n)] \qquad (4.154)$$

Here, $\mathbf{R}_{xx}$ is the $N \times N$ autocorrelation matrix for the input signal, and $\mathbf{R}_{rx}$ is a cross-correlation vector. The resulting filter is called the *Wiener filter* (or estimator) and (4.152) is referred to as the *Wiener–Hopf* equation. This is the best filter in the least squares sense for a stationary signal. Note the similarity between (4.152) and (2.32); for Wiener filtering the autocorrelation term $\mathbf{R}$ in (2.32) has become a cross-correlation term $\mathbf{R}_{rx}$.

4.6.2 *The LMS algorithm*

In practice the adaptive filter in Fig. 4.51 can provide an iterative, approximate solution to (4.152) for real-time problems, *without the need to compute* $\mathbf{R}_{xx}$ *and* $\mathbf{R}_{rx}$. For a stationary input the filter will then quickly move to near optimal coefficients. More significantly, an adaptive filter is applicable when $x(n)$ is *non-stationary* or only locally stationary (as for speech signals) and in this case the filter will first adapt to near optimal coefficients and then track slow time variations in signal statistics. Particularly important adaptive algorithms are the *least mean square (LMS)* algorithm (Widrow *et al*, 1976), and the class of *recursive least squares (RLS)* algorithms (e.g. Cioffi and Kailath, 1984). RLS algorithms generally adapt considerably faster than the LMS algorithm, but require more computation.

As previously discussed, the mmse criterion leads to a concave error 'surface' with a definite minimum corresponding to the case $\mathbf{h} = \mathbf{h}_{opt}$. The LMS algorithm seeks this minimum in an iterative way based upon the *method of steepest descent*. This states that the next value of $\mathbf{h}$ is given by the recursion rule

$$\mathbf{h}(n+1) = \mathbf{h}(n) + \frac{\mu}{2}[-\nabla(n)] \tag{4.155}$$

where μ is a positive constant and $\nabla(n)$ is the *error gradient vector* at time n. This gradient search procedure moves $\mathbf{h}(n)$ in the direction *opposite* to the error gradient and so, intuitively, it should lead to the optimum Wiener solution. The gradient is given by

$$\nabla(n) = \begin{bmatrix} \dfrac{\partial E[e^2(n)]}{\partial h_0} \\ \vdots \\ \dfrac{\partial E[e^2(n)]}{\partial h_{N-1}} \end{bmatrix}_{\mathbf{h}=\mathbf{h}(n)}$$

and the salient point about the LMS algorithm is that it roughly approximates $E[e^2(n)]$ in order to achieve a computationally simple result. In particular, a 'noisy gradient' is derived by using the instantaneous term $e^2(n)$ as an estimate for $E[e^2(n)]$ giving

$$\nabla(n) \approx 2e(n) \begin{bmatrix} \dfrac{\partial e(n)}{\partial h_0} \\ \vdots \\ \dfrac{\partial e(n)}{\partial h_{N-1}} \end{bmatrix}_{\mathbf{h}=\mathbf{h}(n)}$$

Expanding (4.150) it can be seen that

$$\begin{bmatrix} \dfrac{\partial e(n)}{\partial h_0} \\ \vdots \\ \dfrac{\partial e(n)}{\partial h_{N-1}} \end{bmatrix}_{\mathbf{h}=\mathbf{h}(n)} = - \begin{bmatrix} x(n) \\ \vdots \\ x(n-N+1) \end{bmatrix} = -\mathbf{X}(n)$$

Therefore,

$$\nabla(n) \approx -2e(n)\mathbf{X}(n)$$

and so

$$\mathbf{h}(n + 1) = \mathbf{h}(n) + \mu e(n)\mathbf{X}(n) \tag{4.156}$$

Equation (4.156) is the Widrow–Hoff LMS algorithm, and in terms of the actual filter coefficients, we can write it as

$$h_j(n + 1) = h_j(n) + \mu e(n)x(n - j), \quad j = 0, 1, 2, \ldots, N - 1 \tag{4.157}$$

Here the $x(n - j)$ terms correspond to the current data in the filter, and μ is the *step size* or *gain*. Note that $\mathbf{h}(n)$ is adjusted on a sample by sample basis.

Implementation

An adaptive filter based on (4.157) requires just $2N$ multiplications and $2N$ additions for each output sample, and so is simple enough to implement for many real time DSP problems. The algorithm is justified largely by experimental evidence and in many cases works very well. It can start from an arbitrary value of $\mathbf{h}$ and this is usually zero for convenience.

It is important to realize that a transversal filter realization for the LMS algorithm does not guarantee stability since we now have a feedback loop which adjusts the coefficients. Clearly, the rate of adjustment depends upon μ and too high a 'loop gain' can, in fact, lead to instability. In order for $\mathbf{h}(n)$ to converge to some 'steady state' value an *approximate* upper bound on μ can be given as (Widrow and Stearns, 1985)

$$\mu_{\max} \approx \frac{2}{\text{tr}\,[\mathbf{R}_{xx}]} \tag{4.158}$$

(although somewhat larger values of μ will sometimes give a stable system). In (4.158), $\text{tr}\,[\mathbf{R}_{xx}]$ is the trace of the $N \times N$ matrix $\mathbf{R}_{xx}$ and is given by the sum of the diagonal elements of $\mathbf{R}_{xx}$. Expanding (4.153) (or using (2.31)) gives

$$\mathbf{R}_{xx} = E\begin{bmatrix} x^2(n) & x(n)x(n-1) & \cdots & x(n)x(n-N+1) \\ x(n-1)x(n) & x^2(n-1) & \cdots & \\ \vdots & & & \\ x(n-N+1)x(n) & & \cdots & x^2(n-N+1) \end{bmatrix}$$

so that (4.158) can be written

$$\mu_{max} \approx \frac{2}{NE[x^2(n)]} \qquad (4.159)$$

where $E[x^2(n)]$ corresponds to the average input signal power. For example, if the delay line samples in Fig. 4.51 always have value 2 and $N = 15$, then $\mu < 0.033$. Assuming a fixed reference $r = 2$ and that the coefficients are initially zero, $e(n)$ will then converge from 2 towards zero. Some justification for (4.159) can be seen as follows. Suppose we write (4.150) as

$$e(n) = r(n) - \sum_{j=0}^{N-1} h_j(n)x(n - j)$$

If now $\mathbf{h}(n)$ is modified according to (4.157) we obtain a new error value

$$e'(n) = r(n) - \sum_{j=0}^{N-1} h_j'(n)x(n - j)$$

$$= r(n) - \sum_{j=0}^{N-1} [h_j(n) + \mu e(n)x(n - j)]x(n - j)$$

$$= e(n)\left[1 - \mu\sum_{j=0}^{N-1} x^2(n - j)\right]$$

Clearly, the value of μ which forces $e'(n)$ to zero is

$$\mu = \left[\sum_{j=0}^{N-1} x^2(n - j)\right]^{-1}$$

and this is similar to (4.159). Assuming (4.159) is satisfied, the choice of μ is then a compromise between convergence rate and steady state error in $\mathbf{h}(n)$ (in steady state, the coefficients wander randomly around the optimal values). Large values of μ give fast adaption at the expense of greater steady state error and vice-versa.

Sometimes it can be advantageous to implement a *leaky LMS algorithm* of the form

$$\mathbf{h}(n + 1) = (1 - c)\mathbf{h}(n) + \mu e(n)\mathbf{X}(n) \qquad (4.160)$$

where constant $c \ll 1$ and would normally be implemented by bit-shifting. The $1 - c$ term tends to bias each coefficient towards zero. This aids recovery from the effects of transmission errors, and combats the accumu-

Table 4.3. *Interrupt service routines for the LMS algorithm.*

Routine 1	Routine 2
store new sample $x(n)$	store new sample $x(n)$
for all j {	for all j {
$\qquad h_j x(n - j) \Rightarrow accumulator$	$\qquad h_j x(n - j) \Rightarrow accumulator$
$\qquad$ update h_j	
	}
}	$e(n) = r(n) - y(n)$
$e(n) = r(n) - y(n)$	for all j {update h_j}
output $y(n)$	output $y(n)$

lation of roundoff noise when the algorithm is implemented on a fixed wordlength processor. Here, roundoff noise arises due to the need to round a double precision scalar product $\mu e(n)$ and from rounding the scalar-vector product $\mu e(n)\mathbf{X}(n)$. Clearly, such quantization errors can accumulate with time due to the recursion in (4.160) and cause overflow. When the LMS algorithm is implemented on a floating-point processor it can be helpful to normalize the maximum value of the signal in the filter to the order of unity.

Finally, the exact order in which the algorithm is implemented can be important. Assuming DSP chip implementation, two possible pseudo-codes for the interrupt service routine are shown in Table 4.3. Routine 1 updates each coefficient h_j immediately after it has been used for multiplication and results in a very efficient code (minimum number of instructions). On the other hand, in this 'interleaved' code, h_j is updated using an error value computed from the previous sample period whilst the error value *could* be computed from the current sample period. In practice better performance may therefore be obtained by using a more straightforward but less efficient code (Routine 2).

4.6.3 Adaptive prediction

Linear prediction was discussed in Section 2.2.3 under DPCM and we saw that, ignoring quantization, the basic Nth-order predictor is an N-tap transversal filter holding previous samples and whose output is a prediction of the current input sample. A general *adaptive* predictor for a random signal might therefore take the form shown in Fig. 4.50(c), where the filter input is simply a delayed version of the desired or reference input

$r(n)$. Note that, just as in Section 2.2.3, the filter coefficients are effectively optimized using an mmse criterion (Equation (4.151)). Besides its obvious application to ADPCM systems, adaptive prediction is also applicable to the LPC of speech signals (Section 2.5). A basic speech predictor is shown in Fig. 2.32, and so the coefficients could be found adaptively rather than via matrix inversion by using Fig. 4.50(c) with $M = 1$. Clearly, this will update the coefficients on a sample-by-sample basis compared to a frame-by-frame basis when using matrix inversion. Fig. 4.52 illustrates the use of Fig. 4.50(c) and the LMS algorithm for voiced speech.

Another application of the adaptive predictor is that of noise cancellation. The classical noise cancellation system (Widrow *et al*, 1975) had two separate inputs as shown in Fig. 4.53, and noise cancellation is possible provided noise components n_1 and n_2 are correlated in some way, and are uncorrelated with the signal s. This idea (but with some modification for practical systems) is used for adaptive cancellation of acoustic noise in aircraft, vehicles and industrial environments. However, where a

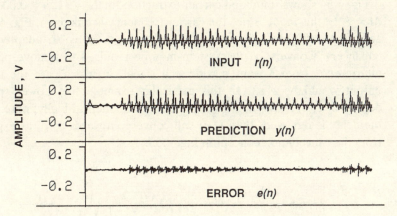

Fig. 4.52 LMS adaptive prediction for 250 ms of voiced speech ($N = 11$, $\mu = 0.5$, $M = 1$).

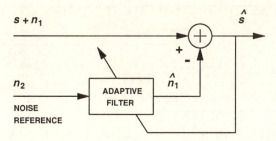

Fig. 4.53 Classical two-input noise cancellation system.

separate noise reference is unavailable, it is still possible to achieve significant noise cancellation using the single input adaptive predictor in Fig. 4.50(c).

Example 4.21

Suppose the input signal in Fig. 4.50(c) is $r(n) = S(n) + N(n)$ where S denotes a sinusoidal carrier of *unknown* frequency and N denotes wideband Gaussian noise of similar amplitude. A typical noisy carrier might appear as in Fig. 4.54 and the task could be to extract it from the noise. Following the argument for Fig. 4.53, the delay z^{-M} must be large enough to ensure that the components of N at the input and output of the delay are uncorrelated. This will prevent the adaptive filter from cancelling the noise component in $r(n)$. On the other hand, the signal component of $r(n)$ will still be correlated to that of $x(n)$ due to periodicity, and so $y(n)$ approximates to $S(n)$. Clearly, this noise cancellation approach is only possible provided S and N have significantly different spectral densities.

Fig. 4.54 shows simulated carrier extraction for $N = 31$, $\mu = 0.0001$ and $M = 5$. In this mode, since the carrier frequency is unknown, Fig. 4.50(c) acts as a *self-tuning filter* (it could also be used as an adaptive line enhancer). Conversely, the system may be required to remove periodic interference from a wanted wideband signal, in which case the system output would be $e(n)$. In this particular example, the power spectral density of $e(n)$ shows negligible periodic component. Finally, note that if the delay is removed, then $e(n)$ will converge towards zero, as expected from the cancellation concept in Fig. 4.53.

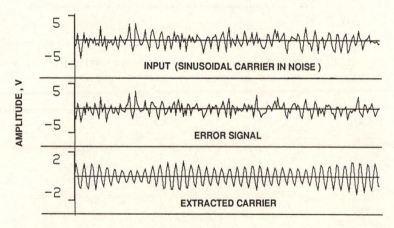

Fig. 4.54 Detection and extraction of an unknown carrier in noise using LMS adaptive prediction.

4.6.4 Adaptive equalization

Here we are concerned with the adaptive filter configuration in Fig. 4.50(*a*). The unknown system, typically a telephone channel, is modelled as a linear channel of unknown characteristics followed by additive Gaussian noise, and the received sampled signal $x(n)$ will comprise the ideal sample amplitude, $s(n)$, plus ISI terms and noise. In data modems, signal $x(n)$ is usually applied to an equalizer (filter) in order to minimise the ISI and noise terms prior to data detection.

Although modems use digital modulation techniques such as PSK and QAM, the concept of equalization is best described using a baseband pulse amplitude modification (PAM) model, as shown in Fig. 4.55. Here, $s(n)$ represents the transmitted quantized pulse amplitude at the *n*th sample instant, and the output of the receive filter will be the superposition of the responses of the analogue system to each transmitted symbol, plus noise. As indicated in Fig. 4.55, since the channel is unknown and time-varying, it is usual to use *adaptive* equalization. For most high-speed modems this is based on the LMS algorithm (and FIR structure in Fig. 4.51) because of its robustness in the presence of noise and large amounts of ISI. The equalizer output is then

$$y(n) = \mathbf{h}^{\mathrm{T}}(n)\mathbf{X}(n)$$

and, in general, it will deviate from the transmitted pulse amplitude $s(n)$ (hence the decision device). Sometimes (but not always) a *training*

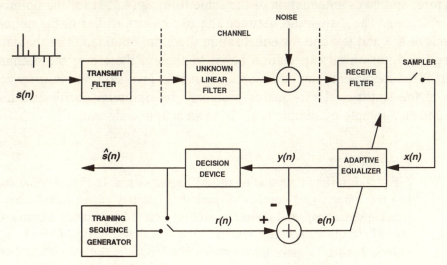

Fig. 4.55 Baseband PAM model for adaptive equalization.

sequence is sent prior to regular data transmission and a *synchronized* version of this signal is generated in the receiver. For example, in the context of ghost cancellation, the training sequence could be a sampled and stored version of the '2T pulse' used to test video systems. After this initial training phase, the error signal is derived as

$$e(n) = \hat{s}(n) - y(n)$$

where, now, $\hat{s}(n)$ is assumed to be a 'usually correct' estimate of $s(n)$. This *decision-directed* mode enables the equalizer to track slow changes in the channel characteristics.

Finally, it is worth noting the theoretical basis of Fig. 4.55. In Sections 4.6.1–4.6.3 we concentrated on *random* signals and the *statistical* least-squares problem. Effectively, the idea was to minimize the (statistical) mean square error (Equation (4.151)) between the filter output $y(n)$ and a delayed version of the transmitted *random* signal $s(n)$. The resulting Wiener filter has coefficients given by (4.152), but, as we have seen, these are usually found recursively using an adaptive algorithm. In contrast, we might regard the equalization training phase as minimizing a sum of squared errors between $y(n)$ and a *known* training signal. In other words, the filter is optimized for a specific (deterministic) sequence and we are dealing with a *deterministic* least-squares problem. In this case the optimality criterion becomes

$$\text{minimize } \sum e^2(n) \qquad (4.161)$$

It can be shown that this leads to a paraboloid-type error surface, as before, and gives an equation of the same form as (4.152) for the optimal coefficients. The distinction between the two equations lies in the definitions of $\mathbf{R}_{xx}$ and $\mathbf{R}_{rx}$ and for equalization these are obtained from the data itself and not on an expectation basis. In theory, therefore, the optimal equalizer coefficients can be found using matrix inversion, but in practice, as in the statistical least-squares problem, the optimum coefficients are found on a sample-by-sample basis using an adaptive algorithm.

Example 4.22

Fig. 4.56 is a simple simulation of LMS equalization. In Figs. 4.56(a) and (b) the input signal consists of periodic, isolated pulses (simulating a training sequence), and these are suitably delayed to provide a correctly phased reference. A single reference pulse is shown in Fig. 4.56(b) and, in fact, its shape closely approximates to a Nyquist pulse with 0% roll-off and a first zero value at $T = 4$ samples (see Fig. 3.35(a)). In contrast, the lowpass channel gives significant amplitude and phase distortion resulting in an asymmetrical pulse and ISI at the equalizer input (Fig. 4.56(b)). However, after 'convergence', the pulse symmetry is restored as $y(n)$ approximates to $r(n)$. Fig. 4.56(c) shows corresponding eye diagrams for random data.

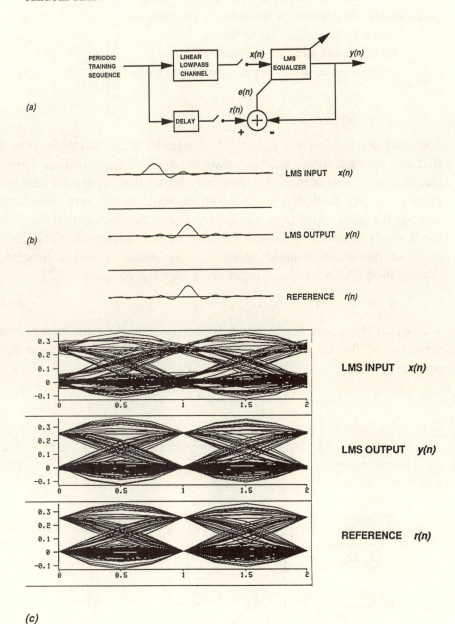

Fig. 4.56 Simulation of LMS equalization $N = 21$, $\mu = 0.5$: (a)(b) training mode; (c) corresponding eye diagrams.

4.7 Finite wordlength effects

It is apparent from the previous sections that established filter design techniques assume infinite precision in the input signal and within the filter itself, and it is left to the designer to assess separately the effects of quantization. In general, it is necessary to consider:

quantization of the input signal;
quantization of filter coefficients;
quantization of arithmetic operations within the filter.

4.7.1 Input quantization

For most practical signals (i.e. random signals) it is reasonable to assume that the wanted signal and the associated ADC quantization error are uncorrelated. This means that they may be treated separately and so the error is usually modelled as an additive, white noise, zero mean source $e_i(n)$ at the filter input (Fig. 4.57(a)). In turn, this means that we can use the linearity property of a linear filter to compute separately the behaviour of input quantization noise. Assuming the usual *rounding* process for quantization, the normalized input noise power is (Section 1.5.2)

$$\sigma_{ni}^2 = E[e_i^2(n)] = \Delta^2/12 \tag{4.162}$$

where Δ is the quantizing interval. By considering the noise power density spectrum at the filter output, it can be shown that the average output noise power is

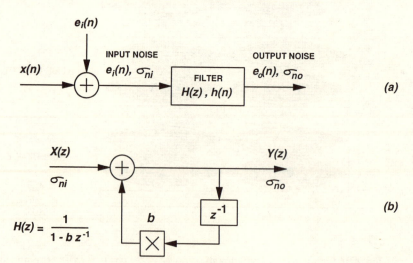

Fig. 4.57 Modelling input quantization noise.

$$\sigma_{no}^2 = E[e_0^2(n)] = \frac{\Delta^2}{12} \sum_{n=0}^{\infty} h^2(n) \tag{4.163}$$

Example 4.23

The impulse response for the simple filter in Fig. 4.57(*b*) is

$$h(n) = b^n, \quad n = 0, 1, 2, \ldots$$

$$\therefore \sigma_{no}^2/\sigma_{ni}^2 = \sum_{n=0}^{\infty} h^2(n)$$

$$= 1 + (b^1)^2 + (b^2)^2 + (b^3)^2 + \ldots$$

$$= 1/(1 - b^2) \tag{4.164}$$

If the pole is close to the unit circle, i.e. $b = 1 - \delta$, $\delta \ll 1$ then

$$\sigma_{no}^2/\sigma_{ni}^2 \approx 1/2\delta \tag{4.165}$$

and so the output noise power is proportional to δ^{-1}. This implies noise *amplification*. Note, however, that

$$H(e^{j\omega}) = 1/(1 - b\,e^{-j\omega})$$

and so

$$H(e^{j\omega})|_{\omega=0} = 1/\delta$$

This means that the output signal power at low frequencies is proportional to δ^{-2}, and so the output SNR at low frequencies is proportional to δ^{-1}. We might expect this result since the filter becomes more selective as δ decreases.

4.7.2 Coefficient quantization

Consider the second-order section represented by

$$H(z) = \frac{a_0 + a_1 z^{-1} + a_2 z^{-2}}{1 + b_1 z^{-1} + b_2 z^{-2}} \tag{4.166}$$

Quantization of coefficients a_i or b_i merely shifts the zero or pole positions respectively. In non-recursive designs the effect is not usually serious and the resulting deviation from the desired specification is easily computed. Typically, coefficient quantization will increase stopband ripple. In recursive designs, quantization means that the poles will have a finite number of possible positions within the unit circle. More significantly, quantization may move the pole *outside* the circle, and so make the filter unstable.

Example 4.24

The stability of the all-pole second-order section in Fig. 4.21 can be examined from the basic equations (4.60) and (4.61), namely

$$b_1 = -2r \cos \theta_1$$

$$b_2 = r^2, \quad r < 1$$

After a little manipulation of these equations we have

$$\Delta r \approx \Delta b_2 / 2r \tag{4.167}$$

$$\Delta \theta_1 \approx \frac{\Delta b_2}{2r^2 \tan \theta_1} + \frac{\Delta b_1}{2r \sin \theta_1} \tag{4.168}$$

Recall that the angular resonant frequency θ_0 ($= 2\pi f_0 T$) is very close to θ_1 when $r \approx 1$. Therefore, according to (4.168), when θ_1 is small, i.e. $f_0 \ll f_s$, the coefficients must be highly accurate if the error in the resonant frequency is to be small. Also, it is clear from (4.167) that when $r \approx 1$, error in b_2 can take the pole outside the unit circle.

4.7.3 *Product roundoff noise*

Fig. 4.21 is also useful for discussing the inaccuracy introduced by fixed-point addition and multiplication. We will assume that m-bit fractional 2's complement arithmetic (see Appendix B) is used for signals throughout the filter, and so any signal, X, will be in the range $-1.0 \leqslant X < 1.0$. Also, practical filters sometimes incorporate a scaling factor b_0 as discussed in Example 4.9, and so Fig. 4.21 can be represented by its fixed-point equivalent in Fig. 4.58. The notation $\langle m, n, t \rangle$ denotes m-bit words, n integer bits, and 2's complement format, and in Fig. 4.58 both signal wordlength and coefficient wordlength are assumed to be the same. Typically, m will be 16 bits.

Consider first the addition process. The addition of two fixed-point numbers cannot lead to inaccuracy unless *overflow* occurs (Appendix B). However, when overflow *does* occur, the 2's complement overflow characteristic can generate large errors, unless the final sum is within the 2's complement range. The probability of overflow should therefore be minimized, and this could be done by appropriate *scaling*, whilst at the same time avoiding severe degradation of the SNR. In Fig. 4.58 the input signal level could be reduced until fixed-point overflow fails to occur at any point in the circuit. Also, some general purpose DSP chips have an overflow checking facility which permits numbers to saturate at the extremes of the 2's complement range.

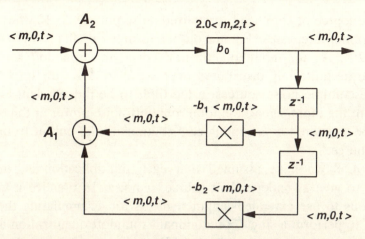

Fig. 4.58 Practical m-bit fractional 2's complement form for Fig. 4.21.

Given adequate overflow handling, addition will not contribute round-off errors in the filter output and the most serious fixed-point arithmetic problems in Fig. 4.58 will be associated with multiplication. In Fig. 4.58, the signal wordlength can be considered to be $m = k + 1$ bits, where k excludes the sign bit, and so multiplication of two such numbers will yield a $2k + 1$ bit product. It is apparent from the figure that this size of product is *not* allowed for and that the least significant k bits have been removed. This is not unreasonable. Clearly, for a recursive structure like Fig. 4.58, larger and larger wordlengths are required as more and more samples are computed, and we are *forced* to limit the wordlength at some point since we are using finite length accumulators and registers. There are several ways of removing the least significant k bits and the usual way is via rounding (Appendix B). Rounding may also be necessary in non-recursive structures, although here wordlength increase is always finite.

Roundoff noise model
The loss of k bits from each multiplier product in Fig. 4.58 is equivalent to assuming a quantizer at each multiplier output. This will generate quant-ization noise sources within the filter and their net effect on the filter output can be estimated by using a roundoff noise model.

The usual assumptions for roundoff noise analysis are that: (1) the filter is properly scaled so that overflow never or very rarely occurs; (2) the filter input signal is 'sufficiently random' to give uncorrelated noise sources; and (3) quantization is achieved by rounding. This being the case, each multiplier is modelled as an ideal multiplier followed by an additive white noise source of variance (normalized noise power) $\Delta^2/12$ where Δ is the quantization step size. For an output format $\langle m, 0, t \rangle$, $m = k + 1$, then $\Delta = 2^{-k}$. As for input quantization noise, this roundoff source is assumed to be uniformly distributed over $-\Delta/2$ to $\Delta/2$ with zero mean. Taking the multiple noise sources in the filter to be independent of each other and of the input signal, the total roundoff noise power at the output can then be estimated by summing contributions at the output from each roundoff source.

In Fig. 4.58 we have assumed that each multiplication is rounded separately to give an independent source of noise. In practice it can be advantageous to use extended length registers for accumulating the products, and to perform a single, predominant roundoff quantization after a final accumulation. For example, in Fig. 4.58, a single predominant quantization could be performed at the output of adder A_2 and before the finite length registers. If this is done, the only significant roundoff source effectively appears at the filter input and so the output noise power due to roundoff is simply given by (4.163).

A roundoff noise model for the general second-order canonical structure is shown in Fig. 4.59, and it is readily shown that the roundoff noise component at the output is

$$Y_n(z) = H(z)[E_3(z) + E_4(z)] + E_0(z) + E_1(z) + E_2(z)$$

$$(4.169)$$

where $H(z)$ is the filter transfer function. In this case, noise sources $E_3(z)$ and $E_4(z)$ are filtered just like any input signal whilst the other sources

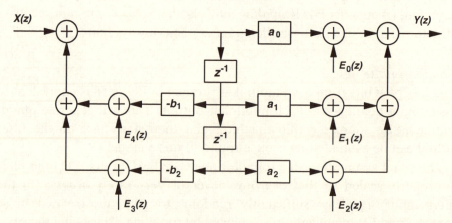

Fig. 4.59 A roundoff noise model for the canonical direct form second-order filter. ($E(z)$ represents a roundoff noise source.)

appear unfiltered at the output. The significant point is that, in general, *the effect of a noise source is spectrally shaped by the transfer function from the point where the noise source is introduced, to the output*, and so the output noise power will be highly dependent upon the filter structure. Two structures which are equivalent in terms of signal transfer function, e.g. a direct form structure and its transpose, can have significantly different roundoff noise performance.

Computation of roundoff noise
As just stated, the effect of a roundoff noise source on the filter output depends upon the transfer function (or impulse response) from the point at which the noise is introduced to the output. Consider the filter signal flow graph in Fig. 4.60. This has two roundoff noise sources e_a and e_b entering node A, corresponding to multipliers a and b, and so the pertinent impulse response is that from node A to the output. The multiplier c contributes a noise power $\Delta^2/12$ directly at output node B. Assuming uncorrelated white noise sources, and following the principle in (4.163), it follows that the total output roundoff noise power for Fig. 4.60 is

$$\sigma_{no}^2 = \frac{\Delta^2}{12}\left[2\sum_{k=0}^{\infty} g_A^2(k) + 1\right] \tag{4.170}$$

where g_A is the impulse response from node A to the output node. Generalizing this for an $n+1$ node system, where node $n+1$ is the output node

$$\sigma_{no}^2 = \frac{\Delta^2}{12}\left[\sum_{i=1}^{n} v_i \sum_{k=0}^{\infty} g_i^2(k) + v_{n+1}\right] \tag{4.171}$$

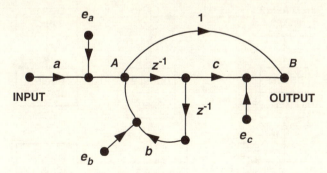

Fig. 4.60 Signal flow graph roundoff noise model for a second-order filter (e_a, e_b and e_c are roundoff noise sources).

Here, v_i multipliers enter node i, and $\mathbf{g}_i$ is the impulse response from node i to the output. Equation (4.171) could be evaluated directly, or, alternatively, the term

$$\sum_k g_i^2(k), \quad i = 1, 2, \ldots, n$$

could be computed using Cauchy's residue theorem. Either way, this type of analysis rapidly becomes tedious as the filter structure becomes more complex. Also, the assumption that the noise sources are independent of each other and of the filter signal itself can be erroneous when the input signal is not sufficiently random, e.g. when it is constant or zero for a period of time, or when it is periodic. Digitized speech, for example, can generate this effect during pauses in conversation. Consequently, especially for complex filters, it may well be quicker and safer to simulate the quantized filter using DSP software.

A typical simulation approach is shown in Fig. 4.61. Here, the reference filter uses double precision, floating-point arithmetic (thereby generating a near ideal output), whilst the practical filter uses m-bit 2's complement arithmetic. Note that, although the input signal to the reference filter is converted to floating-point format, it still retains the m-bit quantization of the source. Excitation is sinusoidal and the results of the simulation are stored in various signal files or 'sinks'. The noise analysis could be an estimation of the PSD of the error signal $e(n)$, or simply a measurement of the average noise power, $\sum_n e^2(n)$.

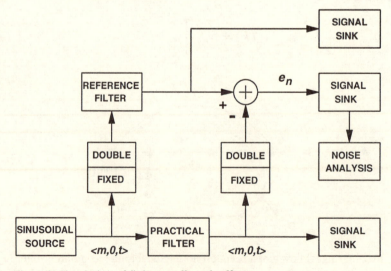

Fig. 4.61 Simulation of finite wordlength effects.

Example 4.25

A particular frequency shift keying (FSK) data modem uses the second-order filter in Fig. 4.58 as part of a tone detector. This filter is cascaded with a similar second-order filter to realize a bandpass response centred at 2100 Hz (one of the modem tone frequencies). The filter in Fig. 4.58 must resonate at 2082 Hz, and since the modem samples at 9600 Hz, the normalized resonant frequency f_0 is 0.216875 Hz. The sinusoidal excitation could be set at this frequency and the amplitude of the sinewave set to maximize the signal in the filter without overflow (all signals within the filter will have magnitude less than unity). Also, for this simulation, the sinusoidal source is quantized to 16-bit fractional 2's complement format.

The required coefficients can be computed from (4.60), (4.61) and (4.63), bearing in mind that the scaling factor $b_0 = 2$. For $Q \approx 100$ we have

$$r = 0.993127766$$

$$b_1 = -0.205211103$$

$$b_2 = 0.493151379$$

Methods for computing the noise PSD are discussed in Section 5.5.7. An estimate of the PSD $P(k)$ could be found directly using the *periodogram*, i.e.

$$\hat{P}(k) = 20 \log_{10} |E(k)| \text{ dB} \tag{4.172}$$

where $E(k)$ $k = 0, 1, \ldots, N - 1$ is the N-point FFT of a windowed version of error signal $e(n)$. Alternatively, here we adopt the indirect *autocorrelation-FFT* method illustrated in Fig. 5.31(b), and use a Hanning window on estimates of the autocorrelation function.

Fig. 4.62 shows the (smoothed) noise PSD for various quantization strategies relative to the signal power density from the reference filter (graph (a)). Graphs (b) and (c) correspond to $m = 12$ and $m = 16$ respectively. Graphs (d) and (e) imply that there is probably little point in increasing the wordlength beyond 16-bits since the error due to 16-bit coefficient quantization starts to dominate beyond this point. Note that this type of simulation can also give a rapid assessment of the effect of different accumulator and register wordlengths.

Example 4.26

Periods of constant input level to a fixed-point filter can generate spurious steady state signals at the filter output. These so-called 'deadband' effects arise from non-linearity within the filter. Suppose the signal $w(n)$ in Fig.

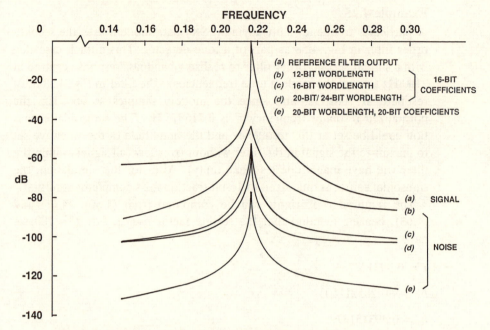

Fig. 4.62 Simulation of signal and quantization noise PSDs.

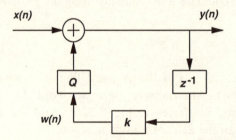

Fig. 4.63 Simple filter to illustrate deadband effects (quantizer Q has interval Δ).

4.63 is quantized by rounding to the nearest level, and let the initial conditions be

$$x(n) = 0 \quad n > 0$$

$$y(0) = 9\Delta$$

where Δ is the quantizing interval. Depending upon the coefficient, k, the filter output can then tend to a constant 'deadband error', or to a spurious periodic signal known as a 'limit cycle'. Table 4.4 illustrates both these effects for the above initial conditions, and it is apparent from the table that the effects can be minimized by increasing the wordlength. Limit cycles can also be caused by overflow in addition (Appendix B).

Table 4.4. *Illustration of deadband effects.*

	$k = 0.91$		$k = -0.91$	
n	$w(n)$	$y(n)$	$w(n)$	$y(n)$
0		9Δ		+9Δ
1	8.19Δ	8Δ	−8.19Δ	−8Δ
2	7.28Δ	7Δ	+7.28Δ	+7Δ
3	6.37Δ	6Δ	−6.37Δ	−6Δ
4	5.46Δ	5Δ	+5.46Δ	+5Δ
5	4.55Δ	5Δ	−4.55Δ	−5Δ
6	4.55Δ	5Δ	+4.55Δ	+5Δ
7	4.55Δ	5Δ	−4.55Δ	−5Δ

4.8 Filters for image processing

In this section we examine some video-rate applications of the classical linear digital filter theory discussed earlier in the chapter. Here, since video signals are 'multidimensional', it is necessary to introduce the concept of filtering in various dimensions. We also examine a class of linear filters used for *image enhancement* applications, either in real time or non-real time. These filters tend to use heuristic algorithms rather than formal linear filter theory and are commonly referred to as '$N \times N$' *convolvers* (also called operators, templates, or masks). Finally, we examine the important concept of *non-linear* filters, these being particularly useful in video and image processing.

4.8.1 Filtering in different dimensions

In video systems we have the option of filtering in different dimensions, depending upon the problem. Consider the lowpass filter in Fig. 4.64(*a*). From Example 4.4, this filter has the amplitude response

$$H^*(e^{j\omega T_o}) = 1 + \cos(\omega T_o) \tag{4.173}$$

Note that the response depends upon T_o and *not* upon the video sampling period, T. If the video signal is sampled at an integer multiple of the horizontal (or line) frequency f_H, we can imagine the samples in the filter as part of the orthogonal *pixel* arrays in Fig. 4.64(*b*). An orthogonal sampling pattern is not essential, but it facilitates a simple explanation of multidimensional processing. In this case we can select the operating dimension by simply selecting the filter delay T_o to be equal to the video

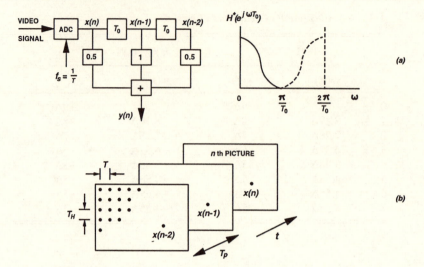

Fig. 4.64 Filtering in different dimensions.

sample period, T, or the horizontal period, T_H, or the picture (frame) period T_p. Consider these three cases, assuming that $f_s \approx 10$ MHz.

(1) $T_o = T$: In this case the filter holds samples along the scan line and so is termed a *horizontal* filter. It has a zero in the amplitude response at the video frequency

$$f = 1/2T \text{ Hz}$$

or approximately 5 MHz, and so it will blur high-frequency detail along the line. However, it will not affect vertical detail.

(2) $T_o = T_H$: Each filter delay is now one *linestore*. This *vertical* filter holds vertical samples from the pixel array, and the first zero in the filter response will be at $f = 1/2T_H$ (typically about 8 kHz). The filter is actually acting as a comb filter with zeros at the video frequencies

$$f = \left(m + \frac{1}{2} \right) f_H, \qquad m = 0, 1, 2, \ldots$$

This type of response could be useful for removing unwanted spectra between line harmonics, as in sub-Nyquist sampling (Section 1.4.3). The effect of this filter upon the displayed image will be to attenuate rapid changes in a column of pixels, i.e. changes down the picture, but it will not affect changes along a line.

(3) $T_o = T_p$: Each filter delay is now one *framestore*, and the samples in the filter are corresponding samples from successive pictures

(see Fig. 4.64(*b*)) – giving rise to the term *temporal* filter. The first zero is now at one half picture frequency (typically at $f = 12.5$ Hz), and the comb response repeats at the picture frequency. Clearly, any change in a particular pixel from picture to picture will tend to be attenuated, and so the filter will blur moving objects. On the other hand, if the image is static, the filter will attenuate random (time-varying) noise, whilst leaving the horizontal and vertical resolution unimpaired.

Example 4.27

Horizontal and vertical filters are sometimes cascaded to realize a video-rate 2-D *spatial* filter. Fig. 4.65 shows a first-order bandpass vertical filter

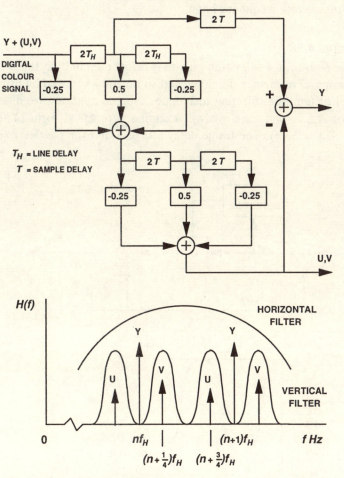

Fig. 4.65 First-order 2-D filter for the separation of luminance and chrominance components in a PAL colour television signal.

followed by a first-order bandpass horizontal filter. Taking $N = 3$ and $T \equiv 2T_H$ in (4.31), the amplitude response of the vertical filter is

$$H^*(e^{j\omega}) = \tfrac{1}{2}[1 - \cos(4\pi f/f_H)] \qquad (4.174)$$

This comb response has peaks at

$$f = \tfrac{1}{4}f_H, \quad \tfrac{3}{4}f_H, \quad \tfrac{5}{4}f_H, \quad \ldots \quad \text{Hz}$$

which means that the filter can be used to select out the chrominance components (U, V) from the luminance components (Y) in a PAL colour television signal, as shown in Fig. 4.65. The horizontal filter has the response

$$H^*(e^{j\omega}) = \tfrac{1}{2}[1 - \cos(4\pi f/f_s)] \qquad (4.175)$$

and is required in order to limit the comb response of the vertical filter to the chrominance frequency band.

Example 4.28

Random noise in a television picture is distributed in both the spatial and the temporal frequency domains, and so it can be reduced by using either spatial or temporal filtering techniques. Spatial methods are discussed in Section 4.8.2, and here we will examine a practical form of temporal filter, Fig. 4.66(a). The frame delay acts as a unit sample delay (z^{-1}) and

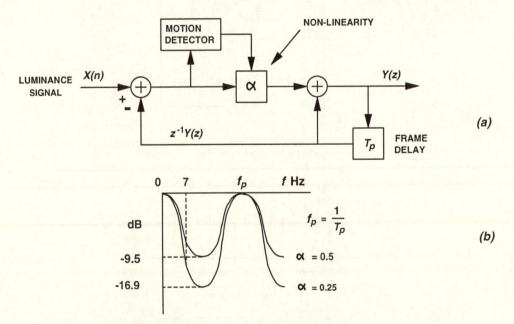

Fig. 4.66 Temporal filter for random noise reduction.

the motion detector varies α over the range 0–1. The z-domain difference equation for this filter is therefore

$$Y(z) = \alpha X(z) - \alpha z^{-1}Y(z) + z^{-1}Y(z) \tag{4.176}$$

giving

$$H(z) = \frac{\alpha}{1 - (1 - \alpha)z^{-1}} \tag{4.177}$$

The corresponding frequency response is

$$H(e^{j\omega}) = \frac{\alpha}{1 - (1 - \alpha)e^{-j\omega T_P}} \tag{4.178}$$

and this reduces to

$$|H(e^{j\omega})|^2 = \frac{\alpha^2}{1 + (1 - \alpha)^2 - 2(1 - \alpha)\cos(\omega T_P)} \tag{4.179}$$

The transfer function is sketched in Fig. 4.66(*b*). To be effective, the filter should give significant attenuation at the peak of the eye's temporal sensitivity characteristic (about 7 Hz). The practical filter must also be *adaptive* if moving objects are not to be blurred, and the role of the motion detector is to increase α towards unity for large difference signals. In this case there is little or no filtering action. Since $H(z)$ is now a function of the input signal, we could also regard the filter as *non-linear* (see Section 4.8.3).

4.8.2 N × N *Convolution*

A digitized image f is a rectangular array of pixels generated by some image formation system from the original object plane F, as in Fig. 4.67. Typically, f is a 512×512 array of 8-bit pixels. Once in this form, *image enhancement* algorithms can be used to generate a processed image, g, and here we are interested in algorithms for image sharpening, smoothing, edge detection, line detection, point detection, etc. These particular operations are basically linear filtering or convolutional processes and can be achieved by passing a template over the pixel array.

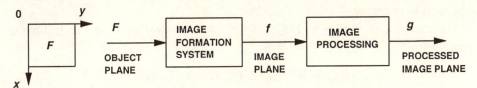

Fig. 4.67 Defining object, image and processed image planes.

Suppose, for example, that a template W covers three pixels of the digitized image and that the template is centred on pixel $f(n)$ (which for convenience will also be assumed to have grey level *value* $f(n)$) i.e.

$$W = \boxed{w(2)} \boxed{w(1)} \boxed{w(0)}$$

$$f = \boxed{f(n-1)} \boxed{f(n)} \boxed{f(n+1)}$$

Forming the inner or dot product of f and W gives the processed output sample

$$g(n) = \sum_{i=0}^{2} w(i)f(n+1-i) \tag{4.180}$$

Clearly, this is a convolutional operation (see (4.7)) with its output centred on $f(n)$, i.e. pixel $f(n)$ would be replaced by $g(n)$ in the processed image. Looked at another way, (4.180) is a linear combination of weighted pixel values and so represents a linear filtering process. Normally, template W would be moved one pixel at a time to process the entire image. In practice, W is usually an $N \times N$ array and filtering becomes a 2-D convolution between W and an $N \times N$ array of pixels. Dedicated $N \times N$ convolvers working at video rates and above are readily available. For $N = 3$, convolution of the general template and image segment in Fig. 4.68(a) gives the output

$$g(x, y) = \sum_{i=1}^{3}\sum_{j=1}^{3} f_{ij} w_{ij} \tag{4.181}$$

$$= (f, W) \tag{4.182}$$

where W is centred on pixel $f(x, y)$. The elements of W define the type of linear filter.

Laplacian operator
This operator is shown in Fig. 4.68(b) and is often used in the following context:

$$g(x, y) = \begin{cases} |(f, W)| & \text{if} \quad |(f, W)| > T \\ 0 & \text{otherwise} \end{cases} \tag{4.183}$$

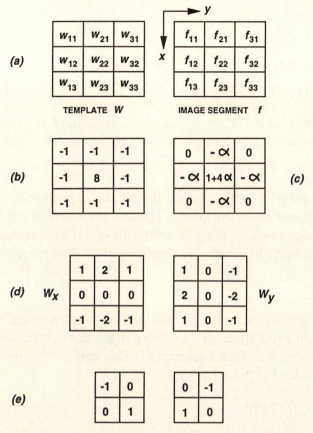

Fig. 4.68 Some convolutional operators: (*a*) general convolution; (*b*) Laplacian; (*c*) high-emphasis; (*d*) Sobel; (*e*) Roberts.

where T is a threshold (the introduction of a threshold actually changes the linear filtering operation to a *non-linear* operation). This means that pixel $f(x, y)$ would be replaced with value $g(x, y)$. The Laplacian operator is particularly sensitive to points, and we could assume a point has been detected if

$$|(f, W)| > T$$

On the other hand, the operator is also sensitive to lines and edges, and it is sometimes used as an edge detector.

High-emphasis operator
Assuming that the general linear operator in Fig. 4.68(*c*) is centred on pixel $f(x, y)$, the corresponding output pixel will be

$$g(x, y) = (f, W)$$

$$= (1 + 4\alpha)f(x, y) - \alpha[f(x + 1, y) + f(x - 1, y)$$

$$+ f(x, y - 1) + f(x, y + 1)]$$

$$= f(x, y) + 5\alpha[f(x, y) - \overline{f(x, y)}] \tag{4.184}$$

where α is a constant and $\overline{f(x, y)}$ is the pixel mean over W, including pixel $f(x, y)$. Extending the operator to a complete image, we could write

$$g = f + 5\alpha[f - \bar{f}] \tag{4.185}$$

i.e. the enhanced image g is the original image plus a proportion of the high-frequency components in the image. This has the effect of enhancing high-frequency components or sharpening the image. The constant α is usually restricted to 1 or 2 to avoid over emphasis of random noise.

Edge and line detection
The objective here is to extract the edge information from an image and to use it for subsequent processing, or to display it for the human observer as, say, 'bright' levels on a black background. This could be achieved by displaying the *gradient G* at point (x, y), i.e. let

$$g(x, y) = G[f(x, y)]$$

$$= \sqrt{\left[\left(\frac{\partial f}{\partial x}\right)^2 + \left(\frac{\partial f}{\partial y}\right)^2\right]} \tag{4.186}$$

The gradient is ideal because it is a true *isotropic* operator and so has equal sensitivity to all edges, irrespective of direction. In practice, simple approximations to G suffice, even though they are not isotropic. Considering the image segment f in Fig. 4.68(a), we could write

$$\frac{\partial f}{\partial x} \approx f_{11} + 2f_{21} + f_{31} - (f_{13} + 2f_{23} + f_{33})$$

$$= (f, W_x) \tag{4.187}$$

$$\frac{\partial f}{\partial y} \approx f_{11} + 2f_{12} + f_{13} - (f_{31} + 2f_{32} + f_{33})$$

$$= (f, W_y) \tag{4.188}$$

Equations (4.187) and (4.188) represent the outputs of linear edge detection filters, and edge detection operators W_x and W_y are shown in Fig.

4.68(d). These approximate differentials can now be substituted into a simplified version of (4.186) to give

$$g(x, y) \approx \left| \frac{\partial f}{\partial x} \right| + \left| \frac{\partial f}{\partial y} \right|$$

$$= |(f, W_x)| + |(f, W_y)| \tag{4.189}$$

Adding these vertical and horizontal filtering operations as in (4.189) yields a simplified form of the *Sobel* edge detection filter. When (4.189) is applied directly (without thresholding) then the enhanced image g is comprised of high grey scale values where there is a strong edge, and virtually zero values (a black image) in quasi-flat areas of f. Changing the factor 2 in the Sobel detector to unity gives the *Prewitt* edge detector, and this can be preferable in some cases. When it is necessary to reduce computation to an absolute minimum, edge detection can be achieved using the 2×2 *Roberts* operator in Fig. 4.68(e). This operator tends to generate thinner lines than the 3×3 operators.

Another approach to edge detection is to define a set of operators each sensitive to a particular edge orientation, as in Fig. 4.69(a) (the factor 1.5 accounting for the fact that within a 3×3 window the diagonal distance between pixels is a factor $\sqrt{2} \approx 1.5$ greater than the vertical or horizontal distance between pixels). The basic detection algorithm is simply

$$\text{find} \quad \max (f, W_i) \quad i = 1, \ldots, 4$$

$$\text{if} \quad \max (f, W_i) > T$$

$$\text{then an edge of orientation } i \text{ is present} \tag{4.190}$$

where T is a threshold. By noting positive and negative signs, the four operators in Fig. 4.69(a) can define eight edge directions, and Fig. 4.69(b) shows the elements of a video-rate DSP chip based on this principle. The four operators are implemented using four FIR filters working in parallel. Since the chip uses 3×3 operators, its three inputs are essentially separated by one line delays.

The multiple template concept can be extended to *line detection*, as in Fig. 4.70(a), and in this case the detector output could be expressed as

$$g(x, y) = |\max (f, W_i)| \quad i = 1, \ldots, 4 \tag{4.191}$$

These operators are most sensitive to a line one pixel wide and the detection of a thin, horizontal line is illustrated in Fig. 4.70(b).

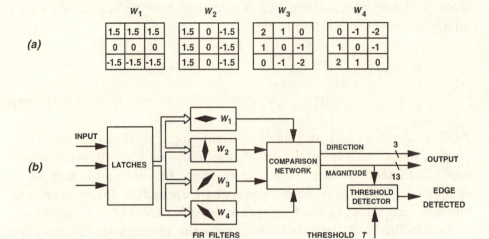

Fig. 4.69 Alternative edge detection algorithm: (*a*) operators; (*b*) DSP implementation (courtesy GEC-Plessey Semiconductors).

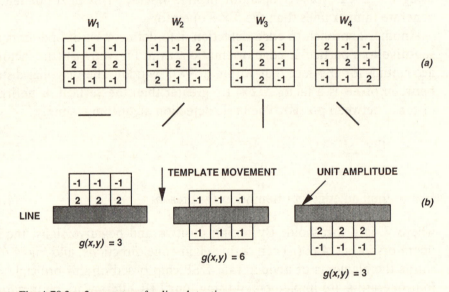

Fig. 4.70 3 × 3 operators for line detection.

Valley detection

It is important to note that edge detectors are not necessarily the best operators for detecting local changes in picture grey level. Consider the problem of detecting the principal features in a human face in order to transmit a simple image down a low bit-rate communications system (Pearson and Robinson, 1985). This type of image is likely to generate local 'dips' or *valleys* in grey level (Fig. 4.71(*a*)), as well as sharp

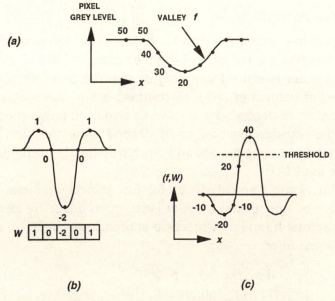

Fig. 4.71 Valley detection: (*a*) grey level valley, f; (*b*) simple valley operator, W; (*c*) $f*W$.

luminance edges. Unfortunately, applying a standard edge detector to a grey level valley will generate *two* lines or edges (since it detects two maximum gradient points) and the resulting image can be misleading. On the other hand, convolving image f with the simple *valley operator W* in Fig. 4.71(*b*) will generate a *single* peak at the centre of the valley, as shown in Fig. 4.71(*c*).

The general form of the valley operator is defined by taking the second derivative of a Gaussian function $G(x)$, i.e.

$$\text{valley operator} \equiv \partial^2 G/\partial x^2$$

It can be shown that the convolution of this operator with an image is equivalent to taking the second derivative of a *smoothed version* of the image (Charniak and McDermott, 1985). For example, when it is convolved with a true grey level edge (as opposed to a valley), any noise is first smoothed and the result will be the second derivative of the smoothed edge. The reader should show that the edge can be detected (albeit with slight error in its position) by thresholding this second derivative. Thus, the valley operator will have a large response to a grey level valley, but also some response to an edge. Alternatively, the operator could be used to noise-smooth edges and take the second derivative simultaneously, and the edge will be accurately located at the zero crossing.

4.8.3 Non-linear filters

So far this chapter has concentrated on linear filters, i.e. filters which have a transfer function which is independent of the nature or level of the input signal. This statement is equivalent to saying that the principle of super-position applies, as defined in (4.4). In contrast, a *non-linear filter* has a transfer function which depends upon its input signal. In image processing for example, the transfer function could depend upon the local image segment covered by a template W and we have already encountered a simple example in (4.183).

Non-linear filters are particularly useful for reducing random noise. Suppose we wish to smooth the random noise in an image by using 2-D convolution or spatial filtering. A practical approach would be to use the following non-linear filter

$$g(x, y) = \begin{cases} \bar{p} & \text{if} \quad |f(x, y) - \bar{p}| > T \\ f(x, y) & \text{otherwise} \end{cases} \qquad (4.192)$$

where $\bar{p}$ is the mean of all pixels covered by a template W, excluding pixel $f(x, y)$. The significant point is that a non-linear filter is often capable of reducing noise *without significant degradation of signal rise-times*, whereas the use of a linear filter (no thresholding) will usually give unacceptable blurr.

Rank filters

An important class of non-linear filters which can be used for edge-preserving smoothing is that of *rank filters* and Fig. 4.72 illustrates the concept for image smoothing. The N pixels in image segment f are ordered by value, i.e.

$$f_1, f_2, f_3, \ldots, f_N \qquad f_{i-1} \leqslant f_i$$

and the rank filter R_k has an output f_k, $1 \leqslant k \leqslant N$. For instance, in Fig.

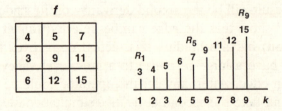

Fig. 4.72 Rank filtering: rank filter R_k replaces the centre pixel with value f_k.

4.72, $R_5 = 7$, and since the median has been selected this filter is referred to as the *median filter*. Put simply, the median filter for Fig. 4.72 will replace the centre pixel (value 9) with the median value, 7.

Example 4.29

Filters with rank positions near the median (N odd) are often used for noise smoothing since, by definition, noise samples are unlikely to occupy the median position. Consider the performance of a three-sample median filter on the following noisy step waveform

filter input	2 2 2 1 3 2 0 2 5 8 9 7 7 8 6 4 7 7 7 7
filter output	2 2 2 2 2 2 2 5 8 8 7 7 7 6 6 7 7 7

It is apparent that the filter preserves edges and reduces edge ripple. Similarly, a median filter applied to a noisy quasi-flat region of an image will reduce the noise variance, and when a sharp edge is encountered it will tend to be preserved.

Rank filters R_1 and R_9 in Fig. 4.72 correspond to the extreme rank positions and are called MIN and MAX filters respectively. These too can be used to create an edge-preserving non-linear filter for random noise reduction. Suppose that we wish to reduce or eliminate the positive and negative noise pulses in Fig. 4.73(a), whilst retaining the signal rise-time. One approach is to assume that the sampled signal comprises legal and illegal n-sample patterns, and to apply transformations to convert illegal patterns to legal patterns. In Fig. 4.73 we could assume that three-sample noise pulses generate illegal patterns.

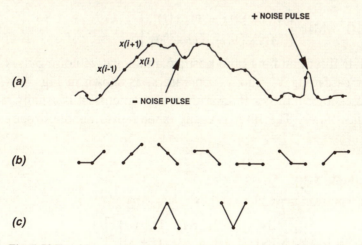

Fig. 4.73 Breaking a noisy signal into three-sample patterns: (a) signal; (b) legal patterns; (c) illegal patterns.

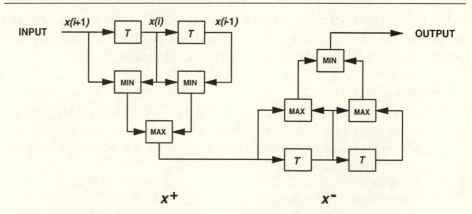

Fig. 4.74 A three-tap logical filter for edge-preserving noise smoothing.

For the positive noise pulse the appropriate three-sample transformation is

$$\{x(i-1), x(i), x(i+1)\} \Rightarrow \{x(i-1), x^+(i), x(i+1)\} \qquad (4.193)$$

where

$$x^+(i) = \text{MAX} \begin{bmatrix} \text{MIN}\,[x(i-1), x(i)], \\ \text{MIN}\,[x(i), x(i+1)] \end{bmatrix} \qquad (4.194)$$

This transformation will force the three-sample positive noise pattern into a legal pattern, but it will have no effect upon legal patterns (highlighting the non-linearity of the filter). Similarly, the negative noise pulse can be transformed into a legal pattern by replacing the centre pixel with

$$x^-(i) = \text{MIN} \begin{bmatrix} \text{MAX}\,[x(i-1), x(i)], \\ \text{MAX}\,[x(i), x(i+1)] \end{bmatrix} \qquad (4.195)$$

A simple 1-D filter handling both positive and negative noise pulses can be realized by cascading x^+ and x^- operators, as shown in Fig. 4.74. Since this filter uses logical rather than arithmetic elements, it is usually referred to as a *logical filter*. The filter is easily extended to handle broader noise pulses.

Example 4.30
For a positive noise pulse and five-sample patterns

$$x^+(i) = \text{MAX} \begin{bmatrix} \text{MIN}\,[x(i-2), x(i-1), x(i)], \\ \text{MIN}\,[x(i-1), x(i), x(i+1)], \\ \text{MIN}\,[x(i), x(i+1), x(i+2)] \end{bmatrix} \qquad (4.196)$$

Using the same 1-D input sequence as in Example 4.29, the sequential
filtering operations are as follows:

filter input	2 2 2 1 3 2 0 2 5 8 9 7 7 8 6 4 7 7 7 7
$x^+(i)$	2 1 1 1 0 2 5 7 7 7 7 7 6 4 7 7
$x^+(i)x^-(i)$	1 1 1 2 5 7 7 7 7 7 7 7

The last row corresponds to the final filtered waveform. Note that the
performance of the filter is comparable to that of the median filter, and
that the step rise-time tends to be preserved.

Logical filters could be applied to the problems of random noise reduction
and luminance–chrominance separation in television receivers (Okada,
Tanaka and Isobe, 1982; Okada, Hongu and Tanaka, 1985). Note,
however, that any sinusoidal input to the filter should have a frequency
significantly less than one quarter of the sampling frequency, otherwise it
could suffer significant attenuation.

When applied to image processing, MIN and MAX filters are usually
referred to as *erosion* and *dilation* respectively and the resulting non-linear
filters are called *morphological filters*. For example, when the MIN filter
or *structuring element* passes over the edge of a bright feature it will select
a dark element as output, thereby eroding the bright object. When erosion
is followed by dilation (the morphological 'opening' process) the effect is
to remove small bright details from an image whilst the overall grey level
and sharp edges tend to be preserved.

5

Discrete transforms

The reader will be familiar with the classical sinusoidal or exponential Fourier series expansion of a continuous time signal. Essentially, it gives us a different perspective on the signal by representing it in the frequency domain as a succession of harmonically related components. Put another way, each Fourier coefficient defines the amplitude of a particular sinusoidal or exponential time function. In a similar way, a *discrete transform* is a way of obtaining 'Fourier-type' coefficients for a discrete time signal, i.e. the input to a discrete transformer is a sequence of N discrete (digital) samples and its output will be a sequence of N Fourier-type coefficients representing a spectral view of the signal. The use of a digital input signal enables sophisticated DSP to be carried out, and this leads to a number of important applications. Clearly, discrete transforms are fundamental to spectral analysis. What is not so obvious is that they also play a major role in data compression (Section 2.3), signal detection (Section 5.5.1) and digital filtering and correlation (Section 5.6). We will first examine their theoretical basis.

5.1 Nth-order linear transforms

We start by considering a generalization of the Fourier series concept. Recall that the exponential Fourier series expansion for an arbitrary time signal $x(t)$ over a time interval $-T/2$ to $T/2$ is

$$x(t) = \sum_{k=-\infty}^{\infty} a(k)\,e^{jk\omega_0 t}; \quad \omega_0 = 2\pi/T$$

where the Fourier coefficients are given by

$$a(k) = \frac{1}{T} \int_{-T/2}^{T/2} x(t) \, e^{-jk\omega_0 t} \, dt; \quad k = 0, \pm 1, \pm 2, \ldots$$

The significant point here is that $x(t)$ is expanded in terms of a *system of functions* (in this case, complex functions) and that these functions are defined over a specific time interval.

Suppose we broaden the discussion and consider a *general* system of functions $\{g(k, \theta)\}$ where k is an integer denoting some spectral harmonic, and $\theta = t/T$ and represents normalized time. By definition, two functions from this general system are *orthogonal* (and normal) if

$$\int_\theta g(m, \theta) g(n, \theta) \, d\theta = \begin{cases} 1 & m = n \\ 0 & m \neq n \end{cases} \tag{5.1}$$

where the integration is carried out over the range of orthogonality (we will usually take this as $-1/2 \leqslant \theta \leqslant 1/2$). For a system of *complex* functions, i.e.

$$\{g(k, \theta)\} = \{e^{j2\pi k\theta}\}; \quad k = 0, \pm 1, \pm 2, \ldots, \infty \quad -\tfrac{1}{2} \leqslant \theta \leqslant \tfrac{1}{2}$$

the complex conjugate $g^*(n, \theta)$ is used in place of $g(n, \theta)$ in (5.1), and the reader should show that this particular system is indeed orthogonal over $-\tfrac{1}{2} \leqslant \theta \leqslant \tfrac{1}{2}$. We shall see later that orthogonal complex functions are fundamental to some signal detection systems.

It should now be clear that the exponential Fourier series just considered is actually a linear combination of mutually orthogonal functions. Moreover, the series expansion of a function is not restricted to any one system of orthogonal functions (or, indeed, to functions of time) and a general Fourier series expansion could be written

$$x(\theta) = \sum_{k=0}^\infty a(k) g(k, \theta) \tag{5.2}$$

where the Fourier coefficients are given by

$$a(k) = \int_{-1/2}^{1/2} x(\theta) g(k, \theta) \, d\theta \tag{5.3}$$

For example, letting $\{g(k, \theta)\}$ correspond to a set of Legendre polynomials would give a Legendre Fourier series, and letting $\{g(k, \theta)\} = \{1, \sqrt{2}\sin(2\pi k\theta), \sqrt{2}\cos(2\pi k\theta)\}$ (see Fig. 5.1(a)) would give the trigonometric Fourier series. Alternatively, we could expand in terms of the set of orthogonal rectangular functions in Fig. 5.1(b). In the following discussion we are more concerned with the Fourier coefficients, $a(k)$, rather than

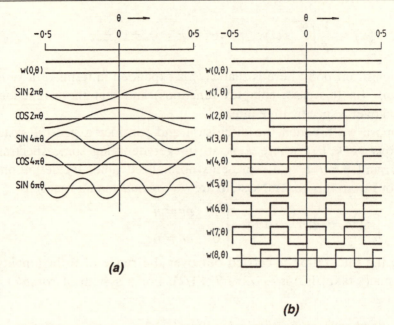

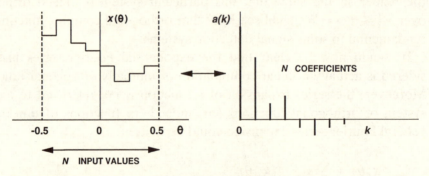

Fig. 5.1 Two orthogonal systems: (*a*) sine and cosine functions; (*b*) Walsh functions.

Fig. 5.2 Generalized Fourier analysis (discrete transformation).

with the Fourier series expansion. This is because (5.3) represents a generalization of the familiar sinusoidal spectral analysis of a signal, i.e. terms $a(k)$ are spectral components, and, when expressed in discrete form, it leads naturally to various discrete transforms.

For DSP applications we consider $x(\theta)$ to be a sampled waveform of N discrete values, as in Fig. 5.2. Here $N = 8$ and spectral analysis is particularly simple if we use the rectangular Walsh functions in Fig. 5.1(*b*) for $\{g(k, \theta)\}$. In this case the integration in (5.3) becomes a summation, and it is readily shown that (5.3) reduces to

$$a(k) = \frac{1}{N} \sum_{n=0}^{N-1} x_n w(k, n); \quad N = 8 \tag{5.4}$$

where x_n is a sample value of $x(\theta)$ and $w(k, n)$ is a sample value of the kth Walsh function $w(k, \theta)$. For example, taking the Walsh function amplitudes in Fig. 5.1(b) as ± 1,

$$\{w(4, n)\} = +1 \ -1 \ -1 \ +1 \ +1 \ -1 \ -1 \ +1$$

and

$$\{w(5, n)\} = +1 \ -1 \ -1 \ +1 \ -1 \ +1 \ +1 \ -1$$

Equation (5.4) is of some practical significance (Pratt, Kane and Andrews, 1969; Clarke, 1975) and is referred to as the discrete Walsh transform (DWT). It transforms a sequence of N samples in time or space into a set of N samples or coefficients in the spectral domain (Fig. 5.2(b)). To return to the time domain we use the *inverse* DWT

$$x_n = \sum_{k=0}^{N-1} a(k)w(k, n) \tag{5.5}$$

The DWT has fast algorithms, e.g. the FWHT in Section 2.3.1, and has been used for its computational simplicity (only additions are required). However, at least for data compression and spectral analysis, other discrete transforms are generally more attractive and so it is helpful to generalize (5.4) to the Nth-order linear transform

$$G(k) = \sum_{n=0}^{N-1} x_n g(k, n); \quad k = 0, 1, \ldots, N - 1 \tag{5.6}$$

where $g(k, n)$ is a general *forward transformation kernel*. By regarding $g(k, n)$ as an element of an $N \times N$ matrix, $\mathbf{A}$, we can also write the general forward and inverse transformation as

$$\mathbf{G} = \mathbf{A}\mathbf{X}; \quad \mathbf{X} = \mathbf{A}^{-1}\mathbf{G} \tag{5.7}$$

where $\mathbf{G}$ and $\mathbf{X}$ are column vectors (see Section 2.3.1).

Clearly, the type of discrete transform depends upon the choice of kernel, and, in practice, it is often advantageous to use a set of band-limited functions rather than rectangular functions. The *discrete cosine transform* (DCT) is particularly useful for data compression (Section 2.3) and can be defined by letting

$$g(k, n) = \frac{2}{N} a_k \cos\left[\frac{(2n + 1)k\pi}{2N}\right] \tag{5.8}$$

where

$$a_k = \begin{cases} \dfrac{1}{\sqrt{2}} & k = 0 \\ 1 & k = 1, 2, \ldots, N - 1 \end{cases} \tag{5.9}$$

The 1-D DCT $C(k)$ can then be written

$$C(k) = \frac{2}{N} a_k \sum_{n=0}^{N-1} x_n \cos\left[\frac{(2n + 1)k\pi}{2N}\right], \quad k = 0, 1, \ldots, N - 1 \tag{5.10}$$

This means that sequence x_n is represented in the spectral domain by a set of cosine functions with special phase and amplitude characteristics. The *inverse* DCT, corresponding to (5.2), is simply

$$x_n = \sum_{k=0}^{N-1} a_k C(k) \cos\left[\frac{(2n + 1)k\pi}{2N}\right], \quad n = 0, 1, \ldots, N - 1 \tag{5.11}$$

As for the DWT, several fast algorithms (FDCT) exist for the DCT, although, in practice, it is not always essential to use such an algorithm (see, for example, Section 5.8.2).

Another discrete transform which has attracted some attention is the *discrete Hartley transform* or DHT (Hartley, 1942). In this case the forward and inverse expressions are

$$H(k) = \frac{1}{N} \sum_{n=0}^{N-1} x_n \operatorname{cas}\left(\frac{2\pi kn}{N}\right), \quad k = 0, 1, \ldots, N - 1 \tag{5.12}$$

and

$$x_n = \sum_{k=0}^{N-1} H(k) \operatorname{cas}\left(\frac{2\pi kn}{N}\right), \quad n = 0, 1, \ldots, N - 1 \tag{5.13}$$

where $\operatorname{cas} \theta = \cos \theta + \sin \theta$. The kernel, $\operatorname{cas} \theta$, is identical to that of the well-known *DFT* (Section 5.2) except that the sine term is real. In fact, the complex DFT coefficients $F(k)$, $k = 0, 1, \ldots, N - 1$ can be computed from the DHT using the relationships

$$F_r(k) = \tfrac{1}{2}[H(N - k) + H(k)]$$

$$F_i(k) = \tfrac{1}{2}[H(N - k) - H(k)]$$

(5.14)

where $F(k) = F_r(k) + jF_i(k)$. It then follows from Section 5.5.7 that an estimate of the signal *power spectrum* can be computed directly from the DHT as

$$\hat{P}(k) = \frac{1}{2N}[H^2(k) + H^2(N - k)]$$

(5.15)

Interest in the DHT increased after the introduction of a fast algorithm – the fast Hartley transform or FHT (Bracewell, 1984). Also, Boussakta and Holt (1988) have described a very fast DSP implementation of the DHT. It appears attractive since, like the DWT and the DCT, the DHT maps a *real* function of time to a *real* function in the spectral domain. Hence, it uses only real arithmetic and it turns out that the central computational element within the FHT (the FHT *butterfly*) is twice as fast as that for the corresponding fast DFT algorithm (the FFT); in fact, the FHT butterfly requires two real multiplications and three real additions (Bold, 1985) whilst the FFT butterfly requires four real multiplications and six real additions (Section 5.3.3). This implies that the DFT of a real sequence found via the FHT (Equation (5.14)) is almost twice as fast as the corresponding *complex input* FFT (Section 5.3).

On the other hand, it is shown in Section 5.3.5 that an efficient form of the FFT for *real* data needs only about half the arithmetic operations of the normal complex input FFT. In other words, an FFT algorithm optim- ized for a *real* input sequence is just as fast as the FHT. Also, a multidimensional FFT is easily computed by separating the kernel (Section 5.8.1), but this is not possible for a multidimensional DHT (Gudvangen and Holt, 1992).

5.2 The discrete Fourier transform (DFT)

The most common discrete transform uses an exponential set of ortho- gonal functions defined by the kernel

$$g(k, n) = e^{-j2\pi kn/N}$$

(5.16)

Using (5.6) this gives the forward transform

$$\text{DFT:} \quad F(k) = \sum_{n=0}^{N-1} x_n e^{-j2\pi kn/N}, \quad k = 0, 1, \ldots, N - 1$$

(5.17)

where

$$\text{IDFT:} \quad x_n = \frac{1}{N} \sum_{k=0}^{N-1} F(k) \, e^{j2\pi kn/N}, \quad n = 0, 1, \ldots, N-1 \quad (5.18)$$

Note that the scaling factor N^{-1} can be associated with either (5.17) or (5.18), or it can be equally distributed between both equations; FFT algorithms usually scale the IDFT, as in (5.18). Also note that, in contrast to the DWT, DCT, and DHT, the DFT is *complex* and so requires more computation and memory. On the other hand, the DFT can deliver a *phase* spectrum.

So far the DFT has been discussed from a Fourier series viewpoint, but it is also helpful to visualize it graphically. Fig. 5.3(a) simply represents the Fourier transform $X_s(f)$ of a sampled signal $x_s(t)$ (Section 1.3). The aperiodic sequence $x_s(t)$ is then windowed to obtain a finite duration sequence $x(nT)$, $n = 0, 1, \ldots, N-1$. The corresponding operation in the frequency domain is a convolution and this somewhat distorts the original spectrum $X_s(f)$ to give spectrum $R(f)$, Fig. 5.3(c).

Now suppose we generate a *periodic* discrete time sequence $x_p(nT)$ by convolving $x(nT)$ with a periodic impulse train $s(t)$ (period $1/\Delta k$). In the frequency domain this process corresponds to multiplication and results in the sampled version of $R(f)$ shown in Fig. 5.3(e). Thus, *a periodic discrete-time sequence $x_p(nT)$ has a periodic discrete spectrum $F_p(k)$*, and we can represent $x_p(nT)$ by the Fourier series ($T = 1$)

$$x_p(n) = \frac{1}{N} \sum_{k=-\infty}^{\infty} F_p(k) \, e^{j2\pi kn/N} \quad (5.19)$$

Expanding (5.19) we find that the terms repeat and that there are actually only N distinct terms. To avoid $x_p(n) \Rightarrow \infty$ we therefore restrict summation to $k = 0, 1, \ldots, N-1$ and state that $x_p(n)$ is completely defined by summation over one period. Equation (5.19) is then written as (5.18). Similarly, $F_p(k)$ is completely defined by one period of $x_p(n)$. When only one period of each periodic sequence is considered we refer to these Fourier series representations as the IDFT and DFT, as indicated in Fig. 5.3(e). Finally, it is usual to simplify the graphical representation of the DFT to that in Fig. 5.3(f), where the absolute frequency of the kth DFT component is given by

$$f = \frac{k}{NT} = \frac{k}{N} f_s \quad (5.20)$$

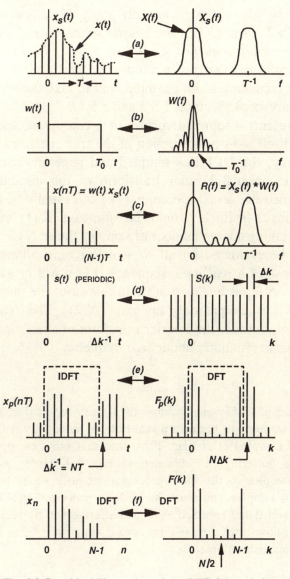

Fig. 5.3 Graphical illustration of the DFT (phase information is omitted for clarity).

and f_s is the sampling frequency. The interval f_s/N is referred to as the *frequency bin* separation.

The foregoing discussion has highlighted several fundamental points concerning the DFT:

(1) Whenever we use (5.17) or (5.18) we must remember that they are really representing the periodic sequences $F_p(k)$ and $x_p(n)$ respect-

ively. In other words, although we only apply N distinct sample values to the DFT, as in (5.17), *the Fourier series basis of the DFT assumes that the input sequence* x_n, $n = 0, 1, \ldots, N - 1$ *is periodic!* It is essential to bear this in mind when performing spectral analysis or fast convolution, for example, and to take the appropriate precautionary steps (Sections 5.5.3 and 5.6.1).

(2) The DFT represents a sampled version of a continuous spectrum $R(f)$ which is itself a distorted version of the true spectrum $X_s(f)$. Put another way, the DFT is a sampled and generally corrupted estimate of the original Fourier transform of the discrete time signal. In practice the actual spectrum $F(k)$ can usually be improved by selecting a suitable *windowing function* $W(f)$ (Section 5.5.5), and by taking a large number of samples (large N).

(3) The DFT can be evaluated for all N points or bins, although for the common case of a *real* input sequence $|F(k)|$ will be symmetrical about $k = N/2$, as in Fig. 5.3(f). In this case, the magnitude spectrum $|F(k)|$ is unique only up to $|F(N/2)|$. The symmetry arises, of course, because the Fourier transform of a real signal has complex conjugate symmetry about zero frequency.

Example 5.1

Direct evaluation of (5.17) can be impractical even for modest applications. Suppose we need to perform a real-time $N = 1024$ point DFT on an audio signal sampled at 50 kHz. First note that real-time operation requires double buffering, as indicated in Fig. 5.4. Here, buffer A supplies 1024 samples to the DFT processor memory whilst buffer B collects 1024 new samples, and when the DFT has processed 1024 samples the roles of A and B are reversed. Similar buffering will be required at the output. It follows that the DFT processing time must be at least as fast as the data acquisition time, i.e.

$$t_{DFT} \leqslant N/f_s = 20.48 \text{ ms}$$

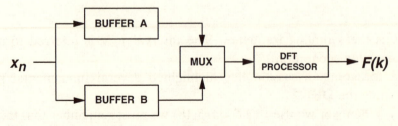

Fig. 5.4 Double buffering for real-time operation.

For a single value of k the DFT in (5.17) can be expressed as

$$F(k) = x_0 \cos(\cdot) + x_1 \cos(\cdot) + \ldots + x_{N-1} \cos(\cdot)$$
$$- j[x_0 \sin(\cdot) + x_1 \sin(\cdot) + \ldots + x_{N-1} \sin(\cdot)]$$

which requires $2N$ real multiplications and $2N - 2$ real additions, i.e. approximately $4N$ real arithmetic operations. The complete DFT therefore requires some $4N^2$ real operations, and these must be performed within 20.48 ms. This allows approximately 5 ns for each addition or multiplication and is clearly a very impractical approach.

5.3 The fast Fourier transform (FFT)

Algorithms for fast computation of the DFT were first brought to prominence by Cooley and Tukey (1965) and since then many FFT algorithms have been developed. Here we will examine the simplest and best known algorithm.

5.3.1 Decimation in time (DIT)

To simplify notation it is usual to write (5.17) as

$$F(k) = \sum_{n=0}^{N-1} x_n W_N^{nk}, \quad W_N = e^{-j2\pi/N} \tag{5.21}$$

Suppose we now decimate x_n into odd and even sequences to give two $N/2$ point DFTs:

$$F(k) = \sum_{n=0}^{N/2-1} x_{2n} W_N^{2nk} + \sum_{n=0}^{N/2-1} x_{2n+1} W_N^{(2n+1)k} \tag{5.22}$$

$$= \sum_{n=0}^{N/2-1} x_{2n} W_{N/2}^{nk} + W_N^k \sum_{n=0}^{N/2-1} x_{2n+1} W_{N/2}^{nk} \tag{5.23}$$

$$= F_e(k) + W_N^k F_o(k), \quad k = 0, 1, \ldots, N - 1 \tag{5.24}$$

In (5.24), $F_e(k)$ and $F_o(k)$ are $N/2$ point DFTs operating upon even and odd time sequences respectively. The important step in deriving an FFT algorithm is to use the periodicity in W_N. In particular,

$$W_{N/2}^{nk} = W_{N/2}^{n(k+N/2)} \tag{5.25}$$

and so (5.23) can be written

$$F(k + N/2) = \sum_{n=0}^{N/2-1} x_{2n} W_{N/2}^{nk} + W_N^{k+N/2} \sum_{n=0}^{N/2-1} x_{2n+1} W_{N/2}^{nk}$$

$$= F_e(k) + W_N^{k+N/2} F_o(k) \tag{5.26}$$

Also, since

$$W_N^{k+N/2} = e^{-j2\pi k/N} e^{-j\pi} = -W_N^k$$

then (5.26) reduces to

$$F(k + N/2) = F_e(k) - W_N^k F_o(k) \tag{5.27}$$

Combining (5.24) and (5.27) gives a pair of equations covering all k values i.e.

$$\left. \begin{array}{l} F(k) = F_e(k) + W_N^k F_o(k) \\[2ex] F(k + N/2) = F_e(k) - W_N^k F_o(k) \end{array} \right\} k = 0, 1, \ldots, (N/2 - 1) \tag{5.28}$$

The significant point is that (5.28) need only be evaluated over the range $k = 0, 1, \ldots, (N/2 - 1)$, as indicated, whereas (5.24) requires evaluation over $k = 0, 1, \ldots, N - 1$. *This means that we have approximately halved the number of complex multiplications*. Fig. 5.5 shows how (5.28) can be utilized for $N = 8$, e.g.

$$F(1) = F_e(1) + W_N^1 F_o(1)$$

$$F(5) = F_e(1) - W_N^1 F_o(1)$$

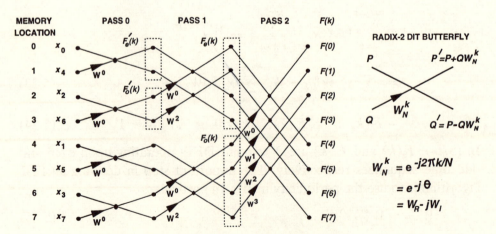

Fig. 5.5 The 8-point DIT radix-2 in-place FFT algorithm.

and two values of $F(k)$ are generated for only one complex multiplication. The flowgraph representing this operation resembles a *butterfly* and coefficients W_N^k are often called 'twiddle factors'!

The foregoing decimation process can be repeated to obtain $F_e(k)$ and $F_o(k)$ in terms of $N/4$ point DFTs. For instance, a little manipulation shows that

$$
\left.
\begin{aligned}
F_e(k) &= F_e'(k) + W_N^{2k} F_o'(k) \\[2ex]
F_e(k + N/4) &= F_e'(k) - W_N^{2k} F_o'(k)
\end{aligned}
\right\} k = 0, 1, \ldots, (N/4 - 1)
\tag{5.29}
$$

and this process can be repeated until the problem has been reduced to 2-point DFTs. Note that $F_e(k)$ operates on the even sequence x_{2n} and (5.29) means that this sequence is itself split into odd and even sequences. Continuing this decimation process down to 2-point DFTs means that the input sequence x_n must be *shuffled* into specific FFT memory locations prior to performing the FFT, as indicated in Fig. 5.5. In practice this is usually done as input samples are loaded into the FFT data RAM, and the basic idea is illustrated in Fig. 5.6 for $N = 8$. Here all bits of the RAM address counter are reversed so that, for example, input sample x_1 will be loaded into RAM address 100 or location 4. The equivalent effect can be achieved in software when using certain DSP devices (Section 5.4.3). Conversely, if the input sequence is not shuffled, the output will be in bit-reversed order.

At this point it is helpful to clarify some of the terminology in Fig. 5.5:

(1) Since this particular algorithm repeatedly decimates odd and even

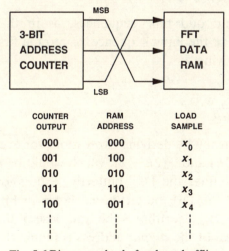

COUNTER OUTPUT	RAM ADDRESS	LOAD SAMPLE
000	000	x_0
001	100	x_1
010	010	x_2
011	110	x_3
100	001	x_4

Fig. 5.6 Bit-reversal rule for data shuffling.

time sequences it is referred to as the *decimation in time (DIT) algorithm.*

(2) The algorithm in Fig. 5.5 is termed a *radix-2* algorithm since N is restricted to the form $N = 2^m$, where m is an integer. In this case, repeated decimation by two will eventually yield 2-point DFTs.

(3) The algorithm is referred to as *in-place* since the output of each butterfly is loaded back into the same memory locations used to supply the butterfly input, i.e. points on the same horizontal line in Fig. 5.5 correspond to the same memory location. Clearly, an in-place or single memory algorithm is very efficient from a hardware point of view since only N complex words of memory are required to transform N data points. On the other hand, all in-place algorithms require some form of shuffling, although this is not a significant problem.

An alternative to the in-place algorithm is to use some form of *double memory* algorithm, i.e. two RAMs each of N complex words. Here, one memory sends data to the butterfly and the other receives data from the butterfly, and after each pass the memories switch roles. This approach can improve FFT speed since a RAM need only be accessed once during each butterfly cycle, and it can also simplify memory addressing. A popular non-in-place, double memory technique is the so-called *constant geometry* algorithm. Appendix D provides a C-routine for a DIT radix-2 in-place FFT.

5.3.2 Decimation in frequency (DIF)

Rather than decimate even and odd time sequences, as in the DIT algorithm, we could initially break the input sequence x_n into the first $N/2$ points and the last $N/2$ points, giving

$$F(k) = \sum_{n=0}^{N/2-1} x_n W_N^{nk} + \sum_{n=N/2}^{N-1} x_n W_N^{nk} \tag{5.30}$$

We could then *decimate in frequency* by deriving expressions for even and odd *frequency* samples, i.e. $F(2k)$ and $F(2k + 1)$, which would be analogous to (5.28). Just as (5.28) defines the DIT butterfly, expressions for $F(2k)$ and $F(2k + 1)$ will yield the DIF butterfly, as shown in Fig. 5.7. Note that the flowgraph for the DIF algorithm is the transpose of the DIT flowgraph in Fig. 5.5 (bit reversal of the *output* address is required for an ordered output sequence) and so both algorithms have the same computa-

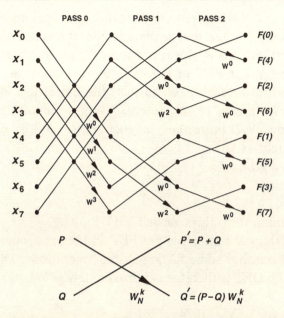

Fig. 5.7 The 8-point DIF radix-2 in-place FFT algorithm.

tional complexity. The choice between the two depends upon the application. Sometimes it may be possible to take advantage of the fact that the P' output of the DIF butterfly can be written to memory whilst Q' is being computed by the multiplier, thereby saving time. On the other hand, when only some of the FFT outputs are required, the DIT algorithm may be preferred since the trivial multiplications at the start of the algorithm can give a computational advantage.

5.3.3 *FFT speed advantage*

Example 5.1 showed that for a real input signal the DFT requires some $4N^2$ real operations (either multiplication or addition). Now consider the radix-2 FFT in Fig. 5.5. Each butterfly input is complex, i.e.

$$P = P_R + jP_I$$

$$Q = Q_R + jQ_I$$

and so the butterfly output samples are

$$P' = [P_R + (W_R Q_R + W_I Q_I)] + j[P_I + (W_R Q_I - W_I Q_R)]$$
$$Q' = [P_R - (W_R Q_R + W_I Q_I)] + j[P_I - (W_R Q_I - W_I Q_R)]$$

(5.31)

It is easily seen that these equations require a total of four real multiplications and six real additions, i.e. ten real arithmetic operations are required for each butterfly operation (this is also true for a DIF radix-2 butterfly). Referring again to the 8-point FFT in Fig. 5.5, we see there are three 'passes' and four butterflies in each pass. In fact, the butterflies fall into 'groups' as can be seen from pass 1. The general N-point radix-2 FFT will have $\log_2 N$ passes and $N/2$ butterflies in each pass, and it will be evaluated from left to right as

> all butterflies in a group
> all groups in a pass
> all passes

In software terms this amounts to three nested 'DO' loops.

Since there are $(N/2)\log_2 N$ butterflies per FFT and ten real operations per butterfly, each FFT requires some $5N \log_2 N$ real operations. The FFT speed advantage over the DFT is therefore approximately

$$\frac{N}{\log_2 N} \approx 100 \quad \text{for} \quad N = 1024$$

Using the FFT for the problem in Example 5.1 will therefore relax the logic speed requirements from 5 ns to 500 ns for each addition or multiplication.

Before leaving the discussion on relative speed it is important to realize that it is not always necessary to use the 'full' FFT. In fact, if only a few frequency bins are needed, the DFT may well be faster than the FFT since the FFT automatically computes all $F(k)$ values. More generally, if only a few bins are required, or if only a few inputs are different from zero, we could try 'pruning' the basic FFT algorithm. A direct approach is to prune the FFT flowgraph so as to disregard redundant paths associated with zero inputs or unwanted outputs.

5.3.4 Inverse FFT (IFFT)

The IDFT in (5.18) can be written

$$x_n = \frac{1}{N} \sum_{k=0}^{N-1} F(k) W_N^{-nk} \tag{5.32}$$

where $W_N = \mathrm{e}^{-\mathrm{j}2\pi/N}$, and taking the complex conjugate gives

$$x_n^* = \frac{1}{N} \sum_{k=0}^{N-1} F^*(k) W_N^{nk} \tag{5.33}$$

The IFFT therefore amounts to performing a normal FFT upon the sequence $F^*(k)$ (including shuffling) and taking the complex conjugate of the result. The significant point is that essentially the same hardware design or software algorithm can be used for both the FFT and the IFFT.

Often x_n is real and so $x_n^* = x_n$. In this case, since the Fourier transform of a real signal has complex conjugate symmetry about zero frequency, then, in general, the FFT of sequence x_n will have complex conjugate symmetry about $k = N/2$, i.e.

$$F^*(k) = F(N - k), \quad k = 0, 1, \ldots, N - 1 \tag{5.34}$$

Example 5.2

Suppose eight samples of a sinewave are extracted from a long sequence using a rectangular window (Fig. 5.8(a)), and then applied to the 8-point FFT in Fig. 5.5. Since the sinewave frequency is $f_o = 1/8T$, the samples are given by

$$x_n = \sin(2\pi n/8), \quad n = 0, 1, \ldots, 7$$

Also, using the general form of the twiddle factor given in Fig. 5.5, we have

$$W^0 = 1 \quad W^1 = e^{-j\pi/4} \quad W^2 = -j \quad W^3 = e^{-j3\pi/4}$$

Inserting these factors into Fig. 5.5 and shuffling the input samples gives the values in the flowgraph shown in Table 5.1. Note the complex conjugate symmetry about $k = N/2$ since x_n is real. In fact, in such cases we are only really interested in the spectrum up to and including $k = N/2 - 1$, and the only nonzero component in this range corresponds to an absolute frequency of

$$f = k/NT = 1/8T$$

as expected. Similar 'exact' spectral analysis is discussed in Example 5.5.

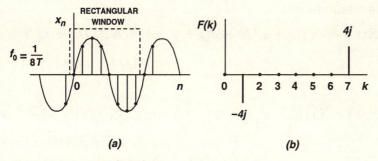

(a)   (b)

Fig. 5.8 8-point FFT (DFT) of a sampled sinewave.

Table 5.1. *FFT computation for* $f_0 = f_s/8$, *N* = 8. *(Example 5.2)*

	Pass 0	Pass 1	Pass 2		
$x_0 = 0$	0	0	0	= $F(0)$
$x_4 = 0$	0	$-2j$	$-4j$	= $F(1)$
$x_2 = 1$	0	0	0	= $F(2)$
$x_6 = -1$	2	$2j$	0	= $F(3)$
$x_1 = 1/\sqrt{2}$	0	0	0	= $F(4)$
$x_5 = -1/\sqrt{2}$	$\sqrt{2}$	$\sqrt{2}(1-j)$	0	= $F(5)$
$x_3 = 1/\sqrt{2}$	0	0	0	= $F(6)$
$x_7 = -1/\sqrt{2}$	$\sqrt{2}$	$\sqrt{2}(1+j)$	$4j$	= $F(7)$

5.3.5 *A real input FFT*

The usual FFT algorithm assumes a complex input signal $x_n = x_{ni} + jx_{nq}$ and this is convenient when the same software or hardware design is to be used to perform the inverse transform, with a complex frequency function as input. A complex input FFT is also essential in some applications, as in Example 5.4. In many cases, however, x_n is real and a complex input FFT will be inefficient since it performs redundant calculations.

Consider $2N$ real input samples. Instead of setting the imaginary part of the data memory to zero at the start of a $2N$-point complex input FFT, it is more efficient to place N real samples in the imaginary memory of an N-point complex input FFT. Odd numbered input samples could be placed in the imaginary memory, thereby generating N complex input points, e.g.

real input x_n: x_0 x_1 x_2 x_3 . . . x_{2N-1}

FFT input array: $x_r[0]$ $x_i[0]$ $x_r[1]$ $x_i[1]$ . . . $x_i[N-1]$

Applying the complex array $x[n]$, $n = 0, 1, \ldots, N-1$ to an N-point complex in-place FFT (which then applies the usual shuffling) generates N complex outputs in array $x[n]$. The required FFT coefficients $F(k) = F_r(k) + jF_i(k)$, $k = 0, 1, \ldots, N-1$ can then be obtained using the following relationships:

$$F_r(k) = \tfrac{1}{2}(x_r[k] + x_r[N-k]) + \tfrac{1}{2}\cos(\pi k/N)(x_i[k] + x_i[N-k])$$

$$- \tfrac{1}{2}\sin(\pi k/N)(x_r[k] - x_r[N-k])$$

$$(5.35)$$

$$F_i(k) = \tfrac{1}{2}(x_i[k] - x_i[N-k]) - \tfrac{1}{2}\sin(\pi k/N)(x_i[k] + x_i[N-k])$$

$$- \tfrac{1}{2}\cos(\pi k/N)(x_r[k] - x_r[N-k])$$

$$(5.36)$$

Note that since $F(k)$ for a real $2N$-point sequence has complex conjugate symmetry about $k = N$, then only N values of $F(k)$ need be computed.

5.4 FFT realization

In this section we examine some practical issues relevant to the hardware and software realization of a radix-2 DIT FFT. The heart of the design problem is the FFT butterfly, and the designer usually tries to minimize the number of machine cycles or instruction cycles needed by the butterfly. In general we also need to ask such questions as:

How many butterfly computation units are required?
Should all butterflies be identical?
How is arithmetic overflow to be avoided?
What wordlength is required for the coefficients (twiddle factors)?
What sort of I/O buffering is required?

5.4.1 Butterfly computation

The general butterfly in Fig. 5.5 has two complex inputs, P and Q, and its two complex outputs are given by (5.31). Noting that

$$W_R = \cos \theta \quad \text{and} \quad W_I = \sin \theta$$

then (5.31) leads to the butterfly computation diagram in Fig. 5.9. Clearly, this computation could be done in either hardware or software, and the

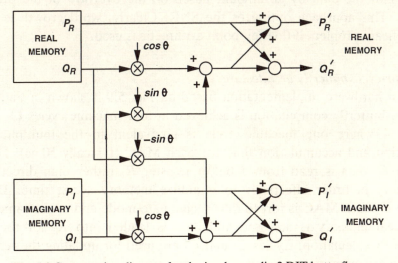

Fig. 5.9 Computation diagram for the in-place radix-2 DIT butterfly.

memories will be repeatedly overwritten. A similar computational diagram can be derived for the DIF butterfly in Fig. 5.7 and, again, it will show that four real multiplications and six real additions are required to complete the computation.

In practice the wordlength in the butterfly can grow from input to output. Consider the butterfly output

$$P'_R = P_R + Q_R \cos\theta + Q_I \sin\theta$$

Assuming unity real and imaginary inputs to the butterfly, then P'_R has a maximum value of

$$1 + \cos(45°) + \sin(45°) = 2.414$$

In this extreme case there is a growth of 2 bits from input to output. Many DSP chips are optimized for fractional 2's complement arithmetic, and so the wordlength growth means there is a real possibility of numbers exceeding unity and generating arithmetic overflow. In practice, for radix-2 algorithms it is common to assume 1 bit of growth per pass or $\log_2 N$ bits of growth for an N-point FFT. We could allow for this by scaling the input data by $\log_2 N$ 'guard' bits, although for large FFTs this can severely restrict the data resolution. For example, for 16-bit output precision the input precision would be only 6 bits for a 1024-point FFT.

A better approach is to incorporate a scaling factor of 0.5 (i.e. a 1-bit shift) into each butterfly computation. Alternatively, using a general purpose DSP chip for realization, overflow handling can be optimized by monitoring the overflow of each pass. Subsequent passes then automatically shift the data by an amount based on the overflow of the previous pass. This approach optimizes the SNR. Clearly, word growth is not a significant problem if floating-point arithmetic is used.

A four-cycle butterfly architecture

A fast hardware implementation based on Fig. 5.9 is shown in Fig. 5.10. Here, butterfly computation is achieved in four machine cycles C_i, $i = 1$, ..., 4 where one machine cycle is equivalent to the multiplication, addition and accumulation time for each MAC (typically 50 ns). During cycle C_1 data is read from a buffer register R_2 rather than direct from memory in order effectively to minimize memory access time. During cycle C_2 each MAC is forced into accumulate mode and at the same time the real memory is accessed and data is loaded into R_2 for the next butterfly calculation. Cycles C_3 and C_4 are used for updating the real and imaginary memories.

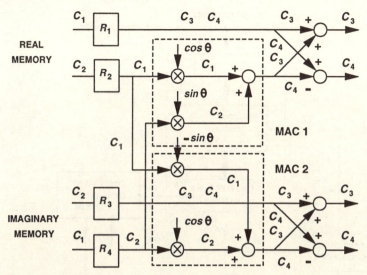

Fig. 5.10 A practical 4-cycle radix-2 DIT butterfly architecture (C_i = machine cycle i).

5.4.2 Dedicated FFT processors

Fig. 5.11 shows the essential elements of a dedicated FFT processor based upon a *single* memory and a *single* butterfly arithmetic unit, e.g. Fig. 5.10. When properly designed, coefficients and memory addresses will be generated without interrupting the actual butterfly computation, and so we can assume that butterflies are computed *contiguously*. In this case, ignoring data I/O, the FFT speed depends only upon the butterfly design (number of machine cycles, C), the machine cycle time, t_c, and on the number of butterflies. For a radix-2 algorithm it can therefore be expressed as

$$t_{FFT} = Ct_c\frac{N}{2}\log_2 N \qquad (5.37)$$

The machine cycle time is the longest operation time, and for Fig. 5.10 corresponds to the multiply, add and accumulate time of each MAC. Taking $C = 4$ and $t_c = 50$ ns we find that a 1024-point FFT takes about 1 ms using this particular architecture. The total number of machine (clock) cycles (20 480 in this case) is, in fact, a good measure of FFT efficiency. Faster transforms can be achieved by using, say, one butterfly computation unit for each pass. Clearly, for a radix-2 algorithm this pipelining technique will reduce the number of cycles/FFT by a factor $\log_2 N$, at the expense of increased hardware and control circuit complexity.

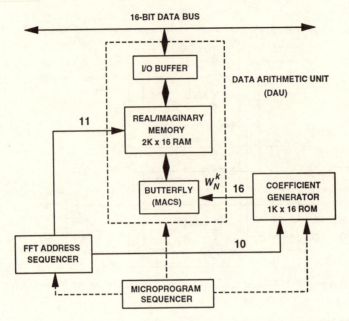

Fig. 5.11 Essential hardware for an in-place 1024-point FFT processor.

Example 5.3

Suppose that a real-time complex-input 128-point FFT is to be performed upon a signal sampled at $f_s = 1$ MHz, and that the FFT uses a single butterfly circuit based upon the 4-cycle architecture in Fig. 5.10. An estimate of the maximum permissible multiply-accumulate time (machine cycle time t_c) can be made by assuming that enough I/O buffering is available to enable the butterflies to be performed contiguously. In this case

$$\frac{N}{f_s} \geqslant Ct_c \frac{N}{2} \log_2 N$$

giving $t_c \leqslant 71$ ns.

Now consider the generation of coefficients W_N^k (the twiddle factors). For an N-point FFT we have, from (5.21)

$$W_N^k = \cos\left(2\pi k/N\right) + \mathrm{j}[-\sin\left(2\pi k/N\right)], \quad k = 0, 1, \ldots, N-1$$

$$(5.38)$$

which may be interpreted as N equally spaced points around a unit circle. In practice only $N/2$ values of k are needed, e.g. for $N = 8$ we only need W^0, W^1, W^2, and W^3, see Fig. 5.5. On the other hand, each twiddle is

Table 5.2. *Twiddle factors for an N-point DIT radix-2 in-place FFT.*

Pass 0	...	Pass $\log_2 N - 4$	Pass $\log_2 N - 3$	Pass $\log_2 N - 2$	Pass $\log_2 N - 1$
$W_N^{Nk/2}$	...	W_N^{8k}	W_N^{4k}	W_N^{2k}	$W_N^k = e^{-j2\pi k/N}$
$k = 0$	...	$k = 0, 1, \ldots, \left(\dfrac{N}{16} - 1\right)$	$k = 0, 1, \ldots, \left(\dfrac{N}{8} - 1\right)$	$k = 0, 1, \ldots, \left(\dfrac{N}{4} - 1\right)$	$k = 0, 1, \ldots, \left(\dfrac{N}{2} - 1\right)$
$W^0 = 1$		W^0	W^0	W^0	W^0
		W^8	W^4	W^2	W^1
	...	W^{16}	W^8	W^4	W^2
		W^{24}	W^{12}	W^6	W^3
		W^{32}	W^{16}	W^8	W^4
		$\vdots$	$\vdots$	$\vdots$	$\vdots$

complex and so requires two words of storage (cos and −sin stores). The *wordlength* for each store depends upon N. For $N = 1024$ we need W^0, $W^1, \ldots, W^{511}$ where

$$W^{511} = \cos\left(\frac{2\pi 511}{1024}\right) + j\left[-\sin\left(\frac{2\pi 511}{1024}\right)\right]$$

Now

$$\cos\left(\frac{2\pi 511}{1024}\right) - \cos\left(\frac{2\pi 510}{1024}\right) = 0.000056$$

and this sort of resolution can be approached using 14 bits ($2^{-14} = 0.000061$). The required resolution can therefore be achieved by using a standard 16-bit wordlength, and the twiddle factors for $N = 1024$ could be stored in a $1 \text{ K} \times 16$-bit ROM, as indicated in Fig. 5.11. The ROM would store 512 words for cos and 512 words for −sin.

Finally, when designing large FFTs it is obviously necessary to know the values and order of the twiddles within the flowgraph. In this respect Table 5.2 can be useful when designing a DIT radix-2 in-place FFT. Each column in the table will give the twiddles for any group within the pass.

Overlapped I/O and data processing

The single memory system in Fig. 5.11 is hardware efficient but it compromises on speed. This is because the FFT processor (butterfly) remains idle while the input data are loaded and the output data are unloaded. In order to maximize system performance, i.e. to achieve contiguous butterfly operation, it is usual to use a *double buffering* technique, as for example in Fig. 5.12(*a*). Here, one buffer captures input data and delivers output data, whilst the other buffer is used for an

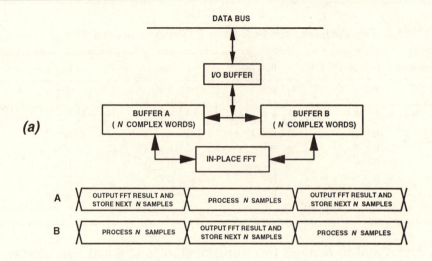

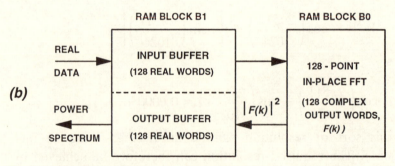

Fig. 5.12 Overlapped I/O and data processing: (*a*) general; (*b*) using the Texas TMS320C25 processor.

in-place FFT. During I/O, the FFT results could be read out and data written in using an alternate read/write process. Once the FFT has been performed, the buffer roles are reversed. Clearly, this type of approach is necessary for *real-time* operation, when the data bus is fed from an ADC, for example.

A similar and practical example of overlapped I/O and data processing is shown in Fig. 5.12(*b*). This diagram illustrates how the on-chip RAM blocks of a DSP chip can be used to compute a 128-point power spectrum of a real-time audio input. While the FFT is being performed in B0 new samples from the ADC are constantly being read into B1 and new spectrum samples are read out of B1. When the input buffer is full the 128-point power spectrum is transferred to B1 and 128 real data samples are transferred to B0.

5.4.3 A DSP-chip based FFT processor

This section describes the implementation of a complex input FFT using a general purpose DSP chip (Wade, 1988). Readers not interested in detail could omit this section.

The TMS320C25 processor comprises a high-speed 16-bit multiplier, a 32-bit ALU and accumulator, eight auxiliary registers (AR0–AR7), and an on-chip RAM and ROM. The on-chip RAM comprises three memories (B0, B1 and B2) and off-chip memory is directly addressable up to 64K words, Fig. 5.13. Suppose we have a 128-point complex input FFT to perform, where the complex points are generated from the I and Q channels of a quadrature demodulation system. The I and Q samples could be available in some external memory (commencing at, say, address 2000 hex) and they could be loaded into B0 as shown in Fig. 5.14. Each external data block (128 samples) could be moved into B0 using the assembler block move instruction BLKD, and by appending the special 'reverse carry' option *BR0 to this instruction the data is also simultaneously shuffled *with no extra time overhead*.

For FFT spectral analysis (signal detection) in communication systems it can be very useful to modify the usual DFT equation by adding a small,

DECIMAL ADDRESS	DATA	PAGE	
0			
	MEMORY-MAPPED REGISTERS		
6		0	
	RESERVED		
96			
	BLOCK B2		
128			
	RESERVED	1-3	
512			
	BLOCK B0 (IN-PLACE DATA MEMORY)	4-5	ON-CHIP
768			
	BLOCK B1 (TWIDDLES)	6-7	
1024			
	EXTERNAL	8-511	

Fig. 5.13 Use of TMS320C25 memory for FFT computation.

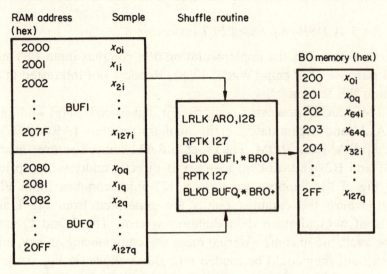

Fig. 5.14 Data shuffle routine for a 128-point complex input FFT.

variable offset, c, to each frequency bin. The modified DFT could be written

$$F(k + c) = \sum_{n=0}^{N-1} x_n W_N^{n(k+c)} \qquad k = 0, 1, \ldots, N - 1 \qquad -0.5 \leqslant c \leqslant 0.5$$

(5.39)

and this permits a DSP system to compensate for frequency offsets in a practical system. This 'offset' FFT (OFFT) requires a larger number of twiddle factors than normal, and for a fixed value of c and $N = 128$ a total of 127 complex twiddles (254 values) are needed. These twiddles are precalculated for a range of values of c and are held in RAM. In order to minimize the butterfly computation time, it is preferable to load the 254 twiddle values for the selected value of c from the external RAM into on-chip RAM B1 prior to FFT computation (see Fig. 5.13).

Butterfly design

Ideally, the DSP implementation of an FFT algorithm should take advantage of trivial twiddle factors in order to simplify some butterflies and so reduce computation time. For example, for a conventional FFT, twiddles can take on simple values such as ± 1 or $\pm j$ and complex multiplication within the butterfly becomes trivial. In such cases a general butterfly macro can be simplified considerably and the use of several simple butterfly macros together with a generalized butterfly macro is therefore quite common.

Unfortunately, for the OFFT the twiddles rarely reduce to a simple value and so a single, generalized butterfly macro is used throughout the algorithm. The code is shown in Fig. 5.15. This butterfly evaluates (5.31),

```
BFLY            $MACRO
* RADIX 2 DIT BUTTERFLY FOR OFFT WHICH USES LOOPED CODE
* SCALES BY 2. ASSUMES PM = 00 AND CURRENT AR IS AR7
* TWIDDLE FACTOR SELECTION VIA AR1
* AR4,AR7 POINT TO DATA IN B1; EACH BUTTERFLY HANDS ON POINTERS TO NEXT BFLY
* EXITS ON ARP = AR1, WITH AR1 RESET TO ITS START VALUE.


        LT    *+,AR1          (QR) TO TREG
        MPY   *+,AR7          WR*QR/2 TO PREG
        LTP   *-,AR1          (QI) TO TREG, PREG TO ACC
        MPY   *-              WI*QI/2 TO PREG
        APAC                  (WR*QR + WI*QI)/2 TO ACC
        MPY   *+,AR7          (TREG) * WR, QI*WR/2 TO PREG
        LT    *               (QR) TO TREG
        SACH  *+,0,AR1        QR = (WR*QR + WI*QI)/2
*
        PAC                   QI*WR/2 TO ACC
        MPY   *-,AR7          QR*WI/2 TO PREG
        SPAC                  (QI*WR - QR*WI)/2 TO
        SACH  *-,0,AR4        QI = (WR*QI - WI*QR)/2
*
        LAC   *,14,AR7        PR/4 TO ACC
        ADD   *,15,AR4        (PR + QR*WR + QI*WI)/4 TO ACC
        SACH  *+,1,AR7        PR = (PR + QR*WR + QI*WI)/2
*
        SUBH  *               (PR - (QR*WR + QI*WI))/4 TO ACC
        SACH  *+,1,AR4        QR = (PR - (QR*WR + QI*WI))/2
*
        LAC   *,14,AR7        PI/4 TO ACC
        ADD   *,15,AR4        (PI + (QI*WR - QR*WI))/4 TO ACC
        SACH  *+,1,AR7        PI = (PI + (QI*WR - QR*WI))/2
*
        SUBH  *               (PI - (QI*WR - QR*WI))/4 TO ACC
        SACH  *+,1,AR1        QI = (PI - (QI*WR - QR*WI))/2
*
        $END
```

Fig. 5.15 TMS320C25 assembly code for a radix-2 in-place DIT butterfly.

Table 5.3. *Twiddle factors for the first three passes of a 128-point OFFT.*

Pass 0			Pass 1			Pass 2		
Address		Twiddle factor	Address		Twiddle factor	Address		Twiddle factor
0	1	$W^0 W^{64c}$	0	2	$W^0 W^{32c}$	0	4	$W^0 W^{16c}$
2	3	$W^0 W^{64c}$	1	3	$W^{32} W^{32c}$	1	5	$W^{16} W^{16c}$
4	5	$W^0 W^{64c}$	4	6	$W^0 W^{32c}$	2	6	$W^{32} W^{16c}$
			5	7	$W^{32} W^{32c}$	3	7	$W^{48} W^{16c}$
	$\vdots$			$\vdots$			$\vdots$	

and uses auxiliary registers AR4 and AR7 to point to the addresses of P and Q respectively in B0. Note that P' and Q' are scaled by 0.5 to minimize overflow and are loaded back into B0, as required for an in-place FFT. Also, within any one butterfly group, each butterfly increments registers AR4 and AR7 by one, ready for the next butterfly. For example, the relative addresses and the corresponding twiddle factor for the first three passes of a 128-point OFFT are shown in Table 5.3. The twiddle factor is initially passed to the butterfly via register AR1 (which points to the twiddle address) and so no macro argument is required. With the selected twiddle factors in on-chip RAM, the macro in Fig. 5.15 requires only 22 machine cycles.

The final code for the OFFT is a compromise between execution time and programme memory space. Generally speaking, a straight-line code (no loops) is fast but requires excessive memory, and in practice a semi-looped code is preferred, Fig. 5.16. The straight-line aspect of the code in Fig. 5.16 lies only in the fact that there is no 'pass counter' and so the overall code is quite compact. Note that, for signal detection, it is usually necessary to compute the power spectrum estimate

$$\hat{P}(k) = |F(k)|^2, \quad k = 0, 1, \ldots, 63$$

Using the TMS320C25, the overall execution time for a 128-point complex-input OFFT, including the loading of B1 and power spectrum computation, is typically 1.3 ms. The total programme memory requirement is typically 0.5K words or about 5% of the program space required for an OFFT written in pure straight-line code.

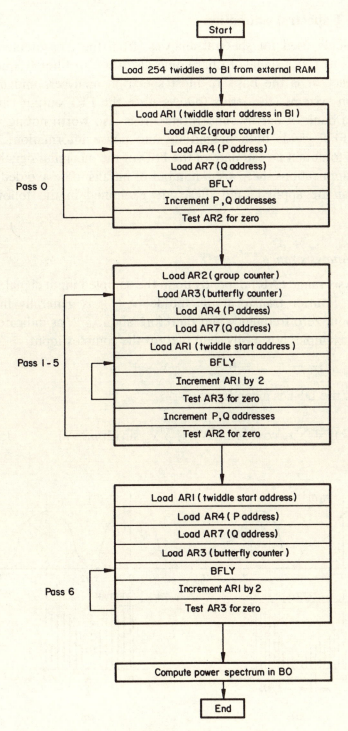

Fig. 5.16 Semi-looped OFFT routine for $N = 128$.

5.5 FFT spectral estimation

The FFT is widely used for spectral analysis, from the measurement of medical signals to radar signals to harmonic distortion to filter frequency response. In fact, it is the basis of most spectrum analysers and many signal detection systems (the latter follows since the FFT output can be equivalent to a bank of narrowband filters). It is also worth noting that, although the FFT yields both magnitude and phase information, very often it is used to obtain an estimate of the PSD of the incoming signal.

In all these applications there are a number of pitfalls to be avoided and 'tricks' that can be applied, and these are examined in the following sections.

5.5.1 Analysis range

The FFT analysis range (AR) depends upon the sampled input signal. If it is complex its Fourier transform magnitude $|X_s(f)|$ is generally highly asymmetric about zero frequency and therefore about $f_s/2$, as indicated in Fig. 5.17. For example, it is easily seen that for the complex input

$$x_n = x_{ni} + jx_{nq}, \qquad n = 0, 1, \ldots, N - 1$$

the real part of the DFT is given by

$$\mathrm{Re}\,[F(k)] = \sum_n x_{ni} \cos{(2\pi nk/N)} + \sum_n x_{nq} \sin{(2\pi nk/N)} \qquad (5.40)$$

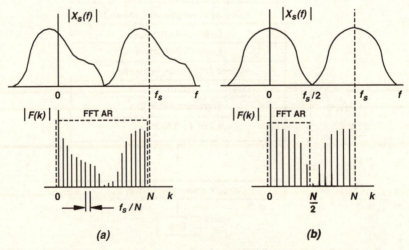

Fig. 5.17 FFT AR for real and complex inputs: (*a*) x_n complex; (*b*) x_n real.

Clearly, this is an even function about $k = 0$, or equivalently about $k = N/2$, only if $x_{nq} = 0$. Similarly, it is readily shown that $\text{Im}[F(k)]$ will be asymmetric about $k = N/2$. For the case of a complex input, $|F(k)|$ will therefore be unique over the AR $0 \leq k \leq N - 1$, as indicated in Fig. 5.17(*a*). On the other hand, when x_n is real its Fourier transform has complex conjugate symmetry about zero frequency which means that its magnitude spectrum is symmetric about $f_s/2$. For real signals the FFT AR is therefore reduced to $0 \leq k < N/2$, as indicated in Fig. 5.17(*b*), and the corresponding frequency range is $0-f_s/2$, or $0-0.5$ on a normalized frequency scale. In practice this may be restricted to typically $0.4f_s$ due to the need to avoid aliasing arising from spectral leakage (see Example 5.7). Also note from Fig. 5.17 that the FFT bins are separated by f_s/N Hz.

Example 5.4

Fig. 5.18(*a*) shows a typical FFT detection system for an Mary FSK signal (M different tones). Quadrature down conversion generates in-phase and quadrature components which form a complex signal $x_n = x_{ni} + jx_{nq}$ at the FFT detector input. Ideally the input signal or tone lies on a frequency bin and we could write

$$x_n = \cos(2\pi nk/N) + j\sin(2\pi nk/N), \qquad k \text{ integer} \qquad (5.41)$$

Here $\cos(2\pi nk/N)$ is applied to the real input and $+\sin(2\pi nk/N)$ to the imaginary input and from the foregoing discussion we expect a unique

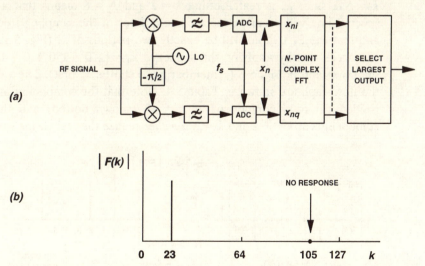

Fig. 5.18 Illustrating the asymmetrical spectrum arising from a complex input signal: (*a*) MFSK detector; (*b*) detector output for $k = 23$, $N = 128$.

spectrum up to $k = N - 1$ since x_n is complex. This also follows since the Fourier transform of a complex tone is asymmetrical about $\omega = 0$, i.e.

$$e^{j\omega_0 t} \Leftrightarrow 2\pi\delta(\omega - \omega_0)$$

For *MFSK*, the normalized signal frequency k/N at any one time could be one of many frequencies. If, for example, $k = 23$ and $N = 128$, then $|F(k)|$ will appear as shown in Fig. 5.18(*b*).

5.5.2 *Self-windowing inputs*

Example 5.4 is a good illustration of exact FFT spectral analysis, i.e. the spectral line occurs exactly where expected. In general, FFT spectral estimation will be exact only when the FFT input sequence x_n, $n = 0, 1,$ $\dots, N - 1$ can be considered to be a complete period or integral number of periods of a discrete periodic sequence. This follows from the Fourier series expansion in (5.19), and from Fig. 5.3(*e*).

Example 5.5

Suppose we simplify the practical illustration in Example 5.4 and assume an input sequence

$$x_n = \cos\left[2\pi\left(\frac{k}{N}f_s\right)nT\right] = \cos\left(2\pi kn/N\right), \qquad k \text{ integer}$$

As before, the signal frequency $f_0 = kf_s/N$ lies on a frequency bin, although now the amplitude spectrum $|F(k)|$ will be *symmetrical* about $k = N/2$ since x_n is real. Letting $k = 2$ and $N = 8$ means that a single spectral line will occur at exactly one quarter of the sampling frequency and that the FFT input will be exactly two periods of x_n (Fig. 5.19). We can check this by applying the sequence $x_n = \{1\ 0\ -1\ 0\ 1\ 0\ -1\ 0\}$ to the flowgraph in Fig. 5.5 (remembering to shuffle x_n first). The values in the flowgraph are shown in Table 5.4. Note that the component at $k = 6$ is identical to a component at $k = -2$ (shown dotted) and that the components at $k = -2$ and $k = 2$ *together* realize the real cosine function.

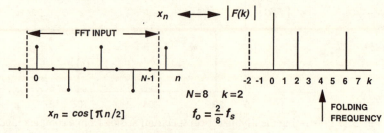

Fig. 5.19 Illustrating exact FFT spectral analysis.

Table 5.4. *FFT computation for* $f_o = 2f_s/8$, $N = 8$. *(Example 5.5)*

		Pass 0	Pass 1	Pass 2	
$x_0 =$	1	2	0	0	$= F(0)$
$x_4 =$	1	0	0	0	$= F(1)$
$x_2 =$	-1	-2	4	4	$= F(2)$
$x_6 =$	-1	0	0	0	$= F(3)$
$x_1 =$	0	0	0	0	$= F(4)$
$x_5 =$	0	0	0	0	$= F(5)$
$x_3 =$	0	0	0	4	$= F(6)$
$x_7 =$	0	0	0	0	$= F(7)$

The FFT of a finite duration sequence

In general the FFT input will not be an integral number of periods of a sinusoidal signal. However, there are many cases where the input can be *assumed* to be periodic without the risk of introducing spectral errors. Typical examples are impulses, shock (transient) responses, sine bursts, chirp bursts and noise bursts. These inputs are called *self-windowing* signals since they do not require processing by any special windowing function prior to FFT spectral analysis (in effect we are using a rectangular window, Section 5.5.5).

Another example is the computation of the frequency response of a FIR digital filter from its impulse response, $h(n)$. Suppose we extend the finite impulse response $h(n)$ of an FIR filter with zeros to form an N-point sequence (which could be assumed to be one period of a periodic signal). Taking the z-transform, we have

$$H(z) = \sum_{n=0}^{N-1} h(n)z^{-n} \tag{5.42}$$

and the continuous frequency response (Fourier transform) is obtained by setting $z = e^{j\omega T}$, i.e.

$$H(e^{j\omega T}) = \sum_{n=0}^{N-1} h(n)e^{-j\omega nT} \tag{5.43}$$

If, on the other hand, we evaluate $H(z)$ at just N equally spaced points on the unit circle then

$$\omega = 2\pi \frac{k}{NT}, \quad z = e^{j2\pi k/N}, \quad k = 0, 1, \ldots, N - 1$$

and the *sampled* frequency response is the DFT relation

$$H(e^{j2\pi k/N}) = \sum_{n=0}^{N-1} h(n) e^{-j2\pi kn/N}, \qquad k = 0, 1, \ldots, N-1 \qquad (5.44)$$

The FFT of $h(n)$ therefore gives an accurate but sampled version of the continuous frequency response. Clearly, providing $h(n) \approx 0$ for $n > N$, the FFT can also be used to estimate the frequency response of an IIR filter.

Example 5.6

The FIR filter in Fig. 5.20(a) has the impulse response in Fig. 5.20(b). To apply an 8-point FFT we can extend $h(n)$ with zeros, and then assume

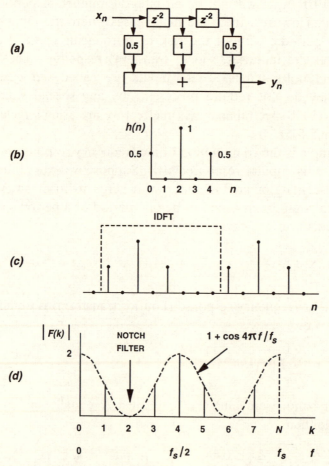

Fig. 5.20 FFT computation of filter frequency response ($N = 8$).

that the extended sequence (Fig. 5.20(c)) is simply the IDFT, as in Fig. 5.3(e). Applying the FFT will then give exact spectral analysis of the IDFT, and hence of $h(n)$.

Applying the 8-point sequence in Fig. 5.20(c) to Fig. 5.5 gives the output in Fig. 5.20(d). This amplitude response is easily confirmed by the digital filter theory in Chapter 4: noting that the impulse response is symmetric and of length 5, the amplitude response is given by (4.31)

$$H^*(e^{j\omega T}) = h(0)_s + 2h(2)_s \cos(2\omega T)$$

$$= 1 + \cos(2\omega T)$$

The filter therefore has a first-order notch response and the 8-point FFT gives an exact but sampled version of this response. Remember, since $h(n)$ is real, the FFT analysis range extends only up to $f_s/2$ or $k = 4$ and frequencies above this are ignored.

5.5.3 Spectral leakage

In general FFT spectral analysis suffers from two fundamental problems:
  (1) a sampled spectrum ('picket-fence' effect);
  (2) spectral leakage.
Example 5.6 illustrates the sampled spectrum quite vividly; it is as though we are looking through a 'picket-fence' and can only see part of the true spectrum. We will discuss ways of reducing this effect in Section 5.5.4. Perhaps the more serious effect is that of 'spectral leakage' and this can be explained with the help of Fig. 5.21. Here, a continuous pure sinewave is multiplied by a rectangular windowing function w_n in order to generate a finite sequence x_n, $n = 0, 1, \ldots, N - 1$. Since multiplication in the time domain corresponds to convolution in the frequency domain, the spectrum $R(\omega)$ of x_n will be a 'spectrally spread' version of $X_s(\omega)$, i.e.

$$R(\omega) = X_s(\omega) * W(\omega) \tag{5.45}$$

where

$$W(\omega) = \left[\frac{\sin(\omega NT/2)}{\sin(\omega T/2)}\right] e^{-j\omega(N-1)T/2} \tag{5.46}$$

(Note that $W(\omega)$ is actually periodic in ω, although only one period is shown in Fig. 5.21 for clarity.) The FFT of x_n is simply a sampled version of $R(\omega)$ and the significant point to note is that the power in the single spectral line at $\omega = \omega_0$ has now been distributed over other frequencies;

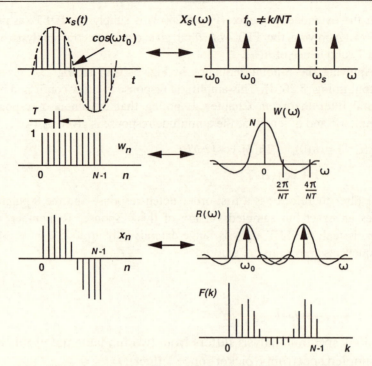

Fig. 5.21 General FFT spectral analysis, illustrating spectral leakage.

this is referred to as spectral leakage. At the outset then, we see that FFT spectral analysis will generally be only an approximation to the true Fourier transform. As discussed in Example 5.5, an exception to this occurs when ω_0 falls on a frequency bin, i.e. when

$$f_0 = \frac{k}{NT} = \frac{k}{N} f_s \qquad (5.47)$$

In this case, provided w_n is rectangular, the tails of $W(\omega)$ pass exactly through bins and spectral leakage disappears.

The usual problem is not so simplistic and corresponds more to the general diagram in Fig. 5.3(*e*). This shows that, whatever form the *N*-point sequence x_n takes, *the FFT assumes that this is replicated throughout time (or space), and it represents the assumed replicated sequence exactly*. Fig. 5.22 illustrates the periodicity assumption made by the FFT for an arbitrary sinusoidal input. It will assume an input waveform with sharp discontinuities and will generate the spectral content of such a waveform. Consequently, very often the FFT derived spectrum will deviate markedly from the expected spectrum unless special precautions are taken.

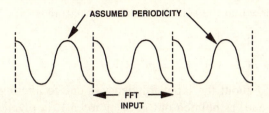

Fig. 5.22 Periodicity assumption of the FFT.

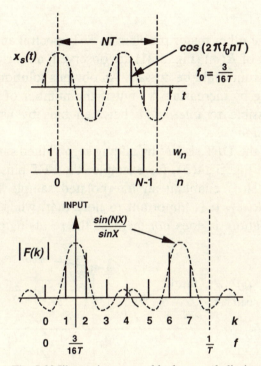

Fig. 5.23 Illustrating spectral leakage and aliasing.

Example 5.7

Consider the sequence

$$x_n = \cos\left[2\pi(3/16T)nT\right] = \cos(3\pi n/8)$$

and assume that eight samples are windowed as shown in Fig. 5.23. Applying these samples to the flowgraph in Fig. 5.5 yields the magnitude spectrum in Fig. 5.23. In this case the FFT input is no longer one or more periods of a discrete periodic sequence and so spectral analysis is far from ideal. Looked at another way, the input frequency (corresponding to $k = 1.5$) falls between bins and so convolutional error becomes apparent.

The reader should show that the actual magnitude spectrum is

k	0	1	2	3	4	5	6	7
$\lvert F(k) \rvert$	1.0	2.4	2.8	1.2	1.0	1.2	2.8	2.4

The lack of symmetry about the input frequency is due to *aliasing* from the repeat spectrum and, as pointed out in Section 5.5.1, in practice this can force a reduction in the FFT AR.

5.5.4 Zero extension

Besides spectral leakage, the other major problem in FFT spectral analysis is the fact that only *samples* of $R(\omega)$ (Fig. 5.21) are observed, and, clearly, important parts of the spectrum could be missed. An obvious solution is to reduce the FFT bin spacing by increasing N, but if the number of input samples is fixed, it is possible to *interpolate* between bins by using a technique called *zero extension*.

Zero extension increases the DFT size N by adding zero-valued samples to the actual input data. In Fig. 5.24(*b*), for example, the FFT bins have been doubled wrt Fig. 5.24(*a*) enabling an interpolated sample to be generated at the peak of $R(\omega)$. It is important to note that, whilst zero extension achieves interpolation, it does *not* increase the resolving power

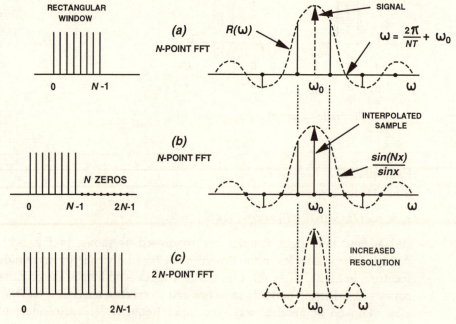

Fig. 5.24 Zero extension.

of the FFT – which can only be increased by taking in more signal samples, as in Fig. 5.24(*c*). Also note that, whenever zero extension is applied, any 'windowing function' (Section 5.5.5) should only span the actual input data and it should not be applied to the added zero valued samples.

Example 5.8
Fig. 5.25 illustrates the advantage of zero extension given only 128 actual input samples. A straight 128-point FFT does not reveal the fact that there are actually just three closely spaced spectral lines of differing amplitudes in the input signal. However, the frequencies and relative amplitudes *can* be identified if we extend the sequence with zeros to generate a 1024-point sequence.

The FFT as a bank of filters
Figs. 5.24(*a*) and (*c*) show that spectral leakage can be 'contained' by increasing the number N of actual input data points. We shall see in Section 5.5.5 that it can also be dramatically reduced by using a good windowing function on the input data. In practice therefore, the N FFT

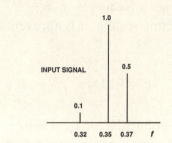

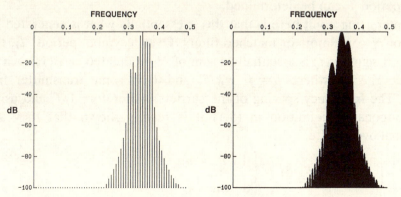

Fig. 5.25 A 128-point FFT and the corresponding 1024-point FFT after zero extension.

outputs $F(k)$, $k = 0, 1, \ldots, N - 1$ should have only a small response either side of the appropriate bin (the actual response is window-dependent) and we could visualize the FFT output as the output of a bank of N *narrowband* filters. In other words, a complex tone input at a frequency $f = kf_s/N$ (k integer, $f_s = 1/T$) will generate only a single, well-defined output at the kth bin, as in Example 5.4.

It is worth pursuing this idea a little further since the FFT filter bank concept, and the detection concept in Fig. 5.18(a), are fundamental to multichannel digital communications systems. In general these systems will use digital carrier modulation (we will assume QPSK) and, *over a transmitted symbol period*, T_s, the complex input to the FFT in Fig. 5.18(a) can be represented (with some simplification) by

$$x_n = \sum_{k=0}^{N-1} Y_k e^{j2\pi nk/N} \tag{5.48}$$

Here we are assuming that a set of N complex, modulated tones or carriers are transmitted and that, over a symbol period, $Y_k = H_k C_k$ where H_k is the complex channel response at this point in time and C_k is a complex number denoting one of the four possible QPSK phases. At the FFT input the kth carrier has a frequency $f_k = k/NT$ where $T = T_s/N$ is the sampling period. Neglecting scaling, it is apparent that x_n has the form of an IDFT where

$$Y_k = \sum_{n=0}^{N-1} x_n e^{-j2\pi nk/N}, \quad k = 0, 1, \ldots, N - 1 \tag{5.49}$$

Therefore, by taking an N-point FFT of x_n, $n = 0, 1, \ldots, N - 1$ we will obtain complex outputs Y_k from which the complex numbers C_k for symbol period T_s can be determined.

The interesting point here is that this is essentially what is generated by a bank of N correlators or matched filters. Over a symbol period, T_s, the received rf signal $x(t)$ is essentially a sum of N modulated carriers each of the form $Y_k e^{j2\pi f_k t}$ where $f_k = f_0 + k/T_s$ and f_0 is some transmitter frequency. The frequency spacing of the carriers is therefore $1/T_s$ and using the orthogonality definition in (5.1) it is readily shown that they are orthogonal over T_s, i.e.

$$\frac{1}{T_s} \int_0^{T_s} e^{j2\pi f_m t} e^{-j2\pi f_n t} \, dt = \begin{cases} 1 & m = n \\ 0 & m \neq n \end{cases}$$

It therefore follows that when the kth correlator multiplies $x(t)$ by the function $e^{-j2\pi f_k t}$ and then integrates over T_s, its output will simply be Y_k. The kth FFT output therefore has a value $H_k C_k$ which is the same as one sample of the *time series* generated by the kth correlator or matched filter. Put another way, the FFT is an efficient way of realizing a bank of N filters each filter being matched to one of the complex carriers. The correlations for all N carriers are performed simultaneously by the FFT and the FFT outputs are the complex amplitudes of the carriers for the corresponding symbol period.

A good example of this form of orthogonal signalling is found in the coded orthogonal frequency division multiplex (COFDM) system (Floch *et al.*, 1989; Shelswell *et al.*, 1991). This system uses many (typically 500) closely spaced orthogonal carriers together with FFT demodulation in order to achieve compact disc quality broadcasting (DAB) to mobile receivers.

5.5.5 *Windowing to reduce spectral leakage*

Fig. 5.23 shows that severe spectral leakage can occur when the FFT input signal is not a complete period or integral number of periods of some periodic sampled signal. For a sinusoidal input this occurs when its frequency f_0 does not fall on a bin, i.e. when $f_0 \neq kf_s/N$. The FFT in Fig. 5.23 actually assumes the input to be the discontinuous waveform in Fig. 5.22, and to obtain a more accurate spectral estimate it is necessary to reduce the discontinuity by forcing the FFT input towards zero at the extremities. This means that, instead of multiplying $x_s(t)$ in Fig. 5.23 by a rectangular window, $w(n)$, it is usual to use a function of the form shown in Fig. 5.26. For a purely sinusoidal input, this window will make the FFT input appear similar to a sampled AM wave – which has a spectrum reasonably close to a single spectral line.

The windowing process is a convolution in the frequency domain and so the shape of $W(\omega)$ is important. A good window $W(\omega)$ has a narrow

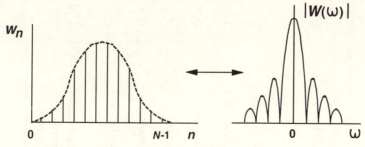

Fig. 5.26 A typical windowing function and its transform.

Table 5.5. *Some useful FFT windows (after F.J. Harris, 1978)* (w_n *is defined over* $n = 0, 1, \ldots, N - 1$).

Window	ENBW (bins)	Highest sidelobe (dB)	Sidelobe roll-off (dB octave)	6 dB bandwidth β (bins)	Function
Rectangular	1	−13	−6	1.21	$w_n = 1$
Hamming	1.36	−43	−6	1.81	$w_n = 0.54 - 0.46 \cos(2\pi n/N)$
Gaussian	$\alpha = 3$: 1.64	−55	−6	2.18	$w_n = \exp\left[-(\alpha^2/2)(2n/N - 1)^2\right]$
$\sin^\alpha(x)$	$\alpha = 3$: 1.73	−39	−24	2.32	$w_n = \sin^\alpha[\pi n/N]$
Hanning	$\alpha = 2$: 1.50	−32	−18	2.00	
Blackman	1.73	−58	−18	2.35	$w_n = 0.42 - 0.5 \cos(2\pi n/N) +$ $0.08 \cos(4\pi n/N)$
Blackman–Harris (3-term)	1.71	−67	−6	1.81	$w_n = 0.42323 -$ $0.49755 \cos(2\pi n/N) +$ $0.07922 \cos(4\pi n/N)$

mainlobe and small and rapidly decreasing sidelobes, i.e. this will achieve good spectral resolution and a high dynamic range. Unfortunately these are conflicting requirements and, generally speaking, a window with a narrow main lobe will have high sidelobes, or vice-versa, as shown in Table 5.5. Here the relative mainlobe width is expressed as the equivalent noise bandwidth (ENBW) of the window since this can be useful for PSD estimation. Note that mathematically these windows are the same as those used for digital filter design (window method) except for a time shift of $N/2$ samples to give an FFT window spanning $0 \leqslant n \leqslant N - 1$.

The choice of window depends upon the application. Generally speaking, the smaller the sidelobes the higher the dynamic range and the greater the ability to detect low-level signals. This is illustrated in Fig. 5.27, where

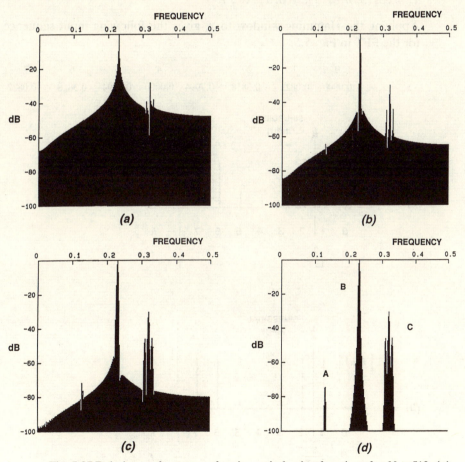

Fig. 5.27 Relative performance of various windowing functions for $N = 512$: (*a*) rectangular; (*b*) Hamming; (*c*) Gaussian ($\alpha = 3$); (*d*) $\sin^3(\cdot)$. Signals: A – low-level carrier; B – high-level carrier; C – AM carrier with sidebands.

it can be seen that only the $\sin^3 x$ window has adequately detected the low-level carrier. In fact, this type of window with $\alpha = 2$ (the Hanning window) is a popular choice for commercial spectrum analysers. Note, however, that the broader main lobe compared to the rectangular window reduces the frequency resolution. The narrow main lobe of the rectangular window is useful for resolving very closely spaced signals of similar strength (conversely, the high leakage of this window could mask weak signals, as shown in Fig. 27(a)). Remember, the rectangular window is also required for self-windowing input signals (Section 5.5.2).

Example 5.8

Consider the same 8-point input sequence used in Example 5.7, i.e.

$$x_n = \cos(3\pi n/8) \quad n = 0, 1, \dots, 7$$

Applying the Hamming window to x_n gives the following input sequence for the FFT in Fig. 5.5:

n	0	1	2	3	4	5	6	7
$x'_n = x_n w_n$	0.0800	0.0822	−0.3818	−0.7994	0.0000	0.7994	0.3818	−0.0822

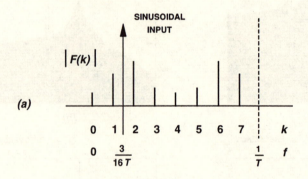

(a)

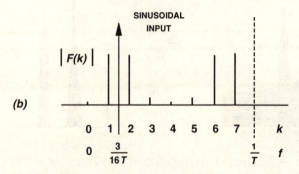

(b)

Fig. 5.28 Reduction of spectral leakage: (*a*) rectangular window; (*b*) Hamming window.

The Hamming window reduces discontinuities at boundaries and so gives a significant reduction in spectral leakage. Fig. 5.28 compares $|F(k)|$ for unwindowed (rectangular window) data x_n and Hamming windowed data x'_n.

5.5.6 Spectral resolution

The limiting resolution of the FFT is equal to the FFT bin width

$$\Delta f = f_s/N \tag{5.50}$$

and so this can be improved by increasing N as shown in Fig. 5.29 (the input signal is identical to that in Fig. 5.27). For a fixed value of N, the convolutional process of windowing spreads each spectral line, or imposes an effective bandwidth on each line, and so the frequency spacing $f_2 - f_1$ between two signals needs to be larger than Δf. A general criterion for

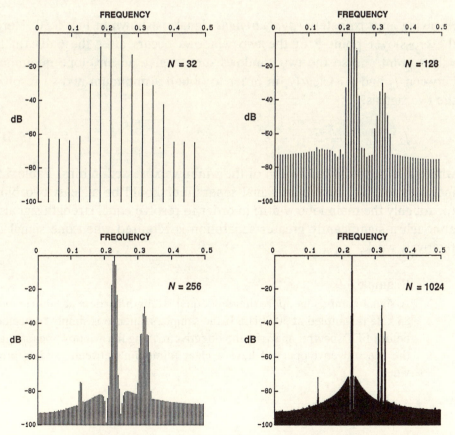

Fig. 5.29 FFT resolution as a function of N (the 3-term Blackman–Harris window is used throughout).

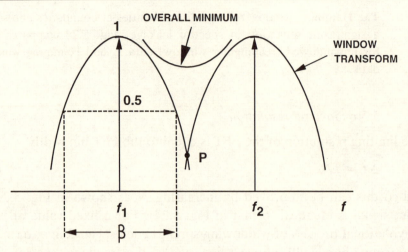

Fig. 5.30 Criterion for resolving equal strength signals.

resolving approximately *equal strength* signals is shown in Fig. 5.30. Here the cross-over point P of the two windows occurs after the 6 dB (half-value) point and so the two windows sum to give an envelope minimum between f_1 and f_2. Clearly, in order to obtain a minimum and so resolve the two signals,

$$f_2 - f_1 > \beta \frac{f_s}{N} \tag{5.51}$$

where β is the 6 dB bandwidth of the window expressed in bins. Table 5.5 indicates that the minimum signal separation should be at least two bins (or roughly the main lobe width) in order to resolve equal strength signals, although a significantly greater separation is required when one signal is relatively weak.

Example 5.9

A signal comprising approximately equal strength carriers at 80 kHz and 85 kHz is sampled at 300 kHz. If the sample sequence is simply truncated before FFT spectral analysis, the effective rectangular window means that the spectral envelope will have a clear minimum between carriers provided

$$N > \frac{2f_s}{f_2 - f_1} = 120$$

A 128-point radix-2 FFT will therefore resolve the two carriers (note the carriers will not fall on frequency bins).

5.5.7 Power spectrum estimation

For simplicity, spectral analysis has been introduced using continuous sinusoidal, i.e. *deterministic*, signals. The real world, of course, comprises either *finite-energy* signals (relatively short duration signals, such as pulses), or *random* (unpredictable) signals, as found in communications systems. For finite-energy signals we can estimate the *energy density* spectrum by invoking Parseval's theorem and using the FFT. For random signals, spectral estimation resorts to a statistical approach, and the two classical methods to be discussed are based upon the autocorrelation function of the signal. In contrast, the more computation intensive *parametric methods* attempt to model the random signal and spectral estimation amounts to a search for the parameters of the model. Typically, an autoregressive (AR) model is assumed. Whichever technique is used, the overall objective for random signals is to obtain a *consistent* estimate of the PSD, $P(\omega)$, together with good spectral resolution.

Suppose our time series x_n (sample period T) can be considered to be a sample function or typical signal from a random process $\{x_n\}$ and that this process has an autocorrelation function R_i. Communication theory (specifically the Wiener–Khinchine theorem) states that the PSD of a *stationary random process* is given by the discrete time Fourier transform of R_i, i.e.

$$P(\omega) = \sum_{i=-\infty}^{\infty} R_i e^{-ji\omega T} \tag{5.52}$$

where R_i is an even function about $i = 0$. In practice we are happy to assume stationarity so that, if required, R_i can be conveniently computed as a time average from a single time series. In this case, R_i is formally defined as an average *over all time* but in practice x_n is known only over a finite interval of, say, N samples. In order to evaluate (5.52) we could therefore define R_i as the estimate

$$\hat{R}_i = \frac{1}{N} \sum_{n=-\infty}^{\infty} x_n^N x_{n+i}^N$$

where

$$x_n^N = \begin{cases} x_n & n = 0, 1, \ldots, N-1 \\ 0 & \text{otherwise} \end{cases}$$

Inserting $\hat{R}_i$ into (5.52), and changing variables where necessary, we find $P(\omega)$ can be reduced to the simple estimate

$$\hat{P}(\omega) = \frac{1}{N}|X(\omega)|^2$$

where $X(\omega)$ is the discrete time Fourier transform of sequence x_n^N, i.e.

$$X(\omega) = \sum_{-\infty}^{\infty} x_n^N e^{-jn\omega T} = \sum_{n=0}^{N-1} x_n e^{-jn\omega T}$$

Clearly, if $X(\omega)$ is reduced to $X(k)$, the DFT of sequence x_n, $n = 0, 1,$ $\ldots, N-1$, then we obtain a sampled estimate of the PSD:

$$\hat{P}(k) = \frac{1}{N}|X(k)|^2 \tag{5.53}$$

Note that this remarkably simple computation for the PSD also applies when the sequence x_n corresponds to a deterministic signal. Equation (5.53) is called the *periodogram* or *power spectrum* and in practice it can be used as follows:

(1) remove any mean value from the signal;
(2) window the input data sequence x_n to minimize leakage (at the expense of broadening the spectral estimate);
(3) if necessary, extend the windowed data sequence $x_n w_n$ with zeros to obtain a sequence length N which is a power of 2;
(4) perform the FFT and find the periodogram;
(5) obtain a consistent estimate of $P(\omega)$ by averaging periodograms.

Frequency domain averaging is particularly important since, in general, the *variance* of $\hat{P}(k)$ does not fall to zero as $N \Rightarrow \infty$, i.e. $\hat{P}(k)$ is not a consistent or reliable estimate of $P(\omega)$. However, a consistent estimate *can* be obtained by simply finding the arithmetic mean of L statistically independent periodograms. A practical approach is therefore to compute

$$\hat{P}(k) = \frac{1}{UNL} \sum_{m=0}^{L-1} |X_m(k)|^2, \quad k = 0, 1, \ldots, \left(\frac{N}{2} - 1\right) \tag{5.54}$$

where

$$U = \frac{1}{N} \sum_{n=0}^{N-1} w_n^2 \tag{5.55}$$

The normalizing factor U provides correction for a bias in the spectral estimate generated by the finite bandwidth of the window transform, $W(\omega)$. An alternative way of accounting for the window function is to use the ENBW of the window. For example, a single (i.e. unaveraged) estimate of the PSD is sometimes computed by spectrum analysers as

$$\hat{P}(k)(\text{dB}) = 10\log|X(k)|^2 - 10\log(\text{ENBW} \cdot \Delta f)$$

where Δf is the analyser frequency resolution (bin width, in Hz) and ENBW is given in Table 5.5.

Provided the periodograms in (5.54) are independent, the variance of the averaged estimate will then be a factor L smaller than that of individual periodograms. In practice it is common to overlap the L sections to avoid missing short events and/or to provide more frequent transforms. Even with as much as 50% overlap (see Fig. 5.31(a)) it is reasonable to assume that the random components in successive periodograms are essentially independent provided a good window is used. Note, however, that if a fixed-length sequence is split into shorter sequences for the purposes of averaging, then, according to (5.50) the spectral resolution will be impaired.

The technique in Fig. 5.31(a) is the classical *direct* approach for estimating $P(\omega)$ and does not require prior computation of R_i. Alternatively, (5.52) suggests that $P(\omega)$ could be found *indirectly* by first estimating R_i and then performing an FFT (Fig. 5.31(b)). Given just N samples of x_n we could estimate R_i using (2.33) (with or without windowing), or, more generally, average over a time window of N samples using

$$\hat{R}_i = \frac{1}{N}\sum_{n=0}^{N-1} x_n x_{n+i}$$

Alternatively, we could use the fast correlation technique discussed in Section 5.6.

Suppose we have found $\hat{R}_i$ for L lags, i.e. $\hat{R}_i$ $i = 0, 1, \ldots, L-1$. Since $\hat{R}_i$ is an even function the FFT input spans a total of $2L-1$ points and so the FFT size must be $N \geqslant 2L-1$, e.g. if $L = 64$ then a 128-point FFT is appropriate. Note that, since the FFT regards its input as periodic, and because we have effectively truncated $\hat{R}_i$, then it is necessary to window the $\hat{R}_i$ samples before applying them to the FFT. Also, since long sequences are required to obtain a relatively few values of $\hat{R}_i$, it will often be necessary to 'zero pad' the windowed FFT input in order to use a larger FFT and so interpolate the PSD. Fig. 5.31(c) shows a typical (short-time) power spectrum for voiced speech computed using the autocorrelation-FFT method. The analysis used $N = 1024$, $L = 512$, and a Hanning window.

Finally, note that periodogram averaging, or the estimation of R_i, both require the availability of long data sequences in order to yield consistent results with good frequency resolution. Where only short data sequences

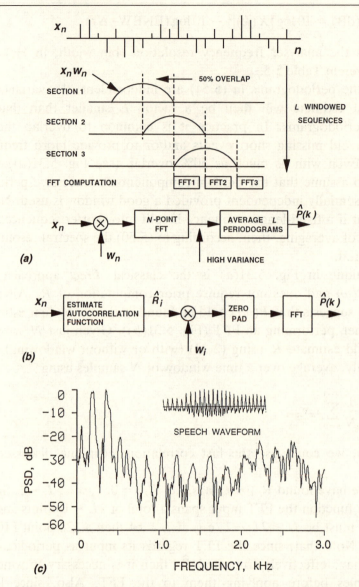

Fig. 5.31 Power spectrum estimation: (*a*) overlapped processing and averaging of periodograms; (*b*) autocorrelation-FFT method; (*c*) typical PSD of voiced speech (autocorrelation-FFT method).

are available, or where higher resolution is required, we could employ a parametric method of estimating the PSD (Proakis and Manolakis, 1988).

Multiband analysis
So far we have assumed that FFT spectral estimation is to be carried out over a single baseband frequency range. However, practical signal ana-

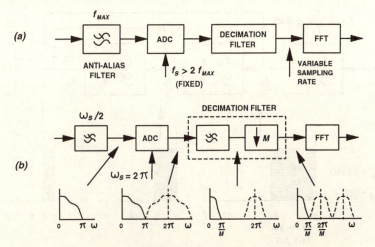

Fig. 5.32 Use of decimation filtering to achieve multiband analysis.

lysers span a wide range of frequencies, e.g. from a minimum span of 1 Hz to a maximum span of 100 kHz, and typically each decade range needs to be covered with at least three spans. This variability could be achieved using multiple analogue anti-alias filters and a variable ADC sampling rate, although a far more efficient and stable solution is shown in Fig. 5.32(*a*). Here the ADC sample rate is fixed at the rate needed for the highest frequency span and only a single anti-alias filter is required. The required frequency span and sampling rate variations are achieved using a decimation filter (Section 1.4.1).

The concept behind the decimation system is shown in Fig. 5.32(*b*), where it is assumed that the highest input band extends to π. Assuming analysis is required over the band $0-\pi/M$, the decimating filter not only restricts the signal to this range, but it also reduces the sample rate by a factor M. The full N-point FFT is now 'zoomed-in' over the range $0-2\pi/M$.

Fig. 5.33 illustrates how identical decimating filters of normalized cutoff frequency f_{cn} could be used to achieve multiband analysis. It is based on the fact that the cutoff frequency f_c of a filter can be changed by controlling its input rate f_s, i.e. $f_c = f_{cn}f_s$. Provided $f_{cn} \leqslant 0.25$ each filter provides sufficient bandlimitation to satisfy Nyquist's sampling theorem at its output, and so a cascade of N identical decimating filters will give an overall cutoff frequency of

$$f_c = f_{cn}f_s/2^{N-1} \tag{5.56}$$

Clearly, if $f_{cn} = 0.25$, then f_c is equivalent to the FFT AR.

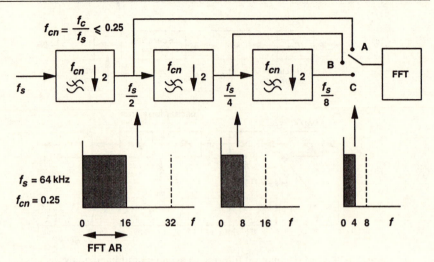

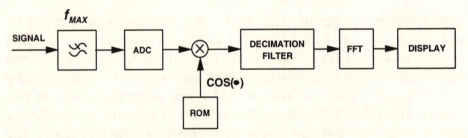

Fig. 5.33 Example of multiband analysis.

Fig. 5.34 General zoom transform in a spectrum analyser.

Example 5.10

Suppose that the input signal in Fig. 5.33 comprises a 1 kHz carrier (level 0 dB) and −60 dB sidebands spaced at 100 Hz from the carrier. It is easily demonstrated that a 1024-point FFT performed directly on Hamming windowed input data sampled at $f_s = 64$ kHz will fail to resolve the sidebands (see also (5.50)), whilst some indication of the sidebands will be found with the switch in position A. On the other hand, switching to position C will give very good resolution of the sidebands.

Similarly, suppose an analyser digitizes a signal at rate $f_s = 50$ kHz and is required to cover the bands 0–6400 Hz and 0–50 Hz. The first band can be covered at the output of the first decimation filter by setting $f_{cn} = 6.4/50 = 0.128$. The output of a second identical decimation filter will therefore cover the band 0–3200 Hz, and, from (5.56), the output from an eighth identical decimating filter will cover 0–50 Hz.

In general, a zoom transform also involves mixing or heterodyning, and this is best carried out after the ADC in order to achieve optimal stability. A basic arrangement is shown in Fig. 5.34.

5.6 Fast convolution and correlation

Sometimes it can be advantageous (faster) to perform the operations of convolution and correlation indirectly using the FFT. Fast or FFT based convolution, for example, is applicable where there are a large number of multiplications to be done, as in high-order FIR filters, or in image filtering. Fast convolution is also used for matched filtering (pulse compression) in radar applications.

5.6.1 Circular convolution

Fig. 5.35(a) shows that digital filtering can be performed either directly, as a time domain convolution, or indirectly by multiplying transforms. The principle of the indirect approach is shown in Fig. 5.35(b). Note that, since $h(n)$ is known, its transform $H(k)$ need be computed only once.

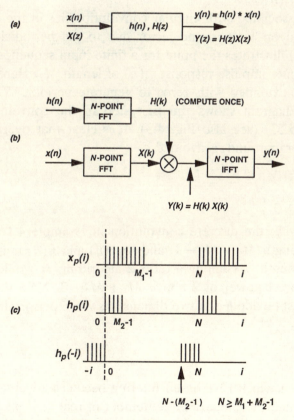

Fig. 5.35 Fast (or circular) convolution: (a) general digital filter; (b) filtering by transform multiplication; (c) avoiding wraparound error.

More significantly, recall that the FFT assumes its input to be periodic, period N, and so we must imagine that two *periodic* sequences $x_p(n)$ and $h_p(n)$ are being convolved, as in Fig. 5.35(c). This process gives rise to the term *circular* or *cyclic* convolution and is formally defined as

$$y_p(n) = \sum_{i=0}^{N-1} x_p(i)h_p(n-i) \qquad (5.57)$$

Sequence $y_p(n)$ is also a periodic sequence, period N. Similarly, the transforms in Fig. 5.35(b) will be periodic (see Fig. 5.3(e)) and could be written

$$Y_p(k) = H_p(k)X_p(k) \qquad (5.58)$$

Therefore, if we multiply the FFTs of two finite duration sequences $x(n)$ and $h(n)$, and then inverse transform the product, the result $y(n)$ will be just one period of $y_p(n)$ and we have effectively convolved *periodic* sequences $x_p(n)$ and $h_p(n)$.

The major pitfall associated with circular convolution lies in the possibility of convolutional or 'wraparound' error due to the periodicity assumption. Fig. 5.35(c) illustrates the point for a finite input sequence $x(n)$ of length M_1 and a finite impulse response $h(n)$ of length M_2. Here both sequences have been extended with zeros to generate periodic N-point sequences, and the diagram shows the first step in the convolutional process defined by (5.57) (see also Fig. 4.3). It is clear that overlap of periodic sequences (wraparound) is avoided if

$$N - M_2 + 1 > M_1 - 1$$

$$N \geq M_1 + M_2 - 1 \qquad (5.59)$$

Looked at another way, the discrete convolution in Example 4.1 shows that $y(n)$ will have length $M_1 + M_2 - 1$, and so $x(n)$ and $h(n)$ must also be extended to this length. Assuming a radix-2 algorithm, N would then be chosen as the nearest power of 2 above $M_1 + M_2 - 2$. Note that the criterion in (5.59) must be applied in two dimensions when performing fast filtering of images.

Speed advantage
The cross-over point at which FFT based filtering becomes advantageous can be roughly estimated by comparing the number of real multiplications required for time and frequency domain convolution. For example, twelve real multiplications are required for the time domain convolution in

Table 5.6. *Approximate comparison of direct and fast convolution in terms of real multiplications.*

$N = 2M$	M^2	K
8	16	128
32	256	768
128	4096	4096
512	65536	20480
1024	262144	45056

Example 4.1, and we could generalize this to $M_1 M_2$ real multiplications for $x(n)$ of length M_1 and $h(n)$ of length M_2. For simplicity, assume $M_1 = M_2 = M$ so that M^2 real multiplications are required for direct convolution.

On the other hand, assuming $H(k)$ is precomputed and stored, fast convolution requires two N-point FFTs plus N complex multiplications. From Section 5.3.3 we note that a radix-2 butterfly requires four real multiplications and that there are $0.5N \log_2 N$ butterflies per FFT. The total number of real multiplications is therefore

$$K = 4N(\log_2 N + 1)$$

Taking $N = 2M$ (Equation (5.59)), Table 5.6 suggests that, on the above assumptions, FFT based filtering is advantageous for FIR filters with impulse response length greater than approximately 64 (more detailed analysis gives a somewhat lower value (Brigham, 1974)). In contrast, the number of multiplications in IIR filters is relatively small and so these are best implemented using direct convolution.

Select–save algorithm for convolution

In practice it is likely that the sum in (5.59) exceeds the capacity of the FFT processor. Suppose for example that the input data sequence is appreciably longer than the available FFT size N. In this case the input sequence $x(n)$ can be sectioned into sets of N points as shown in Fig. 5.36(a), and we start by computing the N-point FFT for section $x_0(n)$. Note that for point by point multiplication in the frequency domain it is necessary to extend $h(n)$ to an N-point sequence by appending zeros, as in Fig. 5.35(c).

The significant point is that, since no attempt has been made to avoid wraparound error, not all values of the FFT sequence will be valid. This

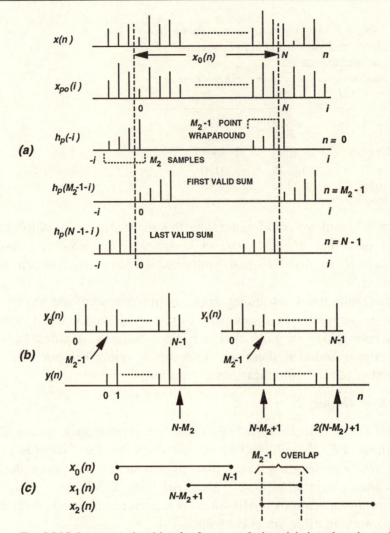

Fig. 5.36 Select–save algorithm for fast convolution: (*a*) time domain equivalent for section x_0; (*b*) generation of $y(n)$ by saving valid samples; (*c*) necessary overlapped processing.

can be seen for section $x_0(n)$ in Fig. 5.36(*a*) where both $x_0(n)$ and $h(n)$ are forced to be periodic by the FFT. The result is an $M_2 - 1$ point wraparound, and so the first $M_2 - 1$ values of sequence $y_0(n)$ are incorrect. In other words, only the last $N - M_2 + 1$ values are selected and saved as part of the output sequence $y(n)$, see Fig. 5.36(*b*). The algorithm now starts on the next N data points, section $x_1(n)$. From Fig. 5.36(*a*) it is clear that the last $M_2 - 1$ samples of section $x_0(n)$ have generated the wraparound and so, in effect, they have been discarded. Section $x_1(n)$

must therefore start with these samples and so it overlaps section $x_0(n)$ by $M_2 - 1$, as shown in Fig. 5.36(c). The last $N - M_2 + 1$ values of $y_1(n)$ are then selected and saved as part of $y(n)$. It is readily shown that, for a given impulse response length M_2, there is an optimum value of N which minimizes the number of real multiplications required for an arbitrary output sequence length (see Problem 34).

5.6.2 Circular correlation

As previously pointed out, the FFT can also be used to perform *fast correlation*. Correlation is useful for detecting signals hidden in noise or for measuring signal delay, or for measuring signal power spectral densities. The peak of the *cross-correlation* function, for example, can indicate the time delay between two signals.

FFT based correlation uses a similar approach to FFT based convolution since both are similar mathematically. For fast convolution we have the DFT pair

$$y(n) = \sum_{i=0}^{N-1} x(i)h(n-i) \Leftrightarrow H(k)X(k) \tag{5.60}$$

whilst for fast or circular cross-correlation we have

$$\hat{R}_{xy}(m) = \frac{1}{N}\sum_{n=0}^{N-1} x(n)y(n-m) \Leftrightarrow \frac{1}{N}X(k)Y^*(k) \tag{5.61}$$

Note that $\hat{R}_{xy}(m)$ is only an *estimate* of the true cross-correlation function $R_{xy}(m)$. This is because $R_{xy}(m)$ can be formally defined as an average *over all time* whilst in (5.61) we are averaging over a finite time window. Even more significantly, the FFT requires all sequences in (5.60) and (5.61) to be understood as periodic (as in (5.57)) and so multiplication of sliding time series can result in convolution or correlation error respectively. For correlation, one solution is to develop a 'select–save' algorithm similar to that for convolution, i.e. we section the data.

Select–save algorithm for correlation
Consider first the direct cross-correlation of two sequences $x(n)$ and $y(n)$:

$$
\begin{array}{cccccccccc}
x_0 & x_1 & x_2 & x_3 & x_4 & x_5 & x_6 & x_7 & x_8 & x_9
\end{array}
$$

y_0 y_1 y_2 y_3 y_4 y_5 y_6 y_7 y_8 y_9 $\quad R(0) = x_0y_0 + x_1y_1 + x_2y_2 + x_3y_3 + \ldots$

$\quad\ y_0$ y_1 y_2 y_3 y_4 y_5 y_6 y_7 y_8 $\quad R(1) = x_1y_0 + x_2y_1 + x_3y_2 + x_4y_3 + \ldots$

$\quad\quad\ y_0$ y_1 y_2 y_3 y_4 y_5 y_6 y_7 $\quad R(2) = x_2y_0 + x_3y_1 + x_4y_2 + x_5y_3 + \ldots$

The mth lag value is therefore

$$R(m) = x_m y_0 + x_{m+1} y_1 + x_{m+2} y_2 + \ldots$$

and this could have been arrived at by sectioning the data. For example, using 50% overlap, the first two sections could be

Section 0:

$x_0(n)$	x_0	x_1	x_2	x_3	x_4	x_5	x_6	x_7	
$y_0(n)$	y_0	y_1	y_2	y_3	0	0	0	0	$R_0(0) = x_0 y_0 + x_1 y_1 + x_2 y_2 + x_3 y_3$
$y_0(n-1)$		y_0	y_1	y_2	y_3	0	0	0	$R_0(1) = x_1 y_0 + x_2 y_1 + x_3 y_2 + x_4 y_3$
$y_0(n-2)$			y_0	y_1	y_2	y_3	0	0	$R_0(2) = x_2 y_0 + x_3 y_1 + x_4 y_2 + x_5 y_3$
$y_0(n-3)$				y_0	y_1	y_2	y_3	0	$R_0(3) = x_3 y_0 + x_4 y_1 + x_5 y_2 + x_6 y_3$
$y_0(n-4)$					y_0	y_1	y_2	y_3	$R_0(4) = x_4 y_0 + x_5 y_1 + x_6 y_2 + x_7 y_3$

Section 1:

$x_1(n)$	x_4	x_5	x_6	x_7	x_8	x_9	x_{10}	x_{11}	
$y_1(n)$	y_4	y_5	y_6	y_7	0	0	0	0	$R_1(0) = x_4 y_4 + x_5 y_5 + x_6 y_6 + x_7 y_7$
$y_1(n-1)$		y_4	y_5	y_6	y_7	0	0	0	$R_1(1) = x_5 y_4 + x_6 y_5 + x_7 y_6 + x_8 y_7$
$y_1(n-2)$			y_4	y_5	y_6	y_7	0	0	$R_1(2) = x_6 y_4 + x_7 y_5 + x_8 y_6 + x_9 y_7$
$y_1(n-3)$				y_4	y_5	y_6	y_7	0	$R_1(3) = x_7 y_4 + x_8 y_5 + x_9 y_6 + x_{10} y_7$
$y_1(n-4)$					y_4	y_5	y_6	y_7	$R_1(4) = x_8 y_4 + x_9 y_5 + x_{10} y_6 + x_{11} y_7$

and, clearly,

$$R(m) = R_0(m) + R_1(m) + R_2(m) + \ldots$$

$$R(m) = \sum_r R_r(m), \quad m = 0, 1, \ldots, 4 \tag{5.62}$$

Now, instead of performing sectioned correlation directly, as in (5.62), we could perform it indirectly in the frequency domain using (5.61). The first step is to transform the x and y samples of the rth N-point section to N-point sequences $X_r(k)$ and $Y_r(k)$ as shown in Fig. 5.37(a).

Taking $N = 8$, the FFT inputs for section 0 would then be the sequences

$$x_0(n) = x_0 \quad x_1 \quad x_2 \quad x_3 \quad x_4 \quad x_5 \quad x_6 \quad x_7$$

$$y_0(n) = y_0 \quad y_1 \quad y_2 \quad y_3 \quad 0 \quad 0 \quad 0 \quad 0$$

Remember, the FFT considers these to be periodic and so Fig. 5.37 will actually be performing the following circular correlations

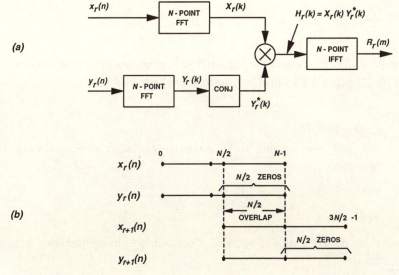

Fig. 5.37 Fast correlation: computation of the rth section (only $R_r(m)$, $m = 0, 1,$ $\ldots$, $N/2$ is valid).

$x_{p0}(n)$	x_6	x_7	x_0	x_1	x_2	x_3	x_4	x_5	x_6	x_7	x_0	x_1	
$y_{p0}(n)$	0	0	y_0	y_1	y_2	y_3	0	0	0	0	y_0	y_1	$\Rightarrow R_0(0)$
.								.					.
.								.					.
$y_{p0}(n-4)$	y_2	y_3	0	0	0	0	y_0	y_1	y_2	y_3	0	0	$\Rightarrow R_0(4)$
$y_{p0}(n-5)$	y_1	y_2	y_3	0	0	0	0	y_0	y_1	y_2	y_3	0	invalid
.								.					.
.								.					.

It is apparent that circular correlation creates invalid products in the same way as circular convolution (Fig. 5.36), and for the above example there are only five valid samples of $R_0(m)$ for a section of length 8. Generalizing, for an N-point FFT we will select and save outputs $R_r(m)$, $m = 0, 1, \ldots, N/2$ and discard the rest, and the sections will overlap by $N/2$ samples as shown in Fig. 5.37(b).

As previously pointed out, in practice summing $R_r(m)$ as in (5.62) will give only an *estimate* of $R(m)$ since our measurement window or input data span is finite. In order to achieve a *consistent* estimate (low variance) it is necessary to restrict the maximum value of m such that $m \ll M$, where M is the total number of input points used for the measurement. Using L 50%-overlapped sections of length N we could therefore rewrite (5.62) as

$$\hat{R}(m) = \sum_{r=0}^{L-1} R_r(m), \qquad m = 0, 1, \ldots, \frac{N}{2} \tag{5.63}$$

where $N \ll M$. For example, given 1024 input points (samples), we could compute 32-point FFTs and estimate $R(m)$ up to $R(16)$.

Example 5.11

We will use 4-point FFTs to perform fast cross-correlation for the sequences

$x(n)$ 7 5 3 2 3 1 0 1 ...

$y(n)$ 6 4 4 2 1 2 1 1 ...

First perform direct, sectioned correlation, using just two sections for brevity:

$r = 0$: $x_0(n)$	7	5	3	2		
$y_0(n)$	6	4	0	0	$R_0(0) = 62$
$y_0(n-1)$		6	4	0	$R_0(1) = 42$
$y_0(n-2)$			6	4	$R_0(2) = 26$

$r = 1$: $x_1(n)$	3	2	3	1		
$y_1(n)$	4	2	0	0	$R_1(0) = 16$
$y_1(n-1)$		4	2	0	$R_1(1) = 14$
$y_1(n-2)$			4	2	$R_1(2) = 14$

Using (5.63) gives

$$\hat{R}(0) = 78, \quad \hat{R}(1) = 56, \quad \hat{R}(2) = 40$$

Now compute $R_0(m)$, $m = 0, 1, 2$ using indirect correlation. Fig. 5.38 shows the detailed computations using radix-2 in-place FFTs. Note that $x_0(n)$ and $y_0(n)$ are shuffled at the flowgraph inputs and that the complex conjugate of $H_0(k)$ is taken in accordance with (5.33) before being applied in a shuffled way to a forward FFT. Also, the three saved outputs of this FFT must be scaled by $1/4$ in accordance with (5.61). Exactly the same procedure is applied for $R_1(m)$, $m = 0, 1, 2$ (i.e. compute $H_1(k)$), yielding

$$R_1(0) = 16, \quad R_1(1) = 14, \quad R_1(2) = 14$$

Summing the two sections gives the same results as found using direct,

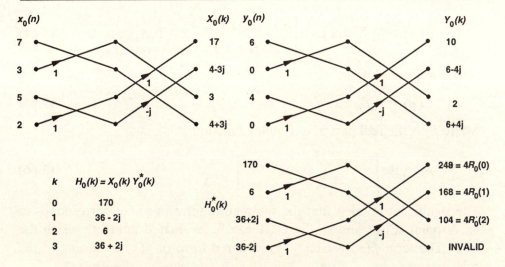

Fig. 5.38 Illustrating fast correlation (Example 5.11).

sectioned correlation. Finally, by expanding (5.63) in terms of IFFTs, it is easily shown that it is more efficient to first sum terms $H_r(k)$, $r = 0, 1$, ..., $L - 1$ and then take a single IFFT, rather than take an IFFT for each section. For this simple example we have

$$\hat{R}(m) = R_0(m) + R_1(m)$$

$$= \frac{1}{N}[\sum H_0(k)W_N^{-nk} + \sum H_1(k)W_N^{-nk}]$$

$$= \frac{1}{N}\sum [H_0(k) + H_1(k)]W_N^{-nk}$$

$$= \frac{1}{N}\sum [H_0(k) + H_1(k)]^* W_N^{nk}$$

and so only five FFTs are required instead of six.

5.7 A fast DCT algorithm

Many fast algorithms for a 1-D DCT have been described in the literature and they could be classified into (*a*) recursive methods (Hou, 1987), (*b*) indirect, FFT based methods (Narasimha and Peterson, 1978) and (*c*) direct, matrix based methods (Suehiro and Hatori, 1986; Brebner and Ritchings, 1988). Here we examine the efficient Narasimha and Peterson algorithm and we will use it in Section 5.8.2 to compute a 2-D DCT. With this in mind, let us change the variables slightly in (5.10), and for clarity ignore the normalization factor $2/N$, giving the 1-D DCT

$$C(k) = a_k \sum_{x=0}^{N-1} f(x) \cos\left[\frac{(2x+1)k\pi}{2N}\right], \quad k = 0, 1, \ldots, N-1 \ (5.64)$$

This can be written

$$C(k) = a_k \beta_k \tag{5.65}$$

where β_k is the real part of a complex function:

$$\beta_k = \text{Re}\left[e^{j\pi k/2N} \sum_{x=0}^{N-1} f_r(x) e^{j2\pi kx/N}\right] \tag{5.66}$$

The significant point is that the summation term in (5.66) corresponds to an N-point IFFT, and so the DCT can be evaluated indirectly using the FFT. The term $f_r(x)$ is simply a reordered form of $f(x)$ according to the rule

$$\left.\begin{array}{l} f_r(x) = f(2x) \\ f_r(N-1-x) = f(2x+1) \end{array}\right\}, \quad x = 0, 1, \ldots, \frac{N}{2}-1 \tag{5.67}$$

For $N = 8$,

$$\{f(x)\} = f(0) \ f(1) \ f(2) \ f(3) \ f(4) \ f(5) \ f(6) \ f(7)$$
$$\{f_r(x)\} = f(0) \ f(2) \ f(4) \ f(6) \ f(7) \ f(5) \ f(3) \ f(1)$$

Similarly, the IDCT in (5.11) can be computed indirectly by first evaluating

$$f(2x) = \text{Re}\left[\sum_{k=0}^{N-1} [a_k C(k) e^{j\pi k/2N}] e^{j2\pi kx/N}\right], \quad x = 0, 1, \ldots, N-1 \tag{5.68}$$

Clearly, (5.68) corresponds to another N-point IFFT and gives an N-point time sequence $f(0), f(2), \ldots, f[2(N-2)], f[2(N-1)]$. The first $N/2$ points of this sequence give the $N/2$ even points of $f(x)$ and the $N/2$ odd points of $f(x)$ are given by

$$f(2x+1) = f[2(N-1-x)], \quad x = 0, 1, \ldots, \frac{N}{2}-1 \tag{5.69}$$

It should be noted that the above algorithm is approximately twice as fast as indirect DCT algorithms based on $2N$-point FFTs, which implies a factor four improvement when used to compute the 2-D DCT (Section 5.8.2).

5.8 Image transforms

It is often necessary to transform an $N \times N$ pixel image to the spectral domain using some type of 2-D discrete transform algorithm. For example, once in this form, the image could be linear filtered to achieve an improved image (as in *image restoration*), or the frequency coefficients could be selectively quantized to achieve data compression (as in transform coding, Section 2.3). We will examine perhaps the two most important image transforms, the 2-D DFT and the 2-D DCT. Fortunately, these two transforms have kernels which are *separable* and so they can be computed in two steps, each involving a 1-D transform; the first step will involve row (column) transforms and the second step will involve column (row) transforms. It is worth noting, however, that *row–column* algorithms are not the only way of performing a 2-D transform. In the case of the 2-D DFT, for example, more computation efficient algorithms such as *vector–radix* algorithms and *polynomial transform* algorithms could be applied (Duhamel and Vetterli, 1990).

5.8.1 2-D FFT

Consider the transform of the $N \times N$ image $f(x, y)$ in Fig. 5.39. Since the input signal is in a 'spatial' domain, the corresponding 2-D transform $F(u, v)$ will be defined in a 'spatial frequency' domain, where u and v are spatial frequencies. The 2-D DFT of $f(x, y)$ can be expressed

$$F(u, v) = \sum_{x=0}^{N-1}\sum_{y=0}^{N-1} f(x, y)\, e^{-j2\pi(ux+vy)/N} \tag{5.70}$$

(where the scaling factor N^{-2} has been associated with the IDFT). Separating the kernel to give a row–column transform gives

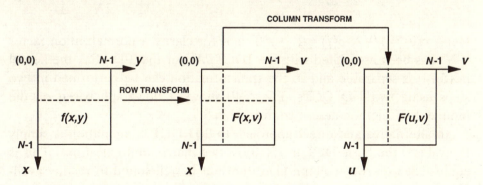

Fig. 5.39 Row–column algorithm for the 2-D FFT.

$$F(u, v) = \sum_{x=0}^{N-1} e^{-j2\pi ux/N} \sum_{y=0}^{N-1} f(x, y) e^{-j2\pi vy/N} \tag{5.71}$$

$$= \sum_{x=0}^{N-1} F(x, v) e^{-j2\pi ux/N} \tag{5.72}$$

Evaluating $F(x, v)$ for a fixed value of x and all values of v amounts in practice to a single 1-D FFT. A row of data in $f(x, y)$ is transformed to a row of data in $F(x, v)$, and so N such row transforms are required to complete $F(x, v)$. Similarly, evaluation of (5.72) for a fixed value of v and all values of u amounts to a single 1-D FFT operating on a column of data in $F(x, v)$. Again, N such column transforms are required to complete $F(u, v)$. In total we require $2N$ 1-D N-point FFTs, or, assuming a radix-2 algorithm, approximately $10N^2 \log_2 N$ real arithmetic operations (addition or multiplication). Note that if fast linear filtering is to be performed on $F(u, v)$ by a function $H(u, v)$, giving

$$G(u, v) = H(u, v)F(u, v) \Leftrightarrow h(x, y)*f(x, y) \tag{5.73}$$

then both $N \times N$ functions $h(x, y)$ and $f(x, y)$ must be assumed periodic with some period M in the x and y directions. To avoid wraparound, $M \geqslant 2N - 1$.

5.8.2 2-D DCT

The 2-D DCT of an $N \times N$ image $f(x, y)$ can be expressed as

$$C(u, v) =$$

$$c(u)c(v) \sum_{x=0}^{N-1} \sum_{y=0}^{N-1} f(x, y) \cos\left[\frac{(2x + 1)u\pi}{2N}\right] \cos\left[\frac{(2y + 1)v\pi}{2N}\right] \tag{5.74}$$

Here, $c(0) = 1/\sqrt{2}$, $c(j) = 1$, $j \neq 0$, and for clarity a normalization factor $4/N^2$ has been associated with the IDCT. As for the 2-D FFT, the kernel in (5.74) is separable and so the transformation can be performed in two steps using two 1-D DCTs. Fig. 5.40 shows a flowgraph based on the indirect FFT method described in Section 5.7.

An alternative and direct approach to 2-D DCT computation is simply to evaluate the basic DCT in (5.10) as two matrix multiplications. This is equivalent to a row–column algorithm but is well suited to readily available hardware. Let (5.10) be expressed in the form

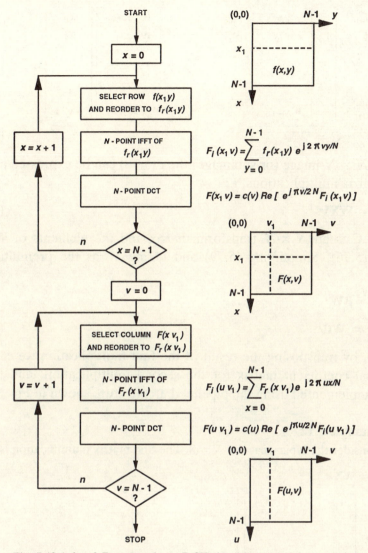

Fig. 5.40 A fast 2-D row–column DCT algorithm based on the *N*-point IFFT.

$$C_k = \sum_{n=0}^{N-1} w_{kn} x_n, \qquad k = 0, 1, \ldots, N - 1 \qquad (5.75)$$

where

$$w_{kn} = \frac{2}{N} a_k \cos \left[\frac{(2n + 1)k\pi}{2N} \right] \qquad (5.76)$$

Putting (5.75) into matrix form

$$\begin{bmatrix} C_0 \\ C_1 \\ \cdot \\ \cdot \\ \cdot \\ C_{N-1} \end{bmatrix} = \begin{bmatrix} w_{00} & w_{01} & \cdots & w_{0N-1} \\ w_{10} & w_{11} & \cdots & w_{1N-1} \\ & & \cdot & \\ & & \cdot & \\ & & \cdot & \\ w_{N-10} & & \cdots & w_{N-1N-1} \end{bmatrix} \begin{bmatrix} x_0 \\ x_1 \\ \cdot \\ \cdot \\ \cdot \\ x_{N-1} \end{bmatrix} \tag{5.77}$$

$$\mathbf{C} = \mathbf{WX} \tag{5.78}$$

If $\mathbf{X}$ is an $N \times N$ image (or subimage) then (5.78) can be expressed as two $N \times N$ matrix multiplications, i.e.

$$\mathbf{C} = \mathbf{WXW}^{\mathrm{T}} \tag{5.79}$$

In (5.79), $\mathbf{C}$ is an $N \times N$ transform matrix and the elements of $\mathbf{W}$ are given by (5.76). Simplifying (5.79) and taking $\mathbf{W}$ as the premultiplier, gives

$$\mathbf{C} = \mathbf{BW}^{\mathrm{T}}$$

$$\mathbf{C}^{\mathrm{T}} = \mathbf{WB}^{\mathrm{T}} \tag{5.80}$$

Therefore, by transposing the result of the first multiplication we can use an identical coefficient matrix for the second multiplication, and (5.80) could be implemented using the pipelined architecture shown in Fig. 5.41.

Example 5.12
Consider the typical case of $N = 8$. The first matrix multiplication is

$$\mathbf{B} = \mathbf{WX}$$

$$= \begin{bmatrix} w_{00} & \cdots & w_{07} \\ & \cdot & \\ & \cdot & \\ & \cdot & \\ w_{70} & \cdots & w_{77} \end{bmatrix} \begin{bmatrix} x_{00} & \cdots & x_{07} \\ & \cdot & \\ & \cdot & \\ & \cdot & \\ x_{70} & \cdots & x_{77} \end{bmatrix}$$

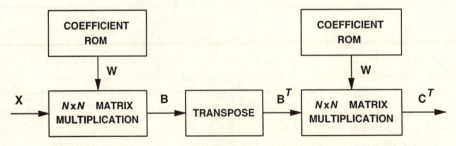

Fig. 5.41 A pipelined $N \times N$ DCT based on direct matrix multiplication.

and so the first of the 64 elements in **B** is found as

$$w_{00}x_{00} + w_{01}x_{10} + \ldots + w_{07}x_{70}$$

This could be computed using a single MAC and eight machine (MAC) cycles. If we used eight MACs to evaluate the eight row elements of **B** in parallel, then **B** could be computed in just 64 cycles. The pipelined architecture in Fig. 5.41 could therefore deliver an 8×8 DCT every 64 cycles or 3.2 μs, assuming a typical MAC cycle time of 50 ns.

Chapter 1

P1 The CCIR 4:2:2 coding standard for component video signals specifies that the colour difference signals should be sampled at 6.75 MHz (and be uniformly quantized to 8-bit PCM). If the maximum colour difference frequency is 1.3 MHz and the PCM signal is subsequently passed through a DAC, what is the maximum $(\sin x)/x$ loss at the output of the anti-imaging filter?

[0.54 dB]

P2 A low-noise monochrome video signal is to be encoded and decoded by the system in Fig. 1.3. The maximum signal excursion (corresponding to a black level to white level change) at the ADC input is set to one-half the (uniform) quantizing range of the ADC. The two lowpass filters cutoff at 5 MHz and the sampling frequency is 12 MHz.

   If the signal to noise ratio (defined as black-to-white range/rms noise) at the output of the anti-alias filter is to be at least 41 dB, what is the minimum digital resolution, n, of the codec?

[$n = 6$]

Chapter 2

P3 A weather map of black (B) pixels on a white (W) background can be modelled as a binary first-order Markov source. The Markov model then comprises state s_1 ($\equiv W$) and state s_2 ($\equiv B$) which state transitions, and in this case (2.7) can be written

$$H(s_i) = -\sum_{j=1}^{2} P(s_j|s_i) \log P(s_j|s_i)$$

If the state transition matrix is

$$T = \begin{bmatrix} p_{11} & p_{12} \\ p_{21} & p_{22} \end{bmatrix} = \begin{bmatrix} 0.99 & 0.3 \\ 0.01 & 0.7 \end{bmatrix}$$

where p_{21} is the probability of a transition from s_1 to s_2, estimate the compression which could be achieved through variable length coding. (Use (2.8) and assume $P(s_1) = p_{12}/(p_{21} + p_{12})$, $P(s_2) = 1 - P(s_1)$.)

$[H_c(A) = 0.1066$ bits/pixel; approximately 10:1]

P4   A discrete source has the following (*a priori*) symbol probabilities:

a_j	a_1	a_2	a_3	a_4	a_5	a_6
$P(a_j)$	0.07	0.25	0.30	0.25	0.08	0.05

Compare the efficiency of a binary Huffman code with that of a comma code for this source.

$[\eta_C/\eta_H = 92\%]$

P5   A ternary Huffman code (i.e. $c = 3$ code symbols) is required for the following discrete source:

a_j	a_1	a_2	a_3	a_4	a_5	a_6	a_7	a_8	a_9	a_{10}
$P(a_j)$	0.025	0.025	0.03	0.16	0.24	0.26	0.14	0.07	0.03	0.02

What is the average length of the code and the average compression ratio when compared with straight ternary coding of the source?

$[\bar{L} = 1.83; 1.64:1]$

P6   State in a few sentences how DPCM achieves data compression.
A monochrome PCM video signal is applied to a 5-bit DPCM coder. The 8-bit PCM input signal has a peak–peak range of 130 quantum intervals and an rms value of one quarter of the peak–peak range. The prediction gain of the DPCM system is 8 dB. Compute the quantizing law for the first four positive output levels (as in Example 2.12).

$[y_k = 0.86, 2.68, 4.62, 6.72; d_k = 1.77, 3.65, 5.67]$

P7   The discrete time sequence $\{x_n\}$ at the input to the DPCM coder in Fig. 2.16 has zero mean and unit variance, and its autocorrelation function R_i has the following values:

$R_0 = 1, R_1 = 0.97, R_2 = 0.91, R_3 = 0.86$

Deduce the coefficients of a 1-D second-order predictor. If the input signal statistics now change such that $R_i = e^{-\alpha i}$ where α is a constant and $R_0 = 1$, $R_1 = 0.94$, what is the prediction gain?

[1.477, -0.523; 9.3 dB]

Chapter 3

P8   An odd parity extended Hamming code is defined by the check matrix

$$\mathbf{H} = \begin{bmatrix} 1 & 1 & 1 & 1 & 0 & 0 & 0 & 0 \\ 1 & 1 & 0 & 0 & 1 & 1 & 0 & 0 \\ 1 & 0 & 1 & 0 & 1 & 0 & 1 & 0 \\ 1 & 1 & 1 & 1 & 1 & 1 & 1 & 1 \end{bmatrix}$$

Three codewords are used to protect decimal numbers in the range 0–999 and for one transmission the received codewords are

01001011 || 11101010 || 01001100

hundreds   tens   units

time $\Rightarrow$

The units of the number are transmitted last. The first digit transmitted in each codeword is the MSB of the message and the message bits are transmitted in descending significance. Noting that the code can detect and reject double errors (see (3.18)), attempt to find the transmitted number.

[probably 4X5, where X is unknown]

P9   Sketch the shift register circuit for multiplying a polynomial $i(x)$ by

$$g(x) = 1 + x^3 + x^4 + x^5 + x^6$$

Compute the circuit output for $i(x) = 1 + x + x^4$ (i.e the data enters as 1 followed by 0011, and is zero thereafter) by examining the shift register state at each clock pulse. Confirm your answer by algebraic multiplication.

[(11011000111) where '111' is generated first]

P10   A BCH(15, 5) coder is used to encode the message polynomial $i(x) = 1 + x^2 + x^3 + x^4$ into a systematic binary codeword $v(x)$, as in (3.25). Errors occur during transmission and the received codeword is

$$r(x) = 1 + x^4 + x^7 + x^8 + x^9 + x^{10} + x^{14}$$

Locate the errors. (Use Table 3.2.)

$[e(x) = x^9 + x^{12} + x^{13}]$

P11 A systematic $(7,4)$ binary cyclic code is generated by the polynomial $g(x) = 1 + x^2 + x^3$. If each systematic codeword $v(x)$ has the form of (3.25) and the message polynomial is $i(x) = 1 + x + x^3$, deduce the corresponding codeword. After transmission the codeword is received as $r(x) = v(x) + x^4$. Find the first syndrome word $s(x)$ generated by a Meggitt decoder (Fig. 3.16) for decoding the first received symbol.

$[v(x) = x^2 + x^3 + x^4 + x^6; s(x) = 1 + x + x^2]$

P12 A BCH$(15, 7)$ coder $(g(x) = 1 + x^4 + x^6 + x^7 + x^8)$ encodes the message $i(x) = 1 + x^2 + x^3$ into a systematic binary codeword (see (3.25)) of the form

$$v(x) = v_0 + v_1 x + v_2 x^2 + \ldots + v_{n-1} x^{n-1}$$

If symbols v_3 and v_{10} are received in error and the decoder uses algebraic decoding, deduce the syndrome vector in the form

$\mathbf{s} = (\alpha^a, \alpha^b, \alpha^c, \alpha^d); a, b, c, d$ integer

where α is an element of the appropriate Galois field (refer to the appropriate Galois field table).

$[\mathbf{s} = (\alpha^{12}, \alpha^9, \alpha^7, \alpha^3)]$

P13 Using the notation in (3.72), an RS$(7, 3)$ codeword is received as $\mathbf{r} = (6\,5\,0\,6\,5\,3\,6)$. Deduce the syndrome vector $\mathbf{s}$ and hence the corrected word. (Use the identities for $GF(2^3)$ given in Example 3.18.)

$[\mathbf{s} = (6\,6\,3\,4); \mathbf{v} = (6\,5\,0\,1\,5\,1\,6)]$

P14 A simple hard decision FEC system comprises a BCH$(15, 11)$ decoder fed from a binary symmetric channel of transition probability $p = 3 \times 10^{-4}$ (see Fig. 3.30(a)). Assuming that, on average, there are 2.7 bits in error for each decoding failure (n-bit block error), estimate by what factor FEC improves the bit error rate, BER.

[176]

P15 The message sequence entering the coder in Fig. 3.26 is

   A

$\{i\} = 1\,0\,1\,1\,0\,0\,0\,1 \ldots$

where symbol A enters first and the sequence is followed by all zeros. If $g(x) = 1 + x^2 + x^5 + x^6$, what is the transmitted parity sequence?

$[\{p\} = 1\,0\,0\,1\,1\,0\,1\,0\,0\,0\,0\,0\,1\,1\ldots]$

P16   The input to the syndrome decoder in Fig. 3.27 is

U

$\{i'\} = 1\,0\,1\,1\,0\,0\,1\,1$

V

$\{p'\} = 1\,0\,0\,0\,1\,1\,0\,0\,0\,1\,0\,1\,0\,1$

where symbols U and V are received first and V is received when U is in the message register, i.e. when $i'_6 = U$. Use (3.101) to deduce a suitable majority logic decision circuit.

Assuming the message and parity sequences are followed by zeros, and that the registers are initially cleared, analyse the syndrome decoder clock by clock in order to determine the corrected output sequence.

[Orthogonal set: $A_1 = s_6$ $A_2 = s_5$ $A_3 = s_2$ $A_4 = s_0$
output sequence $\{\hat{i}\} = 1\,0\,1\,0\,0\,0\,1\,1$]

P17   A convolutional code sequence is generated by the coder and trellis in Fig. 3.25 and is transmitted over a noisy channel. The corresponding received sequence (codeword) at the input to the Viterbi decoder is

$r = (1\,1\,1\,0\,1\,1\,1\,0\,0\,1\,0\,0\,1\,0\,0\,0\,0\,0\ldots)$

Perform hard decision Viterbi decoding on this sequence up to and including the fifth branch and hence estimate the data sequence at the coder input. Assume the decoder starts from the all-zeros state.

$[\{i\} = 1\,0\,1\,1\,0\ldots]$

P18   Sketch linear feedback shift register circuits corresponding to the characteristic polynomials

$\phi(x) = 1 + x + x^2 + x^3 + x^4$

$\phi(x) = 1 + x^3 + x^6$

and show, algebraically, that they have cycle lengths of 5 and 9 respectively.

Chapter 4

P19 Find the transfer function $H(z)$ for the filter in Fig. P19.

$[H(z) = (1 + 0.2z^{-1} - 0.3z^{-2} - 0.18z^{-3})/(1 - 0.9z^{-2})]$

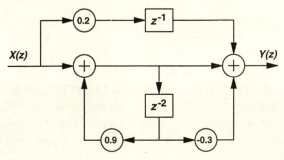

Fig. P19

P20 State the pole-zero locations for the network in Fig. 4.20(b) if $N = 4$ and $g = 1$. What would be the phase shift at normalized frequency $f = 0.00292969$ if $N = 20$ and $g = 0.75$?

[on-circle at $f = 0, 0.25, 0.5, 0.75$; -1.83 radians]

P21 The second-order filter in Fig. 4.21 has complex conjugate poles at $z = 0.8 \angle \pm 30°$. Find the coefficients b_1 and b_2 and the *exact* frequency at which the magnitude of the normalized response is a maximum.

[$b_1 = -1.38564$ $b_2 = 0.64$ $f = 0.0761$]

P22 The circuit in Fig. P22 represents a transversal filter cascaded with two second-order recursive filters. The overall transfer function is to have zeros only at normalized frequencies 0.25, 0.35, and 0.45. Find coefficients b_i, $i = 1, \ldots, 4$.

[$b_1 = -1.90211303$, $b_2 = 1$, $b_3 = -1.17557051$, $b_4 = 1$]

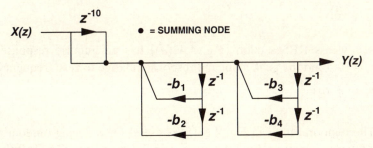

Fig. P22

P23 The section in Fig. 4.24(a) is to realize a second-order highpass Chebyshev filter with a passband ripple of 0.5 dB. The filter is to have unity gain at zero frequency and a normalized cutoff frequency of 0.2. Find the coefficients, assuming $b_0 = 1$.

$$[b_1 = -0.641813 \quad b_2 = 0.327384$$
$$a_0 = 1 \quad a_1 = -2 \quad a_2 = 1 \quad gain = 0.4923]$$

P24 A fourth-order bandpass filter is to be generated from a second-order Butterworth lowpass filter. The lower passband edge frequency is $\omega_l = 1$ and the upper passband edge frequency is $\omega_u = 2$, where $\omega_s = 2\pi$ is the normalized angular sampling frequency. Derive the prototype filter $H_p(s)$. Hence factorize the denominator and generate a canonical cascade form filter as in Fig. 4.12(a).

$$\left[H_p(s) = \frac{1.0223338s^2}{s^4 + 1.4299187s^3 + 2.7239652s^2 + 1.2165973s + 0.7238874} \right.$$

$a_0 = 0.1441189$

$C_1 = 0.4662 \quad D_1 = 0.4925 \quad A_1 = 2 \quad B_1 = 1$

$\left. C_2 = -0.6821 \quad D_2 = 0.5155 \quad A_2 = -2 \quad B_2 = 1 \right]$

P25 The structure in Fig. 4.16 is to be used for a bandstop filter. If the stopband edges are placed at the normalized frequencies 0.25 and 0.35, use the window method to find the unwindowed coefficients.

$[h(0) = -0.139 \quad h(1) = 0.151 \quad h(2) = 0.061 \quad h(3) = 0.8]$

P26 A simple linear phase lowpass FIR filter has impulse response length $N = 5$ and is designed using the window method and a Hamming window (Equation (4.114)). The sampling frequency is 13.5 MHz and the cutoff frequency is 5.5 MHz. What is the filter attenuation at 6 MHz?

[5.1 dB]

P27 A lowpass RRS section (Fig. 4.42(a)) has an impulse response length $k = 10$. Find the peak stopband response relative to zero frequency.

$[-13.1 \text{ dB}]$

P28 The temporal filter in Fig. 4.66(a) is useful for reducing random noise in a television picture. Find the attenuation of the filter at 15 Hz if the frame (picture) delay, T_p, is 1/30 s, and the scaling factor, α, is 1/8.

$[-23.5 \text{ dB}]$

P29 The pixel values (on a 0–255 grey scale) of part of a digitized image are as follows:

$$
\begin{array}{ccccccccc}
195 & 200 & 190 & 40 & 45 & 48 & 185 & 205 & 200 \\
198 & 204 & 192 & 42 & 38 & 46 & 194 & 200 & 202 \\
200 & 200 & 184 & 46 & 35 & 42 & 180 & 202 & 198 \\
204 & 190 & 180 & 42 & 40 & 49 & 182 & 195 & 202 \\
206 & 198 & 182 & 49 & 42 & 52 & 180 & 200 & 204
\end{array}
$$

The sub-image shows a vertical dark feature. Apply a 3×3 median filter to the sub-image (avoiding end effects) and comment upon the image quality.

$$
\begin{bmatrix}
198 & 190 & 45 & 42 & 46 & 185 & 200 \\
198 & 184 & 42 & 42 & 46 & 182 & 198 \\
198 & 182 & 46 & 42 & 49 & 180 & 198
\end{bmatrix}
$$

Chapter 5

P30 A discrete Fourier transformer has a 128-point input time sequence but only five frequency bins are actually required. If the computation time is to be minimized, should the transformer use the DFT directly, or should it use a radix-2 FFT algorithm?

[DFT]

P31 An FFT spectrum analyser has the input signal

$$x(n) = e^{j2\pi fn}; \quad n = 0, 1, 2, 3$$

where f is a normalized frequency, i.e. the sampling frequency f_s is equal to unity. Use an in-place 4-point radix-2 complex input FFT to compute the magnitude spectrum for $f = 0.25$ and explain the spectrum (it may be helpful to refer to Table 5.2).

$[|F(k)| = 0, 4, 0, 0]$

P32 A direct form FIR digital filter has the transfer function

$$H(z) = 0.5 + z^{-1} + 0.5z^{-2}$$

Use an 8-point FFT (e.g. Fig. 5.5) to determine the frequency response of the filter, and hence state the type of filter. Confirm your answer by deriving the continuous response using (4.31).

$[|F(k)| = 2.0, 1.7071, 1.0, 0.2929, 0.0, 0.2929, 1.0, 1.7071; \quad \text{lowpass}]$

P33 A direct form FIR digital filter has the transfer function

$$H(z) = -0.25 + 0.5z^{-1} - 0.25z^{-2}$$

Examine the frequency response of the filter using 4-point and 8-point FFTs and state the essential difference between your two FFTs. Hence state the type of filter and confirm your answer by deriving the continuous response using (4.31).

$[|F(k)| = 0.0, 0.5, 1.0, 0.5$

$|F(k)| = 0.0, 0.1464, 0.5, 0.8535, 1.0, 0.8535, 0.5, 0.1464]$

P34   The select–save algorithm for fast convolution (Section 5.6.1) has an optimum FFT size N if the number of real multiplications is to be minimized for a specific impulse response length M and arbitrary output sequence length. Show that for a radix-2 FFT this is found by minimizing the term

$$\frac{N(\log_2 N + 1)}{N - M + 1}$$

What is the optimum value of N for $M = 64$?

[512]

P35   Estimate the speed of the FDCT algorithm expressed by Equation (5.66) relative to the FWHT algorithm in Table 2.4, for $N = 8$. Assume a radix-2 FFT (Fig. 5.5), and that addition takes about the same time as multiplication. Ignore trivial operations.

[FWHT faster by ≈ 6]

P36   An FFT has to be performed on a 256×256 pixel image. Assuming an in-place radix-2 FFT algorithm, how many butterfly computations have to be performed?

[524, 288]

Transposition of digital filters using signal flow graphs

A signal flow graph (SFG) is a graphical representation of a set of equations which define a system, and by manipulating the graph according to simple rules we can obtain alternative system configurations.

An SFG represents a system by *branches* and *nodes* and a simple example is given in Fig. A1. Note that branches are *directed* and are labelled with a signal operation, such as sample rate expansion, delay, or scaling by a constant gain. Very often the label will be a linear transfer function, H, as indicated, and the SFG represents a connection of sub-systems specified by their transfer function. Nodes define connection points and summing points and are labelled with signals. The signal at each node is equal to the sum of the signals entering the node so that, for example, in Fig. A1 we have

$$W_1(z) = a[X(z) + H_3(z)W_2(z) + H_4(z)Y(z)] \tag{A1}$$

Transposed filters
A particularly useful SFG manipulation is that of *transposition* since it offers alternative filter structures. The rules for obtaining a transposed

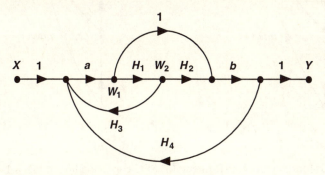

Fig. A1 An SFG for a discrete-time system.

structure can be stated as follows:
 (1) interchange the roles of network input and output;
 (2) reverse the directions of all branches;
 (3) replace all branch operations by their transpose operations. In the case of linear time-invariant operations, such as gains and delays, the branch operation remains unchanged;
 (4) replace connecting nodes by summation nodes and vice-versa.

For linear time-invariant filters the transfer function of the transposed structure will be identical to that of the parent structure. On the other hand, the roundoff noise and dynamic range limitation of a transposed structure can be quite different from the parent structure. This offers an additional degree of freedom in hardware design with respect to register lengths and choice of the point where the predominant roundoff occurs. In addition, transposition can sometimes give a structure more suited to VLSI implementation, as in Example A1.

Example A1

The direct form transversal FIR filter in Fig. 4.14 can be represented by the SFG in Fig. A2(a). Applying the above rules gives the transposed direct form in Fig. A2(b), corresponding to Fig. 4.15. The reader could write expressions for $Y(z)$ in each case in order to demonstrate that the two structures do indeed have the same transfer function.

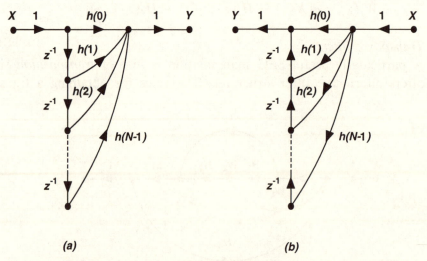

(a)   (b)

Fig. A2 SFGs for direct form FIR filters: (a) SFG corresponding to Fig. 4.14; (b) the transposed form corresponding to Fig. 4.15.

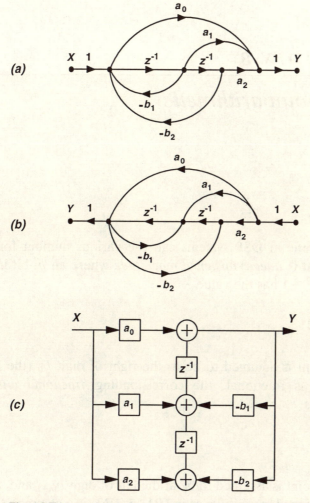

Fig. A3 Transposition of a second-order recursive filter: (a) SFG of parent; (b) transposed SFG; (c) transposed filter.

Example A2

The SFG for the second-order recursive section in Fig. 4.24(a) (but with $b_0 = 1$) is shown in Fig. A3(a). Applying the transposition rules gives the SFG in Fig. A3(b) and the alternative structure in Fig. A3(c). Under certain conditions this can have significant arithmetic advantages over the parent structure.

APPENDIX B

Fixed-point arithmetic

Number formats

Fixed-point arithmetic in DSP systems can use various number formats. The simplest format is *integer unsigned magnitude* where an m-bit integer in the range 0 to $2^m - 1$ has the value

$$X = \sum_{k=0}^{m-1} b_k 2^k \tag{B1}$$

and the binary point is assumed to be to the right of digit b_0 (the LSB). Conversely, if X is fractional, the corresponding *fractional unsigned magnitude* format is

$$X = \sum_{k=1}^{m} b_{m-k} 2^{-k} \tag{B2}$$

where the binary point is assumed to be at the left of digit b_{m-1} and, again, b_0 is the LSB. Unsigned magnitude as in (B1) or (B2) is a common format for dedicated hardware devices such as VLSI filters, multiplier-accumulators and ripple-carry adders. We could generalize the unsigned magnitude format to the form $\langle m, n, u \rangle$ where m is the total number of bits used for representing (varying) the number value, n is the number of integer bits, and u denotes the unsigned format (see Table B1).

Negative numbers could be represented by simply adding a sign bit or 'flag' to the unsigned number, giving the *sign-magnitude* format. In general the number value will be

$$X = (-1)^{b_{m-1}} \sum_{k=0}^{m-2} b_k 2^k \tag{B3}$$

where b_{m-1} corresponds to the sign bit, and this is illustrated in Fig. 1.16

410

Table B1. *Illustrating unsigned magnitude (u) and 2's complement (t) number formats. S is a sign bit.*

Format	Binary form	Range	Resolution
$\langle 4,4,u \rangle$	BBBB.	0 to 15	1
$\langle 4,5,u \rangle$	BBBB0.	0 to 30	2
$\langle 8,0,u \rangle$	.BBBBBBBB	0 to 0.99609375	1/256
$\langle 4,2,t \rangle$	SBB.B	−4 to +3.5	1/2
$\langle 4,5,t \rangle$	SBBB00.	−32 to 28	4
$\langle 8,0,t \rangle$	S.BBBBBBB	−1 to +0.9921875	1/128

for $m = 8$. Note, however, that there are actually *two* representations for zero, i.e. in Fig. 1.16 $+0 = 00000000$ whilst $-0 = 10000000$. This problem is resolved by using the 2's *complement* format since this has a unique representation for zero. Generally speaking, 2's complement also results in simpler addition when compared to sign-magnitude addition. An m-bit number in *integer 2's complement* format will have a value

$$X = -b_{m-1}2^{m-1} + \sum_{k=0}^{m-2} b_k 2^k \tag{B4}$$

where the binary point is to the right of b_0, e.g. 10111. corresponds to -9 decimal. The much used *fractional 2's complement* format is defined by

$$X = -b_0 + \sum_{k=1}^{m-1} b_k 2^{-k}; \quad -1 \leqslant X < 1 \tag{B5}$$

where now the binary point falls between b_0 and b_1, and b_{m-1} is the LSB. The general 2's complement form for m-bit words is simply

$$X = A\left(-b_0 + \sum_{k=1}^{m-1} b_k 2^{-k}\right) \tag{B6}$$

where A is an arbitrary scaling factor (see, for example, Section 1.5.1). In general this will give a 2's complement word with a mix of integer and fractional bits and we can denote this using the notation $\langle m, n, t \rangle$, where, again, n defines the number of integer bits (bits to the left of the binary point, excluding the sign bit), and t denotes 2's complement. Table B1 gives several examples. Different 2's complement formats are often found in a single DSP system, as illustrated in Fig. B1.

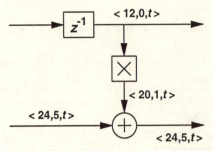

Fig. B1: Illustrating mixed 2's complement formats in a single DSP system.

Example B1

The sum (1.25–6.125) can be performed using the $\langle 7, 3, t \rangle$ format as follows

	-2^3	2^2	2^1	2^0	2^{-1}	2^{-2}	2^{-3}	
	S	B	B	B.	B	B	B	
1.25:	0	0	0	1.	0	1	0	
−6.125:	1	0	0	1.	1	1	1	
	1	0	1	1.	0	0	1	$= -4.875$

Note that, in general, any carry from the sign bit will be neglected.

Rounding and truncation

Consider the multiplication of two fractional 2's complement numbers each $m = k + 1$ bits long, where k excludes the sign bit. In general the $2k + 1$ bit product will need rounding or truncating in order to remove some or all of the least significant k bits. Rounding is the same process used in the quantization of analogue samples, i.e. a sample is assigned to the nearest appropriate level. To round a fractional 2's complement number to t fractional bits we add $2^{-(t+1)}$ to the number and drop the LSBs. This gives a rounding error magnitude $\leqslant 2^{-(t+1)}$ and the final rounded word has a quantization step of 2^{-t}. Usually rounding gives a more accurate result than truncation as shown by the following example.

Example B2

Round the fractional 2's complement number 0.01111 to $t = 3$ bits and compare it with a truncated result:

0.01111
0.00010

0.100   rounded (error $= -0.03125$)
0.011   truncated (error $= 0.09375$)

Overflow handling

The addition of two or more fixed-point numbers cannot lead to inaccuracy unless overflow occurs, i.e. unless the sum exceeds the word-size available. For example, using fractional 2's complement arithmetic (Equation (B5))

$$5/8 - 3/4 \equiv \begin{array}{c} 0.101 \\ 1.010 \\ \hline 1.111 \equiv -1/8 \end{array}$$

but

$$5/8 + 3/4 \equiv \begin{array}{c} 0.101 \\ 0.110 \\ \hline 1.011 \equiv -5/8 \text{ (overflow)} \end{array}$$

The overflow can be represented by the periodic 2's complement overflow characteristic in Fig. B2(a) where the true sum 11/8 is incorrectly represented by $-5/8$. This 'wraparound' characteristic will be followed by DSP systems which fail to check for overflow and, clearly, small overflows can give large errors or noise spikes. In general, overflow can also lead to *overflow oscillation* or a *limit cycle* in an otherwise stable digital filter, and this will be in addition to any limit cycle arising from rounding (Example 4.26). On the other hand, it is worth pointing out that the 2's complement overflow characteristic in Fig. B2(a) has the useful property that indi-

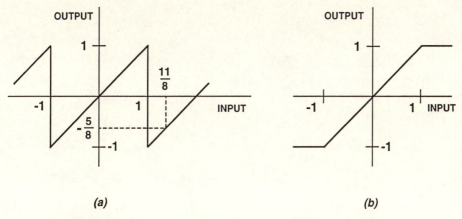

(a)   (b)

Fig. B2: Overflow: (a) 2's complement characteristic; (b) saturation characteristic.

vidual quantized numbers in a sum may be out of range providing the final sum is in-range. In other words, overflow is sometimes permissible.

Figure B2(*b*) corresponds to a more complex but commonly used alternative to the 2's complement overflow characteristic. The idea is to detect overflow (or underflow) and replace the number with the maximum (or minimum) 2's complement value. Provided that the input signal is also *scaled* to minimize the occurrence of overflow then the distortion caused by the saturation characteristic will tend to be small. This characteristic is also attractive compared to that in Fig. B2(*a*) since it eliminates limit cycles arising from overflow.

APPENDIX C

C-routine for FIR filter design

```c
/*      DESIGN OF LINEAR PHASE FIR FILTERS BY THE WINDOW METHOD (N ODD).
 *      Lowpass,highpass,bandpass and bandstop filters.
 *      Sampling frequency fs, lower cutoff frequency f1, upper cutoff
 *      frequency f2. Window: rectangular R, Hamming H, Blackman B,
 *      Blackman-Harris S.
 */
#include <math.h>
main()
{
char window,bs;
int i,j,k,l,m,n,N;
float f,f1,f2,fs,h,hm,ho,pi,r,s1,T,w,wm,x1,x2,xm,y;
double db;
    printf("\n");
    printf("DESIGN OF LINEAR PHASE FIR FILTERS : WINDOW METHOD\n");
    printf("N ODD\n");
    for(n=0;;n++) {
        printf("\n");
        printf("Enter parameters: \n");
        printf("N = ");
        scanf("%d",&N);
        printf("fs = ");
        scanf("%f",&fs);
        printf("f1 = ");
        scanf("%f",&f1);
        printf("f2 = ");
        scanf("%f",&f2);
        printf("frequency grid size = ");
        scanf("%d",&l);
        printf("window = ");
        scanf("%s",&window);
        printf("bandstop design ? (y/n): ");
```

```
        scanf("%s",&bs);
        pi = 3.14159265358979;
        j  = (N-1)/2;
        T  = 1/fs;
        i  = 0;
        ho = 2*(f2-f1)/fs;
        if(bs == 'y')  ho = 1-ho;
        printf("\n       m        h\n\n");
        printf("    %3d     %9.6f\n",i,ho);
        for(k=1; k<l+1; k++) {
            w=pi*fs*k/l;
            r=0;
            for(m=1; m<j+1; m++) {
                x1=m*2*pi*f1*T;
                x2=m*2*pi*f2*T;
                if(x1 == 0.0) s1 = 1.0;
                else s1=sin(x1)/x1;
                hm=2*T*(f2*sin(x2)/x2 - f1*s1);
                if(bs == 'y')  hm = -hm;
                    if(window == 'R') wm = 1;
            else    if(window == 'H')
                wm=0.54 + 0.46*cos(2*pi*m/N);
            else    if(window == 'B')
                wm=0.42 + 0.5*cos(2*pi*m/N) + 0.08*cos(4*pi*m/N);
            else
                wm=0.42323 + 0.49755*cos(2*pi*m/N) + 0.07922*cos(4*pi*m/N);
                h = hm * wm;
                if(k <= 1)  printf("    %3d     %9.6f\n",m,h);
                xm=2*h*cos(m*w*T);
                r = r + xm;
            }
            y = ho + r;
            if(y < 0)  y *= -1.0;
            db = 20.0*log10(y);
            f  = 0.5*fs*k/l;
            if(k <= 1)
            printf("\n      k            f            y            dB\n\n");
            printf("    %3d    %12.4f     %7.4f    %7.2f\n",k,f,y,db);
        }
    }
}
```

C-routine for DIT in-place radix-2 FFT

```c
/*  DIT IN-PLACE RADIX-2 FFT
 *  FFT size N = 2 power M (N <= 1024)
 *  Complex I/O data held in arrays xr[],xi[]
 */
#include <math.h>

float    xr[1024],xi[1024];
int M;

dit(T)
int T;
{
    extern    float xr[1024],xi[1024];
    extern    int  M;
    int       i,ip,j,k,l,le,le1,N,nm1,nv2;
    float     Tr,Ti,Ur[2],Ui[2],Wr,Wi,pi;

    if(T == 0)    {
        N = 1;
        for(i=0; i<M; i++)  N *= 2;
        nv2 = N/2;
        nm1 = N - 1;
        j = 0;
        for(i=0; i<nm1; i++)    {
            if(i < j) {
                Tr = xr[j];
                Ti = xi[j];
                xr[j] = xr[i];
                xi[j] = xi[i];
                xr[i] = Tr;
                xi[i] = Ti;
            }
                k = nv2;
```

```
RE:
            if(k > j)        j += k;
            else {
                    j -= k;
                    k = k/2;
                    goto RE;
            }
        }

        pi = 3.14159265358979;
        for(l=0; l<M; l++)  {
            le = 1;
            for(i=0; i<l+1; i++)       le *= 2;
            le1 = le/2;
            Ur[0] = 1.0;
            Ui[0] = 0.0;
            Wr = cos(pi/le1);
            Wi = -sin(pi/le1);

            for(j=0; j<le1; j++)      {
                i = j;
                while(i < N)    {
                        ip = i + le1;
                        Tr = xr[ip]*Ur[0] - xi[ip]*Ui[0];
                        Ti = xi[ip]*Ur[0] + xr[ip]*Ui[0];
                        xr[ip] = xr[i] - Tr;
                        xi[ip] = xi[i] - Ti;
                        xr[i] = xr[i] + Tr;
                        xi[i] = xi[i] + Ti;
                        i += le;
                }
                Ur[1] = Ur[0]*Wr - Ui[0]*Wi;
                Ui[1] = Ui[0]*Wr + Ur[0]*Wi;
                Ur[0] = Ur[1];
                Ui[0] = Ui[1];
            }
        }
    }
    return;
}
```

C-routine for Meggitt decoder simulation

```
/*  SIMULATION OF CYCLIC CODER AND MEGGITT DECODER
    Double error correcting  BCH(15,7)  code
    User supplied information vector and received vector
    enables random and burst errors to be simulated.
    */

main()
{
    int n,k,i,j,m,corr,err;
    int r[15],e[15],d[7],v[15],vo[15];
    int s0[17],s1[17],s2[17],s3[17],s4[17],s5[17],s6[17],s7[17];
    int p0[15],p1[15],p2[15],p3[15],p4[15],p5[15],p6[15],p7[15];

    n = 15;
    k = 7;
    printf("\nBCH(15,7) CODEC SIMULATION \n");
    printf("\nCyclic coder and Meggitt decoder simulated\n");

    /*    program PROM with syndrome patterns    */

    printf("\nsyndrome patterns to be detected by PROM:\n");
    printf(" s0  s1  s2  s3  s4  s5  s6  s7\n");
    e[n-1] = 1;
    for(j=0; j<n; j++)  {
        for(i=0; i<n-1; i++)    e[i]=0;
        e[n-j-1] = 1;  /* error vector set */

    s0[0]=s1[0]=s2[0]=s3[0]=s4[0]=s5[0]=s6[0]=s7[0]=0;

    /*    find syndrome for jth error vector    */
        for(i=1; i<n+1; i++)    {
            s7[i] = s7[i-1] + s6[i-1];
```

```
                if(s7[i] == 2) s7[i] = 0;
                s6[i] = s7[i-1] + s5[i-1];
                if(s6[i] == 2) s6[i] = 0;
                s5[i] = s4[i-1];
                s4[i] = s7[i-1] + s3[i-1];
                if(s4[i] == 2) s4[i] = 0;
                s3[i] = s2[i-1];
                s2[i] = s1[i-1];
                s1[i] = s0[i-1];
                s0[i] = s7[i-1] + e[n-i];
                if(s0[i] == 2) s0[i] = 0;
        }
        /*   SR now holds syndrome for jth error vector   */
        printf(" %d   %d   %d   %d   %d   %d   %d   %d\n",
                 s0[n],s1[n],s2[n],s3[n],s4[n],s5[n],s6[n],s7[n]);
        p0[j]=s0[n];p1[j]=s1[n];p2[j]=s2[n];p3[j]=s3[n];
        p4[j]=s4[n];p5[j]=s5[n];p6[j]=s6[n];p7[j]=s7[n];
}


/* PROM coding complete */

for(j=0;;j++)  {
err=corr=0;
printf("\nenter 7 information bits, i(k-1) last:\n");
printf(".......\n");
for(i=0; i<k; i++) {
        d[i] = getchar();
        if(d[i] == 48)       d[i] = 0;
        if(d[i] == 49)       d[i] = 1;
        v[n-k+i] = d[i];
}
getchar();      /* get return char */

s0[0]=s1[0]=s2[0]=s3[0]=s4[0]=s5[0]=s6[0]=s7[0]=0;   /* reset SR */

/* generate code vector */
for(i=1; i<k+1; i++)      {
        s7[i] = s6[i-1] + s7[i-1] + d[k-i];
        if(s7[i] == 3) s7[i] = 1;
        if(s7[i] == 2) s7[i] = 0;
        s6[i] = s5[i-1] + s7[i-1] + d[k-i];
        if(s6[i] == 3) s6[i] = 1;
        if(s6[i] == 2) s6[i] = 0;
        s5[i] = s4[i-1];
        s4[i] = s3[i-1] + s7[i-1] + d[k-i];
        if(s4[i] == 3) s4[i] = 1;
```

```
            if(s4[i] == 2) s4[i] = 0;
            s3[i] = s2[i-1];
            s2[i] = s1[i-1];
            s1[i] = s0[i-1];
            s0[i] = s7[i-1] + d[k-i];
            if(s0[i] == 2) s0[i] = 0;
    }


    v[0] = s0[k];
    v[1] = s1[k];
    v[2] = s2[k];
    v[3] = s3[k];
    v[4] = s4[k];
    v[5] = s5[k];
    v[6] = s6[k];
    v[7] = s7[k];


    /*     integer to char conv     */
    for(i=0; i<n; i++)  {
            if(v[i] == 0)  v[i] = 48;
            else v[i] = 49;
    }


    printf("\nsystematic code vector :\n");
    for(i=0; i<n; i++)      putchar(v[i]);
    printf("\n");
    printf("\nenter received vector, r(n-1) last:\n");
    for(i=0; i<n; i++)  {
            r[i] = getchar();
            if(r[i] == 48)       r[i] = 0;
            if(r[i] == 49)       r[i] = 1;
    }
    getchar();     /*  get return char  */
    printf("\n");


    /* Meggitt decoder simulation */


    s0[0]=s1[0]=s2[0]=s3[0]=s4[0]=s5[0]=s6[0]=s7[0]=0;   /* reset SR */


    printf("state of decoder syndrome register :\n");
    printf("\nclock   s0 s1 s2 s3 s4 s5 s6 s7\n");
        for(i=1; i<n+1; i++)       {
                s7[i] = s7[i-1] + s6[i-1];
                if(s7[i] == 2) s7[i] = 0;
                s6[i] = s7[i-1] + s5[i-1];
                if(s6[i] == 2) s6[i] = 0;
```

```
            s5[i] = s4[i-1];
            s4[i] = s7[i-1] + s3[i-1];
            if(s4[i] == 2) s4[i] = 0;
            s3[i] = s2[i-1];
            s2[i] = s1[i-1];
            s1[i] = s0[i-1];
            s0[i] = s7[i-1] + r[n-i];
            if(s0[i] == 2) s0[i] = 0;
            printf(" %2d    %d %d %d %d %d %d %d %d\n",
                  i,s0[i],s1[i],s2[i],s3[i],s4[i],s5[i],s6[i],s7[i]);
    }


/*   received vector now in buffer, initial syndrome computed   */

s0[1] = s0[n]; s1[1] = s1[n]; s2[1] = s2[n]; s3[1] = s3[n];
s4[1] = s4[n]; s5[1] = s5[n]; s6[1] = s6[n]; s7[1] = s7[n];

for(i=1; i<n+1; i++)      {
    m = 15 + i;
    e[n-i] = 0;
    if(s0[i] == 1 || s1[i] == 1 || s2[i] == 1 || s3[i] == 1
    || s4[i] == 1 || s5[i] == 1 || s6[i] == 1 || s7[i] == 1) err = 1;
    for(j=0; j<n; j++) {
    if(s0[i] == p0[j] && s1[i] == p1[j] && s2[i] == p2[j] && s3[i] == p3[j]
    && s4[i] == p4[j] && s5[i] == p5[j] && s6[i] == p6[j] && s7[i] == p7[j])
        { e[n-i] = 1;
          corr = 1;
        }
    }
    vo[n-i] = e[n-i] + r[n-i];   /* error correction */
    if(e[n-i] == 1)         printf(" buffer output complemented\n");
    if(vo[n-i] == 2) vo[n-i]=0;      /* modulo 2 add */
    s7[i+1] = s7[i] + s6[i];
    if(s7[i+1] == 2) s7[i+1]=0;
    s6[i+1] = s7[i] + s5[i];
    if(s6[i+1] == 2) s6[i+1]=0;
    s5[i+1] = s4[i];
    s4[i+1] = s7[i] + s3[i];
    if(s4[i+1] == 2) s4[i+1]=0;
    s3[i+1] = s2[i];
    s2[i+1] = s1[i];
    s1[i+1] = s0[i];
    s0[i+1] = s7[i]; /* no feedback, no data input */
    printf(" %2d    %d %d %d %d %d %d %d %d\n",
    m,s0[i+1],s1[i+1],s2[i+1],s3[i+1],s4[i+1],s5[i+1],s6[i+1],s7[i+1]);
}
```

```
/*    integer to char conv      */
for(i=0; i<n; i++)  {
    if(vo[i] == 0)   vo[i] = 48;
    else vo[i] = 49;  ·
}

if(err == 1 && corr == 0)
printf("\nuncorrectable error pattern detected\n");/* eg burst up to n-k */
if(corr == 1)
printf("\nreceived vector modified\n"); /* not necessarily corrected */
printf("\ndecoded vector.    : ");
for(i=0; i<n; i++)     putchar(vo[i]);
printf("\ntransmitted vector: ");
for(i=0; i<n; i++)     putchar(v[i]);
printf("\n");
}
}
```

REFERENCES

Adams, J.W., & Wilson, Jr., A.N. (1983), A new approach to FIR digital filters with fewer multipliers and reduced sensitivity, *IEEE Trans. Circuits & Systems*, **CAS-30** (5), 277–83.

Adams, J.W., & Wilson, Jr., A.N. (1984), Some efficient digital prefilter structures, *IEEE Trans. Circuits & Systems*, **CAS-31** (3), 260–5.

Agarwal, R.C., and Sudhakar, R. (1983), Multiplier-less design of FIR filters, *IEEE*, **ICASSP**, Boston, 209–12.

Agrawal, B.P., and Shenoi, K. (1983), Design methodology for $\Sigma\Delta M$, *IEEE Trans.* **COM-31**, 360–9.

Alexander, S.T., & Rajala, S.A. (1985), Image compression results using the LMS adaptive algorithm, *IEEE Trans. Acoustics, Speech, and Signal Proc.*, **ASSP-33** (3), 712–14.

Anastassopoulos, V., *et al* (1985), Method for designing half-band delta-modulation FIR filters, *IEE Proc.*, **132** Pt.F (1), 13–17.

Annegarn, M.J., Nillesen, A.H., & Raven, J.G. (1986), Digital signal processing for television receivers, *Philips Tech. Rev.*, **42** (6/7), 183–200.

Ansari, R., and Liu, B. (1983), Efficient sampling rate alteration using recursive (IIR) digital filters, *IEEE Trans. Acoustics, Speech and Signal Proc.* **ASSP-31**, (6), 1366–73.

Atal, B.S., & Schroeder, M.R. (1984), Stochastic coding of speech signals at very low bit rates, *IEEE Proc. ICC'84*, 1610–13.

Avenhaus, E. (1972), On the design of digital filters with coefficients of limited word length, *IEEE Trans. Audio Electroacoustics*, **AU-20** (3), 206–12.

Benvenuto, N. & Daumer W.R. (1987), Waveform coding of 9.6 kb/s voiceband data signals at 32 kb/s, *Proc. IEEE Int. Conf. Communications*, **3**, 1700–3.

Berlekamp, E.R. (1968), *Algebraic Coding Theory*, McGraw-Hill, New York.

Blankenship, P.E., & Hofstetter, E.M. (1975), Digital pulse compression via fast convolution, *IEEE Trans. Acoustics, Speech, and Signal Proc.*, **ASSP-23** (2), 189–201.

Bold, G. (1985), A comparison of the time involved in computing fast Hartley and fast Fourier transforms. *Proc. IEEE*, **73** (12), 1863–4.

Bose, R.C., & Chaudhuri, D.K. (1960), On a class of error correcting binary group codes, *Inf. Control*, **3**, March, 68–79.

Bota, V. (1992), University of Cluj-Napoca, Romania, private communication.

424

Boussakta, S., & Holt, A.G.J. (1988), Calculation of the discrete Hartley transform via the Fermat number transform using a VLSI chip, *IEE Proc.*, **135** Pt.G (3), 101–3.

Bracewell, R.N. (1984), The fast Hartley transform, *Proc. IEEE*, **72** (8), 1010–18.

Brebner, G.E., & Ritchings, R.T. (1988), Image transform coding: A case study involving real time signal processing, *IEE Proc.*, **135** Pt.E (1), 41–8.

Brigham, E.O. (1974), *The Fast Fourier Transform*, Prentice-Hall, Englewood Cliffs, NJ.

Burrus, C.S. (1977), Digital filter structures described by distributed arithmetic, *IEEE Trans. Circuits Syst.*, **CAS-24** (12), 674–80.

Byrne, C.J., *et al* (1963), Systematic jitter in a chain of digital regenerators, *Bell System Tech.*, **42**, 2679–714.

Cabezas, J.C.E, & Diniz, P.S.R. (1990), FIR filters using interpolated prefilters and equalizers, *IEEE Trans. Circuits & Systems*, **37** (1), 17–23.

Campbell Jr., J.P. *et al* (1989), An expandable error-protected 4800 BPS CELP Coder (US Federal Standard 4800 BPS Voice Coder), *Proc. ICASSP, IEEE*, **May**, 735–7.

Cain, G.D., Kale, I., & Yardin, A. (1988), Host windowing employing frequency sampling for painless FIR filter design, *IERE Fifth International Conference on Digital Processing of Signals in Communications*, Loughborough University of Technology, UK, 39–49.

Catchpole, R.J. (1975), Efficient ternary transmission codes, *Electron. Lett.*, **11** (20), 482–4.

CENELEC (1990), Specification of the radio data system (RDS), European Committee for Electrotechnical Standardization, Rue de Stassart 35, B-1050 Brussels, EN 50067.

Charniak, E. & McDermott, D. (1985), *Introduction to Artificial Intelligence*, Addison-Wesley, Reading, Massachusetts.

Chien, R.T. (1964), Cyclic decoding procedures for Bose–Chaudhuri–Hocquenghem codes, *IEEE Trans. Inf. Theory*, **IT-10**, 357–63.

Chow, M. (1971), Variable-length redundancy removal coders for differentially coded video telephone signals, *IEEE Trans. Commun. Technol.*, **COM-19** (6), 923–6.

Church, R. (1935), Tables of irreducible polynomials for the first four prime moduli, *Annals of Mathematics*, **36** (1), 198–209.

Cioffi, J.M., & Kailath, T. (1984), Fast recursive-least-squares transversal filters for adaptive filtering, *IEEE Trans. Acous., Speech and Signal Proc.*, **ASSP-32**, (2), 304–37.

Cioffi, J.M. (1987), Limited-precision effects in adaptive filtering, *IEEE Trans. on Circuits and Systems*, **CAS-34** (7), 821–33.

Cooley, J.W., & Tukey, J.W. (1965), An algorithm for the machine calculation of complex Fourier series, *Mathematics of Computation*, **19**, 297–301.

Claasen, T.A., Mecklenbrauker, W.F., & Peek, J.B. (1976), Effects of quantization and overflow in recursive digital filters, *IEEE Trans. Acoustics, Speech, and Signal Proc.*, **ASSP-24** (6), 517–28.

Clarke, C.K.P. (1975), Hadamard transformation: Walsh spectral analysis of television signals, Report 1975/26, BBC Research Department.

Cohn, D.L., & Melsa, J.L. (1975), The residual encoder – an improved ADPCM system for speech digitization, *IEEE Trans. Communications*, **COM-23** (9), 935–41.

Crochiere, R.E. and Rabiner, L.R. (1983), *Multirate Digital Signal Processing*, Prentice-Hall, Englewood Cliffs, New Jersey.

Croisier, A. (1970), Compatible high-density bipolar codes: An unrestricted transmission plan for PCM carriers, *IEEE Trans. Commun. Technol.*, **COM-18** (3,2), 265–8.

Davis, L. (1991), *Handbook of Genetic Algorithms*, Van Nostrand Reinhold.

Devereux, V.G. (1975), Digital video: differential coding of PAL colour signals using same-line and two-dimensional prediction, Report 1975/20, BBC Research Department.

Devereux, V.G., & Stott, J.H. (1978), Digital video: Sub-Nyquist sampling of PAL colour signals, *Proc. IEE*, **125** (9), 779–86.

Dolivo, F., Hermann, R., & Olcer, S. (1989), Performance and sensitivity analysis of maximum-likelihood sequence detection on magnetic recording channels, *IEEE Trans. Magnetics*, **25** (5), 4072–4.

Drewery, J.O., Storey, R., & Tanton, N.E. (1984), Video noise reduction, Report 1984/7, BBC Research Department.

Duhamel, P., & Vetterli, M. (1990), Fast Fourier transforms: A tutorial review and a state of the art, *Signal Processing*, Elsevier Science Publishers B.V., **19**, 259–99.

Elspas, B. (1959), The theory of autonomous linear sequential networks, *IRE Trans. Circuit Theory*, **CT-6**, 45–60.

ETSI (1991), GSM full rate speech transcoding, *European Telecommunications Standards Institute, Valbonne Cedex, France*, **GSM 06.10** (Version 3.2.0).

Ferebee, I.C., Tait, D.J., & Taylor, D.H. (1989), An ARQ/FEC coding scheme for land mobile communication, *Int. J. Satellite Communications*, **7**, 219–224.

Fire, P. (1959), A class of multiple-error correcting binary codes for non-independent errors, Report RSL-E-2, Sylvania Reconaissance Systems Laboratory, Mountain View, CA.

Flanagan, J.L., *et al* (1979), Speech coding, *IEEE Trans. Communications*, **COM-27** (4), 710–36.

Floch, B., *et al.* (1989), Digital sound broadcasting to mobile receivers, *IEEE Trans. Consumer Electronics*, **35** (3), 493–503.

Franaszek, P.A. (1972), Run-length-limited variable length coding with error propagation limitation, U.S. Patent 3 689 899.

Frei, W., & Chen, C. (1977), Fast boundary detection: A generalization and a new algorithm, *IEEE Trans. Computers*, **C-26** (10), 988–98.

Gilchrist, N.H.C. (1984), Digital sound: Programme-modulated noise levels in sound signals subjected to a number of digital companding processes, Report 1984/11, BBC Research Department.

Gillard, C.H. (1986), Error correction strategy for the new generation of 4:2:2 component digital video tape recorders, *Sixth Conf. on Video, Audio and Data Recording*, IERE, **March**, 165–75.

Golding, L.S., and Garlow, R.K. (1971), Frequency interleaved sampling of a colour television signal, *IEEE Trans. Commun. Technol.*, **COM-19** (6), 972–9.

Goodman, D.J. (1969), The application of delta modulation to analog-to-PCM encoding, *Bell Syst. Tech. J.*, **48**, 321–44.

Gudvangen, S., & Holt, A.G.J. (1992), Computation of 2-dimensional fast transforms and image filtering on transputer networks, *IEE Proc.-E*, **139** (3), 249–56.

Hamming, R.W. (1950), Error detecting and correcting codes, *Bell System Tech. J.*, **29** (2), 147–60.

Harmuth, H.F. (1969), *Transmission of Information by Orthogonal Functions*, Springer-Verlag, New York.

Harris, F.J. (1978), On the use of windows for harmonic analysis with the discrete Fourier transform, *Proc. IEEE*, **66** (1), 51–83.

Hart, J., *et al* (1982), Manipulation of speech sounds, *Philips Tech. Rev.*, **40** (5), 134–45.

Hartley, R.V.L. (1942), A more symmetrical Fourier analysis applied to transmission problems, *Proc. IRE*, **30**, March, 144–50.

Heygster, G. (1982), Rank filters in digital image processing, *Computer Graphics and Image Processing*, **19**, 148–64.

Heuval, A.P., *et al* (1991), A spectrum efficient combined speech and channel coding method providing high voice quality for land mobile radio systems, *IEE Colloquium on Future Mobile Radio Trunking and Data Systems, Savoy Place, February*, Digest **1991/051**.

Hocquenghem, A. (1959), Codes correcteurs d'erreurs', *Chiffres*, **2**, 147–56.

Holland, J. (1975), *Adaption in Natural and Artificial Systems*, University of Michigan Press, Michigan.

Holmes, J.N. (1982), A survey of methods for digitally encoding speech signals, *The Radio & Electronic Engineer*, **52** (6), 267–76.

Hou, H.S. (1987), A fast recursive algorithm for computing the discrete cosine transform, *IEEE Trans. Acoustics, Speech, and Signal Proc.*, **ASSP-35** (10), 1455–61.

Hsiao, M.Y. (1970), A class of optimum odd-weight-column SEC-DED codes, *IBM J. Res. Dev.*, **14**, July.

Huang, T.S. (1977), Coding of two-tone images, *IEEE Trans. Communications*, **COM-25** (11), 1406–24.

Huffman, D.A. (1952), A method for the construction of minimum redundancy codes, *Proc. IRE*, **40**, 1098.

Hunter, R. & Robinson, A.H. (1980), International digital facsimile coding standards, *Proc. IEEE*, **68** (7), 854–67.

Immink, K.A.S. (1989), Coding techniques for the noisy magnetic recording channel: A state-of-the-art report, *IEEE Trans. Communications*, **37** (5), 413–19.

ITU (1982), CCIR Rec.601: Encoding parameters of digital television for studios, *International Telecommunication Union, XVth Plenary Assembly, Geneva*, **XI** (1), Broadcasting Service (TV).

ITU (1988), CCITT Rec.G.721, 32 kbit/s adaptive differential PCM (ADPCM), *International Telecommunication Union, Melbourne*, Fascicle III.4.

ITU (1989), CCITT SGXV, Description of ref. model 8 (TM8), *Specialist Group on Coding for Visual Telephony*, Document 525.

ITU (1990), CCITT Rec.H.261, video codec for audiovisual services at $p \times 64$ kbit/s, *International Telecommunication Union, Geneva*, Study Group XV.

Jain, A.K. (1981), Image data compression: A review, *Proc. IEEE*, **69** (3), 349–81.

Jessop, A., & Waters, D.B. (1970), 4B3T: An efficient code for PCM coaxial line systems, *Proc. 17th Intern. Scientific Congress on Electronics, Rome*, 275.

Justesen, J. (1982), Information rates and power spectra of digital codes, *IEEE Trans. Inform. Theory*, **IT-28** (3).

Kabal, P., & Pasupathy, S. (1975), Partial-response signaling, *IEEE Trans. Communications*, **COM-23** (9), 921–34.

Kallaway, M.J., & Mahadeva, W.A. (1977), CEEFAX: Optimum transmitted pulse-shape, Report 1977/15, BBC Research Department.

Kasai, H., *et al* (1974), PCM jitter suppression by scrambling, *IEEE Trans. Communications*, **COM-22** (8), 1114–22.

Klieber, E.J. (1970), Some difference triangles for constructing self-orthogonal codes, *IEEE Trans. Inform. Theory*, **IT-16**, 237–8.

Kleijn, W.B., *et al* (1990), Fast methods for the CELP coding algorithm, *IEEE Trans. Acoustics, Speech, and Signal Proc.*, **38** (8), 1330–41.

Knee, M.J., & Wells, N.D. (1989), Comparison of DPCM prediction strategies for high-quality digital television bit-rate reduction, Report 1989/8, BBC Research Department.

Kodek, D.M. (1980), Design of optimal finite wordlength FIR digital filters using integer programming techniques, *IEEE Trans. Acoustics, Speech, and Signal Proc.*, **ASSP-28** (3), 304–7.

Kretzmer, E.R. (1952), Statistics of television signals, *Bell System Tech. J.*, **31**, 751–63.

Kroon, P., *et al* (1986), Regular-pulse excitation – a novel approach to effective and efficient multipulse coding of speech, *IEEE Trans. Acoustics, Speech and Signal Proc.*, **ASSP-34** (5), 1054–63.

Kupnicki, R., & Moote, S. (1984), High security television transmission using digital processing, *Milcom'84, IEEE Military Comms. Conf., Los Angeles, Cali.*, 284–9.

Larsen, K.J. (1973), Short convolutional codes with maximum free distance for rates 1/2. 1/3, and 1/4, *IEEE Trans. Inform. Theory*, **IT-19**, 371–2.

Lender, A. (1963), The duobinary technique for high-speed data transmission, *IEEE Trans. Communication and Electronics*, **82**, 214–18.

Lender, A. (1966), Correlative level coding for binary-data transmission, *IEEE Spectrum*, **February**, 104–15.

Lin, S., & Costello, Jr., D.J. (1983), *Error Control Coding*, Prentice-Hall, Englewood Cliffs, NJ.

Lucky, R.W., Salz, J., & Weldon Jr, E.J. (1968), *Principles of Data Communication*, McGraw-Hill.

Lynn, P.A. (1980), FIR digital filters based on difference coefficients: Design improvements and software implementation, *Proc. IEE, Part E*, **127** (6), 253–8.

Makhoul, J. (1975), Linear prediction: A tutorial review, *Proc. IEEE*, **63** (4), 561–80.

Marguinaud, A., & Sorton, G. (1983), The benefits of coding in satellite communication systems, *Space Communication and Broadcasting*, **1**, 293–318.

Martinson, L.W., & Smith, R.J. (1975), Digital matched filtering with pipelined floating point fast Fourier transforms (FFTs), *IEEE Trans. Acoustics, Speech and Signal Proc.*, **ASSP-23** (2), 222–34.

Massey, J.L. (1963), *Threshold Decoding*, MIT Press, Cambridge, Massachusetts.

Max, J. (1960), Quantizing for minimum distortion, *IRE Trans. Inform. Theory*, **IT-6**, 7–12.

McClellan, J.H., Parks, T.W., & Rabiner, L.R. (1973), A computer program for designing optimum FIR linear phase digital filters, *IEEE Trans. Audio Electroacoust.* **AU-21** (6), 506–26.

McEliece, R.J. (1977), *The Theory of Information and Coding*, Addison-Wesley Publishing Company, Reading, Massachusetts.

Meggitt, J.E. (1961), Error correcting codes and their implementation for data transmission systems, *IRE Trans. Inform. Theory*, **October**, 234–44.

Meyr, H., *et al* (1974), Optimum Run Length Codes, *IEEE Trans. Communications*, **COM-22** (6), 826–35.

Musmann, H.G., *et al*, (1985), Advances in picture coding, *Proc. IEEE*, **73**(4), 523–47.

Narasimha, M.J., & Peterson, A.M. (1978), On the computation of the discrete cosine transform, *IEEE Trans. Communications*, **COM-26** (6), 934–6.

Natvig, J.E., (1988), *Pan-European Speech Coding Standard for Digital Mobile Radio*, Elsevier Science Publishers B.V. (North Holland), Amsterdam, 113–23.

Naus, P.J.A., *et al* (1987), A CMOS stereo 16-bit D/A converter for digital audio, *IEEE J. Solid-State Circuits*, **SC-22** (3), 390–4.

Netravali, A.N., & Limb, J.O. (1980), Picture coding: A review, *Proc. IEEE*, **March**, 366–406.

Neuvo, Y., Cheng-Yu, D., & Mitra, S.K. (1984), Interpolated finite impulse response filters, *IEEE Trans. Acoustics, Speech, and Signal Proc.*, **ASSP-32** (3), 563–70.

Nicol, R.C., *et al* (1980), Transmission techniques for picture viewdata, 1980 *International Broadcasting Convention, IEE Conference Publication*, **191**, 109–13.

Nishitani, T., *et al* (1982), A 32 kb/s toll quality ADPCM codec using a single chip signal processor, *Proc. ICASSP*, **April**, 960–3.

Noonan, J.P., & Marquis, D.A. (1989), New algorithm for the design of linear phase FIR filters with $+1$, -1, and 0 coefficients, *Signal Processing*, **17**, 81–5.

Nyquist, H. (1924), Certain factors affecting telegraph speed, *Bell System Tech. J.*, **3**, 324–46.

Nyquist, H. (1928), Certain topics in telegraph transmission theory, *Trans. AIEE*, **47**, 617–44.

Okada, T., Tanaka, Y., & Isobe, T. (1982), New filter technology in picture processing, *IEEE Trans. Consumer Electronics*, **CE-28** (3), 157–66.

Okada, T., Hongu, M., & Tanaka, Y. (1985), Flicker-free non-interlaced receiving system for standard color TV signals, *IEEE Trans. Consumer Electronics*, **CE-31** (3), 240–53.

O'Neal Jr., J.B. (1966), Predictive quantizing systems (differential pulse code modulation) for the transmission of television signals, *Bell System Technical Journal*, **May–June**, 689–721.

Oppenheim, A.V., & Willsky, A.S. (1983), *Signals and Systems*, Prentice-Hall, Inc., Englewood Cliffs, NJ.

Paaske, E. (1974), Short binary convolutional codes with maximal free distance for rates 2/3 and 3/4, *IEEE Trans. Inform. Theory*, **IT-20**, 683–9.

Paez, M.D., & Glisson, T.H. (1972), Minimum mean-squared-error quantization in speech PCM and DPCM systems, *IEEE Trans. Communications*, **April**, 225–30.

Pearson, D.E., & Robinson, J.A. (1985), Visual communication at very low data rates, *Proc. IEEE*, **73** (4), 795–812.

Peterson, W.W. (1960), Encoding and error-correction procedures for the Bose–Chaudhuri codes, *IRE Trans. Inform. Theory*, **IT-6**, 459–70.

Peterson, W.W., & Brown, D.T. (1961), Cyclic codes for error detection, *Proc. IRE*, 228–35.

Pratt, W.K., Kane, J., & Andrews, H.C. (1969), Hadamard transform image coding, *Proc. IEEE*, **57** (1), 58–68.

Proakis, J.G., & Manolakis, D.G. (1988), *Introduction to Digital Signal Processing*, Macmillan Publishing Company, New York.

Qureshi, S.U.H. (1985), Adaptive equalization, *Proc. IEEE*, **73** (9), 1349–87.

Rabiner, L.R., & Gold, B. (1975), *Theory and Application of Digital Signal Processing*, Prentice-Hall, Englewood Cliffs, NJ.

Rabiner, L.R., & Schafer, R.W. (1978), *Digital Processing of Speech Signals*, Prentice-Hall, Englewood Cliffs, NJ.

Rader, C.M., & Gold, B. (1967), Digital filter design techniques in the frequency domain,

Proc. IEEE, **55** (2), 149–70.

Ratliff, P.A. & Stott, J.H. (1983), Digital television transmission: 34 Mbit/s PAL investigation, Report 1983/9, BBC Research Department.

Reed, I.S., & Solomon, G. (1960), Polynomial codes over certain finite fields, *J. SIAM*, **8**, 300–4.

Reid, D.F. (1974), Digital video: Some bit-rate reduction methods which preserve information in broadcast-quality digital video signals, Report 1974/37, BBC Research Department.

Robinson, J.A. (1986), Low-data-rate visual communication using cartoons: A comparison of data compression techniques, *IEE Proc.* **133** Pt F (3), 236–56.

Robinson, J.P., & Bernstein, A.J. (1967), A class of binary recurrent codes with limited error propagation, *IEEE Trans. Inform. Theory*, **IT-13**, 106–13.

Saramaki, T., Neuvo, Y., and Mitra, S.K. (1988), Design of computationally efficient interpolated FIR filters, *IEEE Trans. Circuits & Systems*, **35** (1), 70–87.

Savage, J.E. (1967), Some simple self-synchronizing digital data scramblers, *The Bell System Tech. J.*, **February**, 449–87.

Shannon, C.E. (1949), Communication in the presence of noise, *Proc. IRE*, **37**, 10–21.

Shannon, C.E. (1951), Prediction and entropy of printed English, *Bell System Tech. J.*, **30**, 50.

Sharma, D.K., & Netravali, A.N. (1977), Design of quantizers for DPCM coding of picture signals, *IEEE Trans. Communications*, **COM-25** (11), 1267–74.

Shelswell, P., *et al* (1991), Digital audio broadcasting: The first UK field trial, Report 1991/2, BBC Research Department.

Smithson, P. (1992), University of Plymouth, private communication.

Stott, J.H. (1984), Digital transmission: 68 Mbit/s PAL field trial system, Report 1984/5, BBC Research Department.

Suehiro, N., & Hatori, M. (1986), Fast algorithms for the DFT and other sinusoidal transforms, *IEEE Trans. Acoustics, Speech, and Signal Proc.*, **ASSP-34**, June, 642–4.

Tan, B.S., & Hawkins, G.J. (1981), Speed-optimized microprocessor implementation of a digital filter, *Proc. IEE*, Part E, **128** (3), 85–93.

Tribolet, J.M., & Crochiere, R.E. (1979), Frequency domain coding of speech, *IEEE Trans. Acoustics, Speech, and Signal Proc.*, **ASSP-27** (5), 512–30.

Uchimura, K., *et al* (1988), Oversampling A-to-D and D-to-A converters with multistage noise shaping modulators, *IEEE Trans. Acoustics, Speech, and Signal Proc.*, **36** (12), 1899–905.

Vary, P., *et al*, (1988), A regular-pulse excited linear predictive codec, *Speech Communication*, **7**, 209–15.

Van Gerwen, P.J., Mecklenbrauker, W.F.G., Verhoeckx, N.A.M., Snijders, F.A.M., & Van Essen, H.A. (1975), A new type of digital filter for data transmission, *IEEE Trans. Communications*, **COM-23** (2), 222–33.

Viterbi, A.J. (1967), Error bounds for convolutional codes and an asymptotically optimum decoding algorithm, *IEEE Trans. Inform. Theory*, **IT-13** (2), 260–9.

Viterbi, A.J. (1971), Convolutional codes and their performance in communication systems, *IEEE Trans. Communications*, **COM-19** (5), 751–72.

Viterbi, A.J., & Omura, J.K. (1979), *Principles of Digital Communication and Coding*, McGraw-Hill, New York.

Wade, G. (1975), Crosscorrelation method for analysing phase and gain errors in PCM systems for colour television, *Proc. IEE*, **122** (4), 367–8.

Wade, G. (1988), Offset FFT and its implementation on the TMS320C25 processor, *Microprocessors and Microsystems*, **12** (2), 76–82.

Wade, G., Van-Eetvelt, P., & Darwen, H. (1990), Synthesis of efficient low-order FIR filters from primitive sections, *Proc. IEE*, Part G, **137** (5), 367–72.

Wade, G., & Li, Z., (1991), Compression of Meteosat data for image dissemination, *Computer Communications*, **14** (8), 489–95.

Wade, G., Roberts, A.R., & Williams, G. (1994), Multiplier-less design of FIR filters using a genetic algorithm (accepted for publication, *IEE Proc., Vision, Image and Signal Processing*).

Walker, R., (1974), Hadamard transformation: A real-time transformer for broadcast standard PCM television, Report 1974/7, BBC Research Department.

Weinberg, L. (1962), *Network Analysis and Synthesis*, McGraw-Hill, Inc., New York.

Welch, T.A. (1984), A technique for high-performance data compression, *IEEE Computer*, **June**, 8–19.

Widrow, B., *et al* (1975), Adaptive noise cancelling: Principles and applications, *Proc. IEEE*, **63** (12), 1692–716.

Widrow, B., *et al*, (1976), Stationary and nonstationary learning characteristics of the LMS adaptive filter, *Proc. IEEE*, **64** (8), 1151–62.

Widrow, B., & Stearns, S.D. (1985), *Adaptive Signal Processing*, Prentice-Hall, Inc., Englewood Cliffs, NJ.

Yan, J., & Donaldson, R.W. (1972), Subjective effects of channel transmission errors on PCM and DPCM voice communication systems, *IEEE Trans. Communications*, **June**, 281–90.

Zierler, N. (1959), Linear recurring sequences, *J. Soc. Indust. Appl., Math.*, **7** (1), 31–48.

Ziv, J., & Lempel, A. (1977), A universal algorithm for sequential data compression, *IEEE Trans. Inform. Theory*, **IT-23** (3), 337–43.

INDEX